AF597914

NEUROMETHODS

Series Editor
Wolfgang Walz
University of Saskatchewan
Saskatoon, SK, Canada

For further volumes:
http://www.springer.com/series/7657

Exocytosis Methods

Edited by

Peter Thorn

School of Biomedical Sciences, University of Queensland, St. Lucia, QLD, Australia

Editor
Peter Thorn
School of Biomedical Sciences
University of Queensland
St. Lucia, QLD, Australia

ISSN 0893-2336 ISSN 1940-6045 (electronic)
ISBN 978-1-62703-675-7 ISBN 978-1-62703-676-4 (eBook)
DOI 10.1007/978-1-62703-676-4
Springer New York Heidelberg Dordrecht London

Library of Congress Control Number: 2013951228

Printed on acid-free paper

Cover illustration: A representation of the process of exocytosis shows a granule fusing with the cell membrane.

Humana Press is a brand of Springer
Springer is part of Springer Science+Business Media (www.springer.com)

Series Preface

Under the guidance of its founders Alan Boulton and Glen Baker, the Neuromethods series by Humana Press has been very successful since the first volume appeared in 1985. In about 17 years, 37 volumes have been published. In 2006, Springer Science + Business Media made a renewed commitment to this series. The new program will focus on methods that are either unique to the nervous system and excitable cells or which need special consideration to be applied to the neurosciences. The program will strike a balance between recent and exciting developments like those concerning new animal models of disease, imaging, in vivo methods, and more established techniques. These include immunocytochemistry and electrophysiological technologies. New trainees in neurosciences still need a sound footing in these older methods in order to apply a critical approach to their results. The careful application of methods is probably the most important step in the process of scientific inquiry. In the past, new methodologies led the way in developing new disciplines in the biological and medical sciences. For example, Physiology emerged out of Anatomy in the nineteenth century by harnessing new methods based on the newly discovered phenomenon of electricity. Nowadays, the relationships between disciplines and methods are more complex. Methods are now widely shared between disciplines and research areas. New developments in electronic publishing also make it possible for scientists to download chapters or protocols selectively within a very short time of encountering them. This new approach has been taken into account in the design of individual volumes and chapters in this series.

Wolfgang Walz

Preface

This *Neuromethods* book brings together the techniques and methods that are currently being applied to the study of exocytosis. The book is organised with Part I focussed on the range of techniques being applied to the study of single-vesicle fusion events. These methods are needed to gain insight into the final steps of vesicle fusion. Simple secretion assays can be used and have been used to identify mechanisms in the secretory cycle. However, to discriminate where in the secretory cascade a particular protein or mechanism is engaged requires techniques that can dissect out the components of these pathways. To this end, the methods in Part I are aimed at determining the steps associated with the final stages of vesicle fusion and content loss. In some cases, in some cell types, the vesicles are large enough to be imaged directly with a light microscope. In other cell types, very small vesicles require novel imaging or electrophysiological methods for their detection.

In Part II, the focus is on some of the model systems that are being employed to unravel the complexities of exocytosis. Model systems have been a traditional mainstay of research in this area and have led to many breakthroughs in the field. Exocytosis in most neuronal systems engages very small vesicles and occurs over a very rapid time-course. This has necessitated the use of cell systems where analogous processes of exocytosis occur but where vesicles are larger and fusion is slower. It includes the use of systems that enable access to what would have normally been hidden sites of exocytosis or systems where large quantities of both membranes engaged in exocytosis can be produced, enabling biochemical approaches. A particular recent advance has been methods that transition from the study of single, isolated cells to multicellular organotypic preparations that preserve the native structure of a gland or organ. These methods are now progressing to in vivo methods that enable imaging of exocytosis in a living animal.

This is an exciting area of research into a fundamental process that is essential to functions ranging from protein secretion to hormone release and neurotransmission. With techniques, such as the ones described in this book, we are making rapid progress in understanding the complex control of exocytosis. With the advent of new methods in microscopy and the development of new preparations, we can look forward to many new discoveries in this field.

St. Lucia, QLD, Australia *Peter Thorn*

Contents

Contributors

PRABHODH S. ABBINENI • *Department of Molecular Physiology, Molecular Medicine Research Group, School of Medicine, University of Western Sydney, Penrith South DC, NSW, Australia*

KYOTA AOYAGI • *Department of Biochemistry, Kyorin University School of Medicine, Tokyo, Japan*

BONG-KYU CHOI • *School of Interdisciplinary Bioscience and Bioengineering, POSTECH, Pohang, South Korea*

JENS R. COORSSEN • *Department of Molecular Physiology, Molecular Medicine Research Group, School of Medicine, University of Western Sydney, Penrith South DC, NSW, Australia*

ADOLFO CUADRA • *Department of Microbiology and Physiological Systems and Program in Neuroscience, University of Massachusetts Medical School, Worcester, MA, USA*

EDWARD CUSTER • *Department Microbiology and Physiological Systems and Program Neuroscience, University of Massachusetts Medical School, Worcester, MA, USA*

JAMES A. DANIEL • *Cell Signalling Unit, Children's Medical Research Institute, The University of Sydney, Sydney, NSW, Australia*

PAUL DIETL • *Institute of General Physiology, University of Ulm, Ulm, Germany*

JURIJ DOLENŠEK • *Faculty of Medicine, Institute of Physiology, University of Maribor, Maribor, Slovenia*

MICHAEL D. DUFFIELD • *Department of Human Physiology, Centre for Neuroscience, Flinders University, Adelaide, SA, Australia*

RORY R. DUNCAN • *Institute of Biological Chemistry, Biophysics, and Bioengineering, Heriot Watt University, Edinburgh, UK*

THOMAS HALLER • *Department of Physiology and Medical Physics, Innsbruck Medical University, Innsbruck, Austria*

DAMIEN J. KEATING • *Department of Human Physiology, Centre for Neuroscience, Flinders University, Adelaide, SA, Australia*

MURRAY C. KILLINGSWORTH • *Department of Molecular Physiology, Molecular Medicine Research Group, School of Medicine, University of Western Sydney, Penrith South DC, NSW, Australia; Electron Microscopy Laboratory, NSW Health Pathology, Liverpool BC, NSW, Australia*

JAE-YEOL KIM • *Department of Physics, POSTECH, Pohang, South Korea*

MAŠA SKELIN KLEMEN • *Faculty of Medicine, Institute of Physiology, University of Maribor, Maribor, Slovenia; Centre of Excellence for Integrated Approaches to Chemistry and Biology of Proteins, Maribor, Slovenia*

NAM KI LEE • *Department of Physics, School of Interdisciplinary Bioscience and Bioengineering, POSTECH, Pohang, South Korea*

JOSÉ R. LEMOS • *Department of Microbiology and Physiological Systems and Program Neuroscience, University of Massachusetts Medical School, Worcester, MA, USA*

CHANDRA S. MALLADI • *Department of Molecular Physiology, Molecular Medicine Research Group, School of Medicine, University of Western Sydney, Penrith South DC, NSW, Australia*
HECTOR MARRERO • *Instituto de Neurobiología, UPR-MSC, San Juan, Puerto Rico*
ANDRIUS MASEDUNSKAS • *Intracellular Membrane Trafficking Unit, Oral and Pharyngeal Cancer Branch, National Institute of Dental and Craniofacial Research, National Institutes of Health, Bethesda, MD, USA*
JAMES MCNALLY • *Department of Psychiatry, VA Boston Healthcare System, Harvard Medical School, Brockton, MA, USA*
OLEG MILBERG • *Intracellular Membrane Trafficking Unit, Oral and Pharyngeal Cancer Branch, National Institute of Dental and Craniofacial Research, National Institutes of Health, Bethesda, MD, USA; Department of Chemical and Biochemical Engineering, Rutgers University, Piscataway, NJ, USA; Department of Biomedical Engineering, Rutgers University, Piscataway, NJ, USA*
SHINYA NAGAMATSU • *Department of Biochemistry, Kyorin University School of Medicine, Tokyo, Japan*
MICA OHARA-IMAIZUMI • *Department of Biochemistry, Kyorin University School of Medicine, Tokyo, Japan*
LAURA PARENTE • *Intracellular Membrane Trafficking Unit, Oral and Pharyngeal Cancer Branch, National Institute of Dental and Craniofacial Research, National Institutes of Health, Bethesda, MD, USA*
NATALIE PORAT-SHLIOM • *Intracellular Membrane Trafficking Unit, Oral and Pharyngeal Cancer Branch, National Institute of Dental and Craniofacial Research, National Institutes of Health, Bethesda, MD, USA*
RAVINARAYAN RAGHUPATHI • *Department of Human Physiology, Centre for Neuroscience, Flinders University, Adelaide, SA, Australia*
COLIN RICKMAN • *Institute of Biological Chemistry, Biophysics, and Bioengineering, Heriot Watt University, Edinburgh, UK*
BOŠTJAN RITUPER • *Laboratory of Neuroendocrinology-Molecular Cell Physiology, Faculty of Medicine, Institute of Pathophysiology, University of Ljubljana, Ljubljana, Slovenia*
PHILLIP J. ROBINSON • *Cell Signalling Unit, Children's Medical Research Institute, The University of Sydney, Sydney, NSW, Australia*
TATIANA P. ROGASEVSKAIA • *Department of Chemical and Biological Sciences, Mount Royal University, Calgary, AB, Canada*
MARJAN SLAK RUPNIK • *Faculty of Medicine, Institute of Physiology, University of Maribor, Maribor, Slovenia; Centre of Excellence for Integrated Approaches to Chemistry and Biology of Proteins, Maribor, Slovenia*
YEON-KYUN SHIN • *Biomedical Research Institute, Korea Institute of Science and Technology, Seoul, South Korea; Department of Biochemistry, Biophysics, and Molecular Biology, Iowa State University, Ames, IA, USA*
ANDRAŽ STOŽER • *Faculty of Medicine, Institute of Physiology, University of Maribor, Maribor, Slovenia; Centre for Open Innovations and Research, University of Maribor, Maribor, Slovenia*
MUHIBULLAH TORA • *Intracellular Membrane Trafficking Unit, Oral and Pharyngeal Cancer Branch, National Institute of Dental and Craniofacial Research, National Institutes of Health, Bethesda, MD, USA*

ROBERTO WEIGERT • *Intracellular Membrane Trafficking Unit, Oral and Pharyngeal Cancer Branch, National Institute of Dental and Craniofacial Research, National Institutes of Health, Bethesda, MD, USA*

DIXON WOODBURY • *Department of Physiology and Developmental Biology, Brigham Young University, Provo, UT, USA*

ELISE P. WRIGHT • *Department of Molecular Physiology, Molecular Medicine Research Group, School of Medicine, University of Western Sydney, Penrith South DC, NSW, Australia*

ROBERT ZOREC • *Laboratory of Neuroendocrinology-Molecular Cell Physiology, Faculty of Medicine, Institute of Pathophysiology, University of Ljubljana, Ljubljana, Slovenia; Celica Biomedical Center, Ljubljana, Slovenia*

Part I

Techniques in the Measurement of Exocytosis

Chapter 1

Solution Single-Vesicle Fusion Assay by Single-Molecule Alternating-Laser Excitation

Jae-Yeol Kim, Bong-Kyu Choi, Yeon-Kyun Shin, and Nam Ki Lee

Abstract

Neurons use Ca^{2+}-triggered exocytosis of synaptic vesicles for interneuronal communication. SNARE proteins and fusion regulatory factors play key roles in the exocytosis. In vitro fusion assays using reconstituted proteoliposomes have been widely used for studying the functions of the SNARE proteins as well as those of the regulatory factors. Recently, single-vesicle assays have provided mechanisms of fusion and their dynamic features in unprecedented details normally hidden in the ensemble level studies. Here we describe a new single-vesicle assay in solution named alternating-laser excitation (ALEX). Eliminating the need for surface immobilization of vesicles, ALEX discriminates docked and fused vesicles in solution and measures their kinetics. We describe the principle, the experimental details, and the applications of ALEX to fusion studies.

Key words Single molecule, Fluorescence resonance energy transfer, Alternating-laser excitation, Fusion, Exocytosis

1 Introduction

1.1 Membrane Fusion and SNARE Proteins

Membrane fusion is used broadly for essential biological processes, such as trafficking of proteins and secretion of hormones and neurotransmitters. Membrane fusion is mediated by the SNARE (soluble *N*-ethylmaleimide-sensitive factor attachment protein receptor) proteins: vesicle (v)-SNAREs (VAMP-2 or synaptobrevin-2) locate on the vesicle membranes, while target (t)-SNAREs (the binary complex of syntaxin-1A and SNAP-25) do on plasma membranes [1]. The SNARE proteins form a tight parallel four-helix bundle when they assemble together, which brings two membranes close enough to induce membrane fusion [2–6].

During the process of SNARE complex formation, vesicles undergo several steps before fusion (Fig. 1). Synaptic vesicles primarily come close to the target plasma membrane to form *trans-SNARE* complex, which is the "docking state" [7]. At this state, the lipids of two separate membranes are not mixed together yet.

Peter Thorn (ed.), *Exocytosis Methods*, Neuromethods, vol. 83,
DOI 10.1007/978-1-62703-676-4_1,

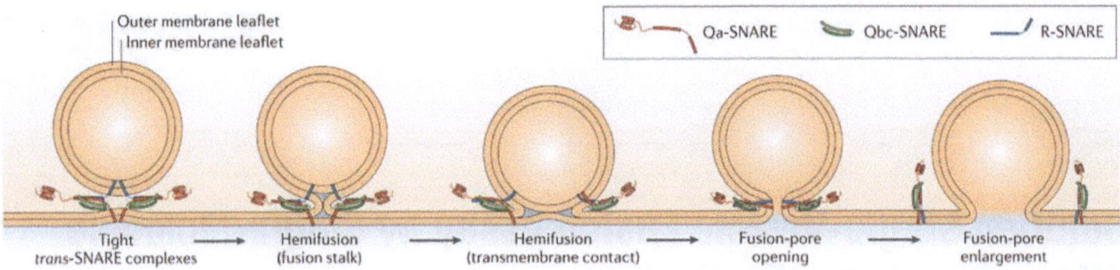

Fig. 1 Multiple steps of SNARE-mediated fusion. During the process of SNARE complex formation, vesicles undergo several steps before fusion, i.e., docking state that synaptic vesicle is tethered on the plasma membrane by forming *trans*-SNARE complex, hemifusion state that membranes in the outer leaflets are fused together but not the membranes in the inner leaflets, fusion-pore opening, and dilation of fusion pore (Reprinted, with permission, from [7])

As C-terminal domains of SNARE proteins form a complex, membranes in the outer leaflets are fused together, prior to the contact between inner leaflets, which is the "hemifusion state" [8]. As SNARE proteins form *cis*-complex, inner leaflets are fused together, which results in the opening of a fusion pore. The fusion pore is expanded further to complete the release of neurotransmitters.

1.2 In Vitro Vesicle Fusion Assays

Rothman and coworkers have demonstrated that vesicle fusion can be achieved only by SNARE proteins, without the help of any other factors, using the in vitro fusion assay of reconstituted proteoliposomes [9]. The reconstituted proteoliposome assay has provided a well-defined fusion system with purified proteins and thus widely used for investigating the functions of SNARE proteins as well as those of fusion regulatory factors, such as lipid molecules, synaptotagmin, complexin, and SM proteins [10–14]. Based on this prototypical in vitro fusion method, various fusion assays have been developed (Fig. 2). For example, a conventional fusion assay is to measure lipid mixing as the indicator of vesicle fusion by using nitro-2-1,3-benzoxadiazole-4-yl (NBD)-rhodamine dye pair (Fig. 2a) [9, 15]. In this assay, both NBD-labeled lipids and rhodamine-labeled lipids are incorporated into one type of vesicles. At this condition, strong fluorescence resonance energy transfer (FRET) from NBD to rhodamine occurs due to the short average distances and frequent collisions between two types of dyes. Thus, the fluorescent signal of NBD dyes is considerably diminished. However, when these dye-labeled vesicles are fused with vesicles containing no lipid dyes, the fluorescent intensity of NBD is increased as the average distance between NBD and rhodamine is increased by the membrane dilution effect (Fig. 2a). This lipid-mixing assay provides a basic platform for studying the functions of the SNARE proteins [9, 15].

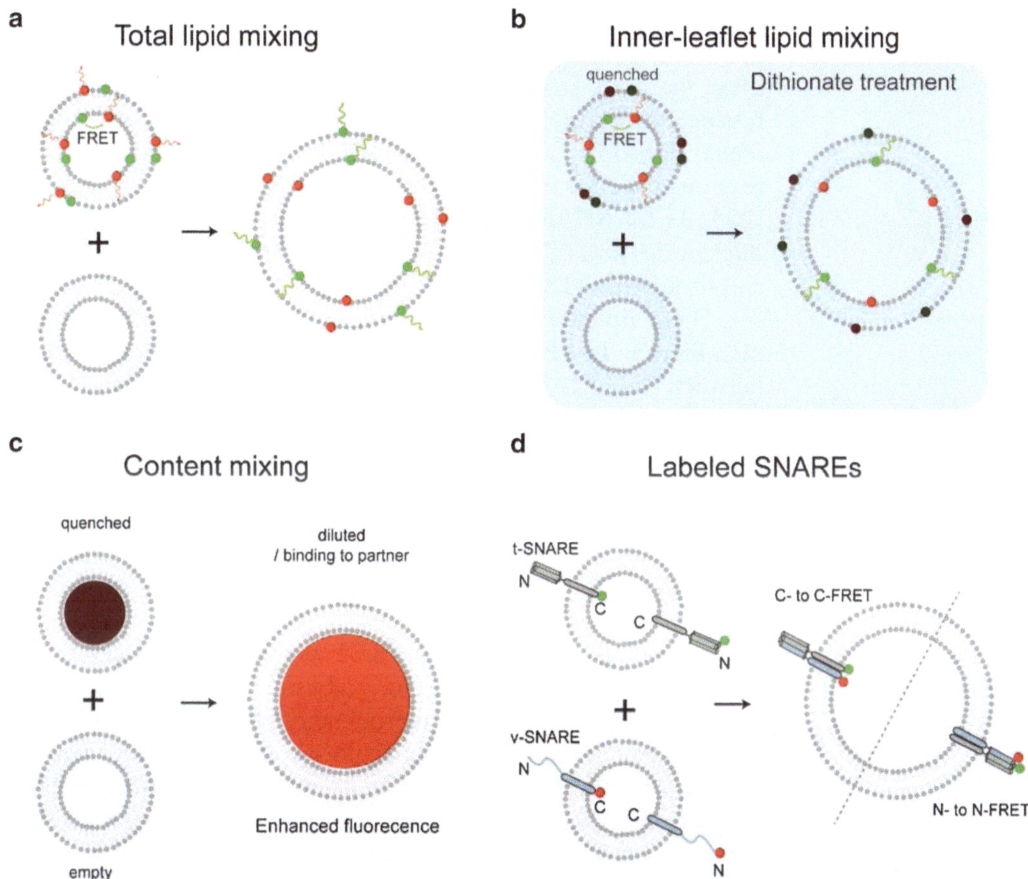

Fig. 2 In vitro fusion assays using reconstituted proteoliposomes. (**a**) Total lipid-mixing assay using fluorescently labeled lipids, such as NBD and rhodamine dye pair. (**b**) Inner-leaflet lipid-mixing assay. (**c**) Content-mixing assay by self-quenching of dyes. (**d**) Direct FRET measurement between the dye-labeled SNARE proteins

One of the weaknesses of this conventional lipid-mixing assay is that it cannot detect hemifusion state. Thus, an inner-leaflet lipid-mixing assay has been developed (Fig. 2b). In this assay, the fluorescent emissions of NBD and rhodamine in the outer-leaflet membrane are quenched by adding dithionate [16], and thus NBD and rhodamine dyes only in the inner-leaflet are fluorescent during fusion reaction. This quenching effect of dithionate has been used frequently for detecting hemifusion state (Fig. 2b) [17].

Compared with the lipid mixing, a more direct indicator of vesicle fusion is observing content mixing (Fig. 2c) [18–22]. Various methods have been developed for quantifying the content mixing, such as the self-quenching of calcein (Fig. 2c) [18, 19], fluorescent enhancement of Tb (terbium) by DPA (dipicolinic acid) [20], quenching of ANTS (1-aminonaphthalene-3,6,8-trisulfonic acid) fluorescence by DPX (*p*-xylene-α,α′-bispyridinium dibromide) [21], and radiolabeled oligonucleotide [22]. Although Tb/DPA and ANTS/DPX assays have been developed early [23], these methods are not suitable for SNARE-mediated membrane fusion. For example, ANTS/DPX is not suitable for small unilamellar liposome [24], and Tb/DPA is possibly interfered by the reagents used for the purification of SNARE-embedded liposomes [22]. As for SNARE-mediated vesicle fusion, radiolabeled oligonucleotides and calcein assays are used for proving content mixing and its kinetics [18, 19, 22].

These lipid- and content-mixing assays reveal the status of membranes. To observe directly the complex formation processes between SNARE proteins, FRET between labeled SNARE proteins has been used (Fig. 2d) [8, 25]. Typically FRET between N-terminally labeled SNAREs (N-N FRET) reports initial SNARE complex formation, such as vesicle docking, while FRET between C-terminally labeled SNAREs (C-C FRET) does a fully zippered state. These FRET assays are useful to obtain the relation between SNARE assembly and vesicle fusion processes [8].

1.3 Single-Molecule FRET

Since Ha et al. have observed FRET at the single-molecule level; single-molecule FRET has been established as a valuable tool for studying the dynamics and heterogeneity of biomolecules, such as the movements of protein machinery, protein conformational change, protein folding, and protein–protein/protein–DNA interactions [26–29]. The energy transfer efficiency ($\boldsymbol{E}$) is inversely proportional to the sixth power of the distance between two probe dyes, $\boldsymbol{E}=(1+(R/R_0)^6)^{-1}$, where R is the distance between donor and acceptor dyes and R_0 is the Förster radius (Fig. 3a). Importantly, R_0 is typically 5–7 nm that FRET can measure the distance in 2–10 nm, which is the optimal distance range for investigating biomolecules' interactions and movements [30].

Two methods have been most frequently used for single-molecule FRET measurements. One is to immobilize molecules on the surface and then observe FRET by total internal reflection microscopy (TIR) (Fig. 3b). This method is powerful to observe the real-time dynamics of individual molecules (Fig. 3c). However, the limitations of TIR are that molecules have to be immobilized on the surface, which may have a surface effect, and its time resolution cannot be better than tens of milliseconds due to using CCD camera for fluorescence detection (Fig. 3c). Another format of single-molecule technique is to detect single molecule freely diffusing in solution, called diffusion-based single-molecule FRET [31]. The in-and-out events of individual molecule at the detection volume are detected as fluorescent bursts (Fig. 3d, e). This method is useful for analyzing the subpopulations of heterogeneous mixture and the dynamics of subpopulation change with a high-throughput caliber. Because the fluorescent burst lasts only a

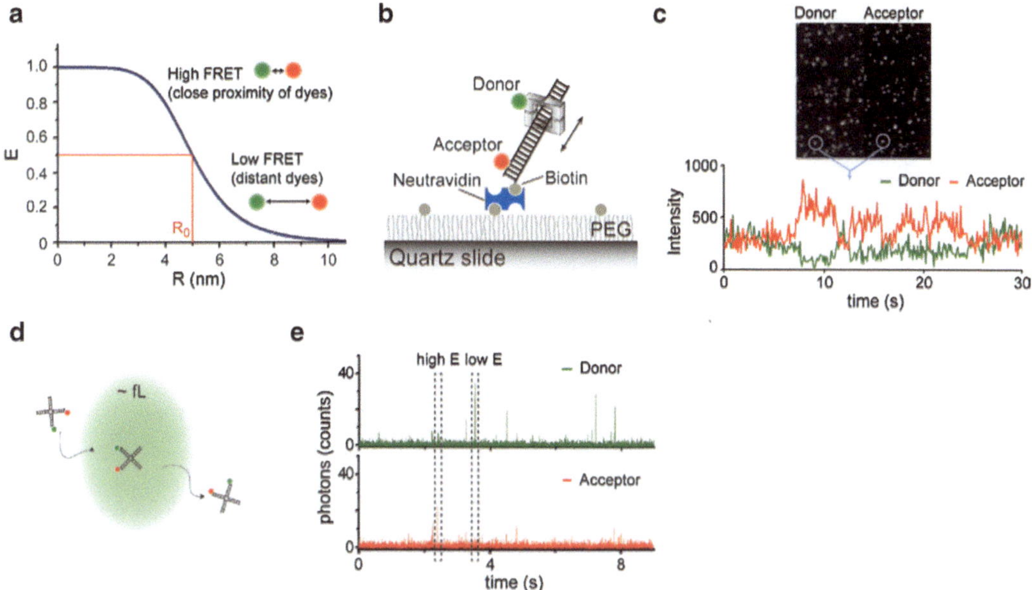

Fig. 3 Two platforms for single-molecule FRET. (**a**) The FRET efficiency (**E**) is inversely proportional to the sixth power of the distance between two probe dyes. (**b**) Immobilization of single molecules on the surface using biotin–NeutrAvidin interactions. (**c**) The fluorescence signals of immobilized molecules are detected by CCD camera. (**d**) Excitation geometry of diffusion-based single-molecule FRET. Excitation laser is tightly focused to have a small confocal excitation volume (-fL). The fluorescently labeled molecule passes through the excitation volume and emits fluorescence during its well time. (**e**) Typical photon stream obtained by diffusion-based single-molecule FRET. *Green* and *red lines* denote donor and acceptor fluorescent emissions, respectively. Fluorescent bursts are originated from the in-and-out events of molecules into the excitation volume

few milliseconds, this method cannot be used for the dynamics (>ms) of single molecule, but can detect fast dynamics shorter than millisecond within a burst (Fig. 3e) [32, 33].

1.4 Single-Vesicle Fusion Assay by TIR

In vitro fusion assays using reconstituted proteoliposomes have successfully recapitulated SNARE-mediated membrane fusion. However, this assay observing many fusion events at a time cannot detect fusion intermediates, such as docked and hemifused vesicles. Single-molecule methods are well suited for observing such fusion intermediates. In the case of a single-vesicle assay based on supported planar membrane (Fig. 4a), fluorescent signal appears abruptly upon the binding of fluorescently labeled vesicle to supported membrane incorporated with SNARE proteins [34–38]. Then the fluorescent intensity decreases gradually as fusion progresses. This method has revealed a minimal number of SNARE complexes for fast vesicle fusion [34, 36] and the productive hemifusion intermediates [37, 38]. Another format of single-vesicle assay is to immobilize liposomes on the surface instead of using supported membrane (Fig. 4b) [39–41]. In this case, vesicles

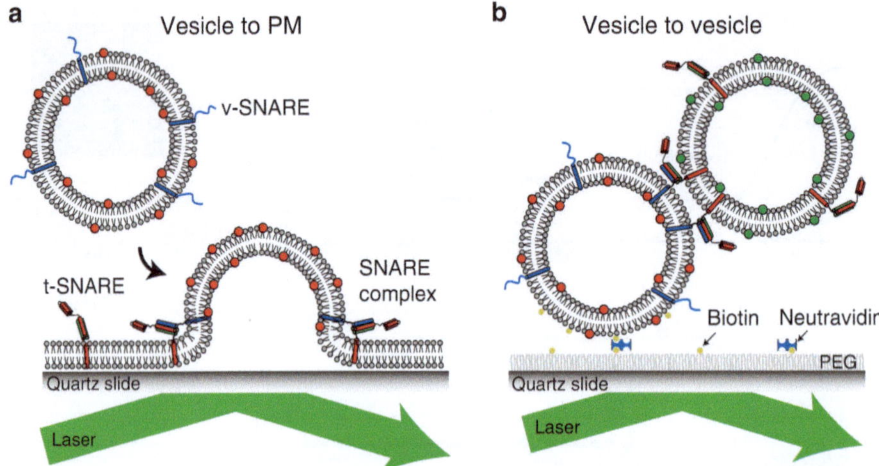

Fig. 4 Single-vesicle fusion assays with surface immobilization. Experimental schemes of single-vesicle fusion assays by the fusion (**a**) between vesicles flow in and support planner membrane (Reprinted, with permission, from [34]) and (**b**) between vesicles immobilized on the surface with vesicles flowed in (Modified from [39])

doped with acceptor dyes are immobilized on the surface by biotinylated-lipid/NeutrAvidin interactions. Vesicles doped with donor dyes afterwards are flowed in, and the fusion progress is monitored by the increment of the FRET signal. This method has revealed the functions of fusion regulatory proteins, such as complexin [41] and synaptotagmin-1 [40].

1.5 Alternating-Laser Excitation FRET for Single-Vesicle Fusion Assay

The alternating-laser excitation (ALEX) technique has been developed for discriminating between low-FRET species and donor-only species since both species have a low-FRET signal and thus are not distinguishable in conventional diffusion-based single-molecule FRET (Fig. 5a) [42]. In addition, acceptor-only species are not detected by the single-laser excitation format (Fig. 5a, c), and thus the conventional diffusion-based single-molecule method is not suitable for studying the bimolecular reactions of protein–protein and protein–DNA interactions. These limitations have been overcome by adding a second laser for acceptor excitation with periodic alternation (Fig. 6) [42]. Using the second laser in ALEX, the S (stoichiometry) parameter is obtained together with the E (FRET efficiency) value of the conventional method (Fig. 5d). This S parameter discriminates low-FRET species from donor-only species in a two-dimensional E–S graph. Furthermore, acceptor-only species are also detected by the second laser, which provides an excellent format for studying bimolecular interactions and detecting all fluorescent species in heterogeneous solution [43]. So far ALEX has been used for single-molecule studies, for example, abortive RNA polymerase initiation [44], the retention of sigma subunit [45], protein conformational change [46],

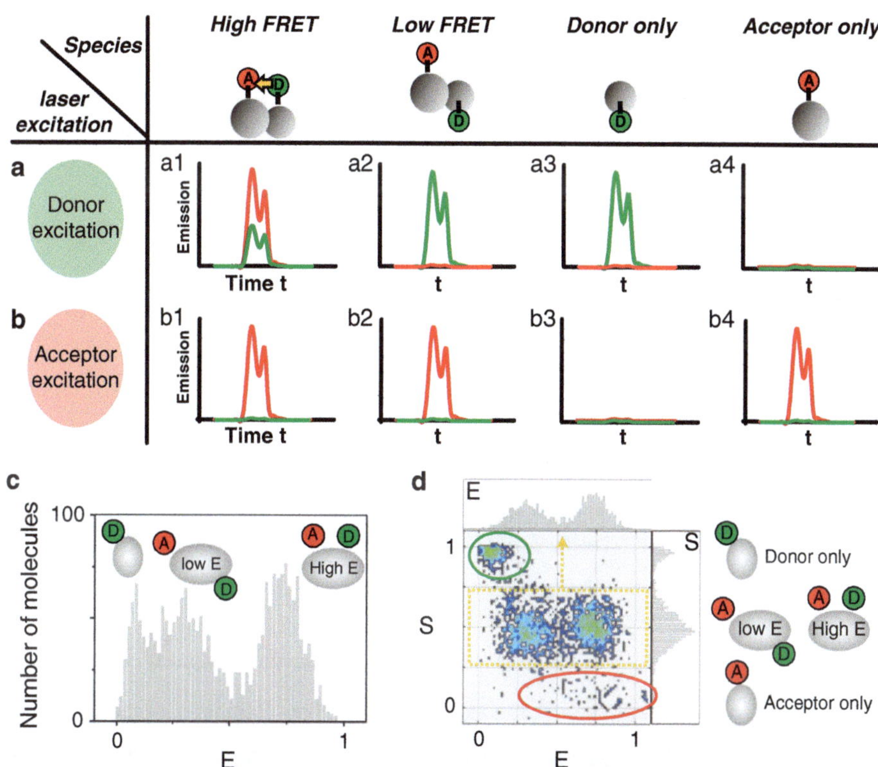

Fig. 5 Comparison between single-laser excitation and alternating-laser excitation (ALEX). (**a**) Schematic descriptions of fluorescent bursts of single-pair FRET species by donor-excitation laser. (*a1*) single-pair FRET species with high FRET, (*a2*) single-pair FRET species with low FRET, (*a3*) donor-only species, and (*a4*) acceptor-only species. (**b**) Schematic descriptions of fluorescent bursts of single-pair FRET species by acceptor-excitation laser. (*b1*) single-pair FRET species with high FRET, (*b2*) single-pair FRET species with low FRET, (*b3*) donor-only species, and (*b4*) acceptor-only species. (**c**) Typical FRET histogram obtained by single donor laser excitation. (**d**) Two-dimensional E–S graph obtained by ALEX (Modified from [42] for (**a**) and (**b**))

accurate FRET measurement [47], multicolor FRET [48], and DNA folding [49]. Recently, the capability of ALEX measuring the stoichiometry has been applied to various single-molecule florescence methods, such as switchable FRET [50] and quenchable FRET [51, 52].

Recently, we have expanded the application of ALEX to vesicle fusion assay [53]. ALEX has the capability of sorting out the subpopulations of fusion products and fusion intermediates in solution [53], which provides an excellent method for studying vesicle docking and their kinetics. In comparison with the surface-immobilization methods, ALEX detects a single vesicle in solution without the need for surface immobilization and directly provides full subpopulations of fusion products. Here we describe the principle of ALEX for vesicle fusion, sample preparation, and experimental details and a perspective for future applications of ALEX on fusion.

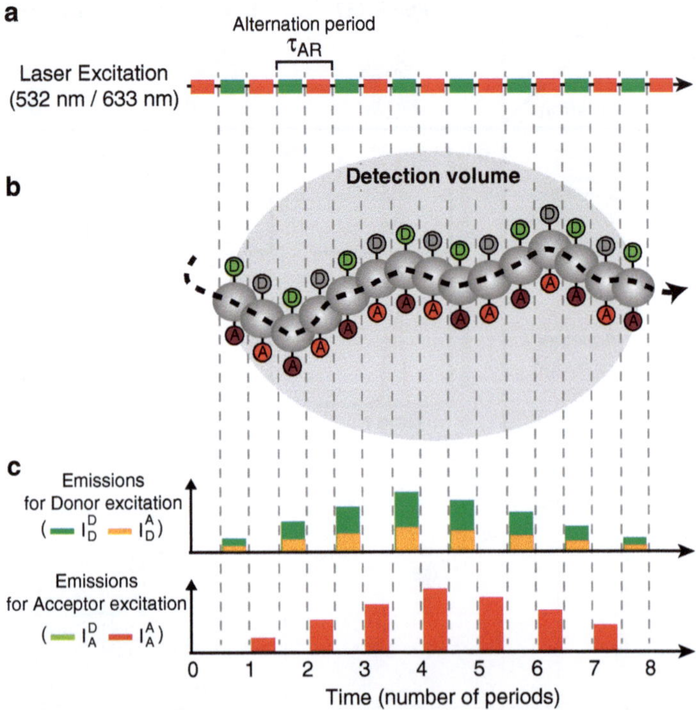

Fig. 6 Data acquisition scheme of alternating-laser excitation. (**a**) Alternating-laser excitation is achieved by modulators with alternation period, τ_{AR}. Typically τ_{AR} is 100 μs for biomolecules, such as protein and DNA. (**b**) Schematic description of molecule's passage through the excitation volume. While the molecule passes through the excitation volume, it is excited alternatively by the two lasers. (**c**) Fluorescent signal registration. According to the source of the excitation laser and fluorescent probes, the fluorescent emissions are divided into four types of signals. These four emissions are used for calculating ***E*** and ***S*** of individual molecules (Modified from [65])

2 Materials and Experimental Setup

2.1 Principle of ALEX

In order to distinguish low-FRET species from donor-only species and to detect acceptor-only species, a second laser for acceptor excitation has been introduced (Fig. 6a) [42]. However, if a continuous second laser is introduced, the FRET signal is not distinguishable from the acceptor emission by acceptor direct excitation [42]. To solve this problem, two lasers are turned on and off alternatively with an alternation period, τ_{Alt}, by modulators (Fig. 6a). In a diffusion-based single-molecule assay, lasers are focused into a small volume, close to 1 fL (Fig. 3d), and then the in-and-out event of a molecule into this excitation volume is detected as a fluorescent burst (Fig. 3e). In ALEX, two lasers are alternated at least five times faster than the average transit time (approximately 1 ms), i.e., the dwell time of a molecule in the excitation volume (Fig. 6b, c). Using these alternating excitations, the emissions from donor and acceptor are collected and divided according to the different excitation laser sources (Fig. 6c). As a result, each single burst contains three types of fluorescent intensities which are $I_D{}^D$, $I_D{}^A$, and $I_A{}^A$ (note that $I_A{}^D$ is a background signal, not used for data analysis), where $I_X{}^Y$ denotes the fluorescent intensity from Y molecule when X molecule-excitation laser is applied. From these three intensities, two parameters, $\boldsymbol{S}$ (a sorting number) and E, are calculated [42]:

$$\boldsymbol{E} = I_D^A / \left(I_D^D + I_D^A \right),$$

$$\boldsymbol{S} = \left(I_D^D + I_D^A \right) / \left(I_D^D + I_D^A + I_A^A \right).$$

In the case of donor-only species, $I_A{}^A = 0$ and then $\boldsymbol{S} = 1$. In the same manner, $I_D{}^D + I_D{}^A$ is close to zero for acceptor-only species so that $\boldsymbol{S} = 0$. When a molecule contains both donor and acceptor, $\boldsymbol{S}$ becomes 0.5. Here E reports the distance between donor and acceptor dyes. As a result, a two-dimensional ***E–S*** graph is obtained (Fig. 5d).

2.2 Instruments for ALEX

The instrumental setup of ALEX has been well described before (Fig. 7) [53]. Two lasers, 532-nm solid-state green laser and 633-nm HeNe laser, are alternated by modulators, such as electro-optical modulator or acousto-optical modulator, before they are coupled by a dichroic mirror (DM1). The coupled light is spatially filtered by a single-mode optical fiber. Typically 100 μs of the alternation period is used for diffusing proteins or DNA molecules since their transit time is about 1 ms. As for the vesicle assay, a 400 μs alternation period is used because the diffusion transit time of vesicle is close to 20 ms. The collimated lasers from optical fiber are entered into the microscope and then focused at 30 μm from the surface of a coverslip by a water-immersion objective (OBJ) after they are reflected by a dichroic mirror (DM2). For vesicles doped with 2 % of DiI (donor) or DiD (acceptor), 100 nW and 50 nW of 532- and 633-nm lasers in alternating mode are used, respectively. The fluorescent emissions collected by the objective pass through a 100 μm pinhole (PH) and then focused again onto silicon avalanche photodiode detectors (APD). The emissions of DiI and DiD are split by a beam splitter (DM3) and then filtered by the band-pass filters (F1 and F2) in front of the APDs. The photons detected by APDs are analyzed by home-built LabVIEW software (National Instruments), which results in the photon time traces (Fig. 3e) and the two-dimensional E–S graph (Fig. 5d).

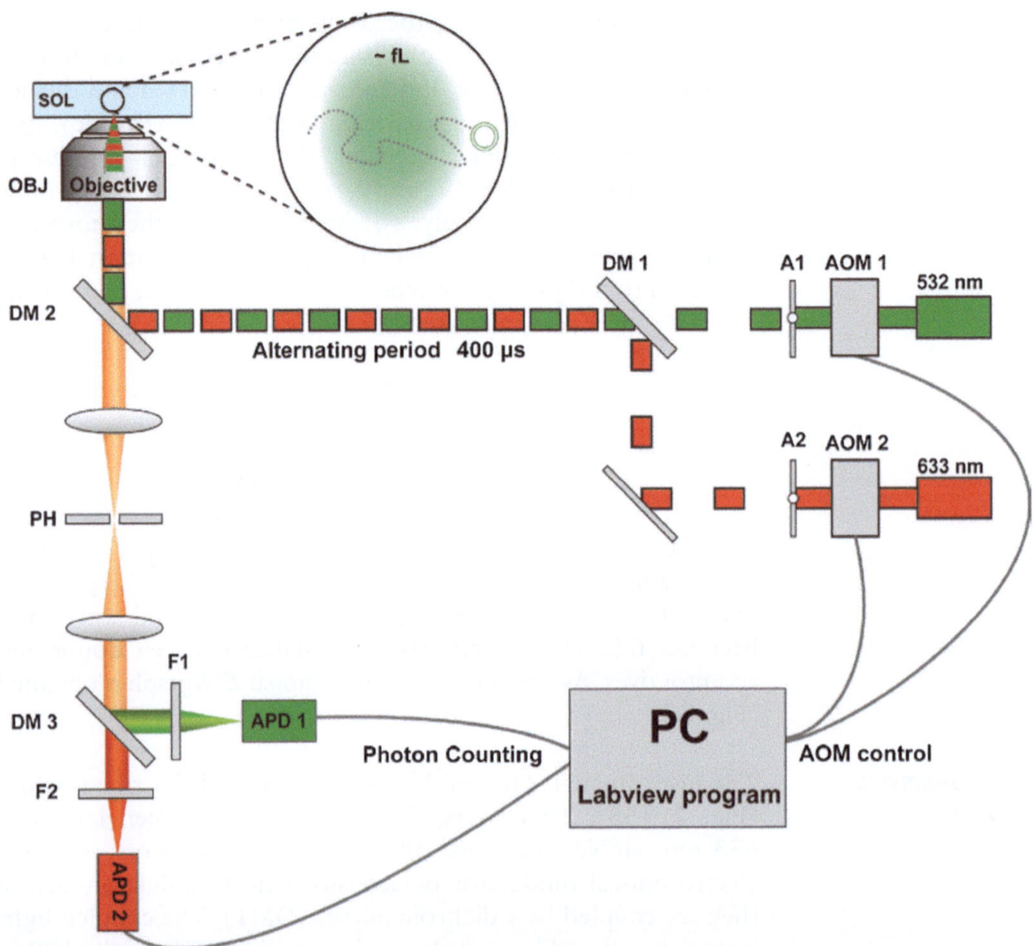

Fig. 7 Experimental setup of ALEX. Two lasers of the 532- and 633-nm excitation wavelengths are alternated by acousto-optical modulators. Since the diffusion transit time of vesicle is approximately to 20 ms, 400 μs alternation period is used. For excitation of vesicles doped with 2 % of DiI (donor) or DiD (acceptor), 100 and 50 nW of 532-nm and 633-nm lasers are used, respectively. Two silicon avalanche photodiode detectors (APDs) detect the emissions of DiI and DiD separately, and the photons detected by APDs are analyzed using home-built LabVIEW software

2.3 Sample Chamber Preparation for ALEX

Samples are put in the sample chamber made by two sandwiched coverslips with silicon gasket. Coverslips are washed with acetone, methanol, and distilled water sequentially and then sonicated with 1 M KOH for 15 min.

2.4 SNARE Proteins and Lipids

SNARE Proteins. The genes of recombinant SNARE proteins syntaxin-1A full-length (1–288, three cysteines are replaced with alanines), SNAP-25 (1–206, four cysteines are replaced with alanines), and VAMP-2 (or synaptobrevin-2, 1–116, one cysteine is replaced with alanine) are cloned into a pGEX-KG vector with GST-tag (glutathione *S*-transferase), respectively. The proteins are expressed by *E. coli* BL21 Rosetta (DE3) pLysS (Novagen). The procedures of the purification are well described elsewhere [25, 54]. His-tagged full-length synaptotagmin-1 (Sytl) is expressed by pET-28a vector and purified by Ni-NTA column.

Lipids. Lipid components, 1-palmitoyl-2-oleoyl-sn-glycero-3-phosphocholine (POPC), 1,2-dioleoyl-sn-glycero-3-[phospho-L-serine] (DOPS), cholesterol (Chol), and 1,2-dioleoyl -sn-glycero-3-phospho-(1′-myo-inositol-4′,5′-bisphosphate) (PIP_2), are purchased from Avanti Polar Lipids. Fluorescently labeled lipids, 1,1′-dioctadecyl-3,3,3′,3′-tetramethylindocarbocyanine perchlorate (DiI, donor dye) and 1,1′-dioctadecyl-3,3,3′,3′-tetramethylindodicarbocyanine perchlorate (DiD, acceptor dye), are purchased from Invitrogen.

3 Methods

3.1 Preparation of SNARE-Embedded Liposomes

The mixtures of lipids with the ratios of POPC:DOPS:Chol:PIP_2:DiI = 62:15:20:1:2 for t-vesicles and POPC:Chol:DiD = 78:20:2 for v-vesicles are prepared, respectively. For no PIP_2 t-vesicle, 1 % PIP_2 is replaced with 1 % POPC. The mixtures are dried and incubated in vacuum for 4 h and then dissolved with vesicle buffer (25 mM pH 7.4 HEPES and 100 mM KCl). To ensure the formation of unilamellar liposomes, the freezing and thawing of the liposome solution is performed ten times using liquid nitrogen. The liposome solution is extruded by a mini extruder (Avanti Polar Lipids) with 100 nm polycarbonate filter (Whatman), which generates monodispersed vesicles with a 100 nm diameter.

For the reconstitution of t-vesicles, the binary t-SNARE, which is formed by incubating syntaxin-1A and SNAP-25 with 1:2 molar ratio at room temperature for 1 h, is mixed with the extruded vesicles doped with DiI and 0.7 % (w/v) octyl-beta-d-glucoside (OG). As for v-vesicles, VAMP-2 and Sytl are mixed with the vesicles doped with DiD and 0.8 % (w/v) OG. Lipid-to-protein molar ratio (L/P) is 500:1 for both t-vesicle (**T**) and v-vesicle (**V**), reconstituted with t-SNARE and VAMP-2, respectively, and 600:1 for v-vesicle with Sytl (**S**). In the case of **SV** containing both VAMP-2 and Sytl, L/P is 500:1 for VAMP-2 and 900:1 for Sytl. Protein–lipid–detergent mixtures are nutated at 4 °C for 10 min, diluted by twofold, and then dialyzed at 4 °C overnight. After dialysis, centrifugation is applied to remove aggregates at 10,000 × *g* for 10 min at 4 °C.

3.2 SNARE-Mediated Membrane Fusion Analyzed by ALEX

Figure 8a–d shows typical time traces of single vesicles in a solution. Unreacted **T**, incorporated with t-SNARE, shows high donor emissions by the donor excitation (Fig. 8a, green line) without any

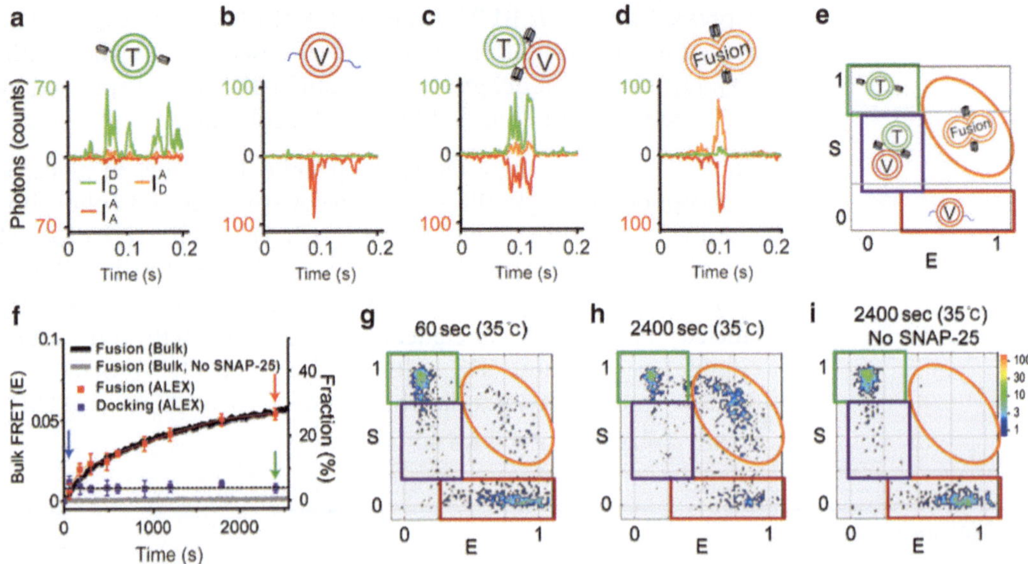

Fig. 8 SNARE-mediated membrane fusion analyzed by ALEX. (**a–d**) Typical time traces of single vesicles in solution obtained by ALEX. I_D^D is the fluorescent emission of donor dyes (DiI) excited by the donor-excitation laser (*green line*); I_D^A is the fluorescent emission of acceptor dyes (DiD) excited by the donor-excitation laser, which is FRET signal (*orange line*); and I_A^A is the fluorescent emission of acceptor dyes excited by the acceptor-excitation laser (*red line*). (**a**) Time trace of unreacted **T**, reconstituted with t-SNARE and doped with DiI. (**b**) Time trace of **V**, reconstituted with VAMP-2 and doped with DiD. Time traces of (**c**) the docked vesicles and (**d**) fused vesicles. (**e**) Two-dimensional E (FRET efficiency) and S (sorting number) graph. The unreacted, docked, and fused vesicles have characteristic values of S and E and thus occupy different areas in the E–S graph. (**f**) Comparison of the fusion kinetics measured by the bulk FRET assay (*black line*) and ALEX (*red square*). In the case of ALEX, the mixture is incubated at 35 °C and diluted to be 3 μM in a lipid concentration at the selected time points, and then ALEX measurement is performed for 5 min. The fractions of the docked and fused vesicles (*purple box* and *orange oblique*, respectively, in **g**, **h**) are obtained. From the single-exponential fitting, the fusion rates are obtained: $k_{fusion,\ bulk} = 1.03 \pm 0.02 \times 10^{-3}\ s^{-1}$ and $k_{fusion,\ ALEX} = 1.19 \pm 0.08 \times 10^{-3}\ s^{-1}$. (**g–i**) The E–S graphs of **T–V** mixture after (**g**) 1 min incubation and (**h**) 40 min incubation. (**i**) No SNAP-25 control after 40 min incubation. Color scale bar indicates the number of vesicles (Reprinted after modification, with permission, from [53])

emission of acceptor by FRET or the direct excitation. On the other hand, unreacted **V**, incorporated with VAMP-2, presents high acceptor emissions by the acceptor direct excitation (Fig. 8b, red line). Docked and fused vesicles show emissions by both donor and acceptor excitations but present different emission patterns. For example, docked vesicles show both high donor emission by donor excitation and acceptor emission by acceptor excitation, but no FRET signal (Fig. 8c, orange line). However, fused vesicles (lipid mixing) present low donor emission (green line) but high FRET signal (Fig. 8d, orange line). As a result, all fusion products and intermediate species of vesicles have different fluorescent emission patterns, thus locating at the different regions in the ***E–S*** graph (Fig. 8e). From the ***E–S*** graph, the populations of each vesicle species are calculated quantitatively.

The fusion kinetics measured by ALEX can be directly compared with the bulk FRET measurement (Fig. 8f). The bulk fusion assay measures the increase of FRET signal continuously after mixing **T** and **V** vesicles (Fig. 8f, black line). The increase of FRET signal indicates the fusion process, but it does not inform the subpopulations of the docked and fused vesicles. For comparison with the bulk assay, ALEX measurements are performed at the selected time points (1, 3, 5, 7, 10, 15, 20, 30, 40 min) after incubating two types of vesicles. In order to ensure the detection of vesicles at the single-vesicle level, the reaction mixture has to be diluted to be a few μM in lipid concentration. After incubating **T–V** mixture for 1 min at 35 °C, the subpopulations of the docked and fused vesicles (purple box and orange oblique, respectively) are rare (Fig. 8g). However, the subpopulation of fused vesicles is considerably increased after 40 min incubation, while the subpopulations of unreacted **T** and **V** vesicles are decreased (Fig. 8h). It is to be noted that the changes of vesicles' subpopulations are captured well by ALEX. When fusion is blocked by removing SNAP-25, almost no fusion is detected (Fig. 8i). The ratio of fused vesicles [(the number of fused vesicles)/(the number of fused vesicles + the number of unreacted **V**s)] obtained from the ***E–S*** graph (Fig. 8g, h) reports the degree of fusion. The fusion kinetics obtained from this ratio is consistent with the bulk FRET assay as shown in Fig. 8h.

3.3 Vesicle Docking and Membrane Protein Interaction Measured by ALEX

Observing vesicle–vesicle interactions, such as vesicle docking or tethering, by protein–protein and protein–lipid interactions is crucial to understanding fusion mechanisms as well as membrane protein interactions. In fluorescence-based bulk fusion assay, vesicle docking is not detectable because this event does not accompany lipid mixing that brings the change in a fluorescent signal. The size-dependent methods, such as dynamic light scattering [55] or turbidity [56], can be used for detecting vesicle aggregation or docking in solution. However, these methods require relatively large size changes as well as an overall population change by the reaction. Above all, these methods are not able to discriminate the aggregated/docked vesicles from fused vesicles. As a result, it is not possible to monitor vesicle-docking and fusion events simultaneously in solution without surface immobilization. Compared with these techniques, ALEX detects both docked vesicles and fused vesicles simultaneously in solution.

The detection ability of docked vesicles makes ALEX an ideal tool for studying vesicle-docking mechanisms. For example, ALEX has been successfully used for investigating the docking by synaptotagmin-1 (Syt1) [53]. Syt1 is a well-known major Ca^{2+} sensor for neuroexocytosis [57]. Besides its Ca^{2+}-sensing role, Syt1 has been proposed to function as a docking factor by several morphological studies using electron microscopy [58–60], and its binding partner has been suggested to be the binary t-SNARE [59].

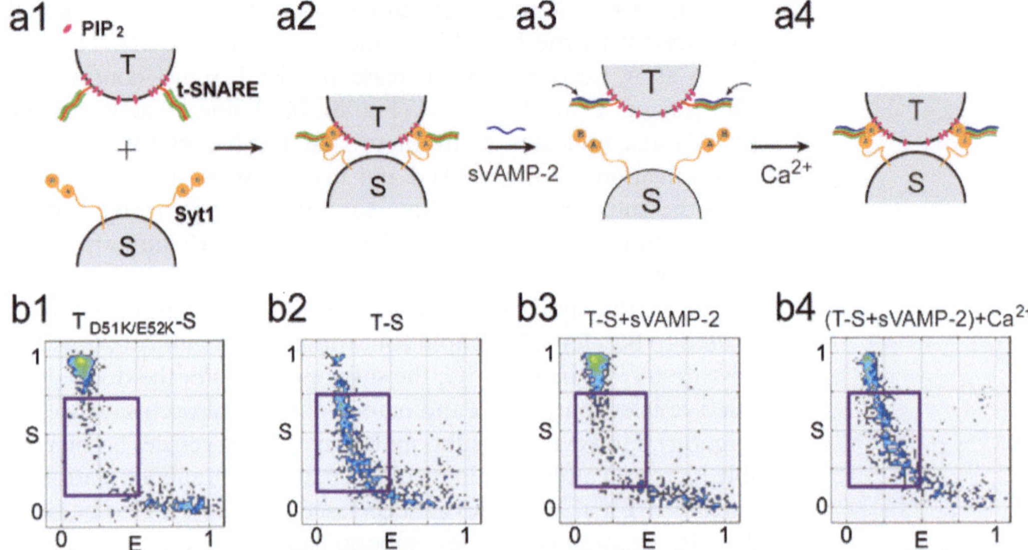

Fig. 9 Vesicle docking and protein-vesicle interaction measured by ALEX. (**a**) Schematic descriptions of the vesicle docking, dissociation, and rebinding by Syt1-SNARE interactions. **S**, v-vesicle with only Syt1. **T**, t-vesicle with t-SNARE and 1 % PIP_2. (*b1*) The mixture of **S** and $\mathbf{T}_{mutant}$ (t-vesicle incorporated with t-SNARE formed with SNAP-25 D51K/E52K mutant). (*b2*) The mixture of **T** and **S** results in vesicle docking (*purple box*). (*b3*) Soluble VAMP-2 is added to the mixture of (*b2*). (*b4*) Ca^{2+} is added to the mixture (*b3*) (Reprinted after modify, with permission, from [53])

On the other hand, an in vitro study using proteoliposomes suggests that Syt1 does not require its specific interaction with t-SNARE for tethering proteoliposomes in the absence of Ca^{2+} [19]. These controversial issues can be resolved by ALEX using reconstituted proteoliposomes.

A simple and direct method to prove whether Syt1 acts as a docking factor is to put only Syt1 on v-vesicles (**S**) (Fig. 9a1). When **S** is incubated with t-SNARE-containing vesicles (**T**) (Fig. 9a2), a considerable amount of docked vesicles is observed by ALEX (purple box, Fig. 9b2). However, when mutants, such as the Syt1 mutant which is unable to interact with t-SNARE and the SNAP-25 mutant that cannot bind to Syt1, are used, no docking is observed (Fig. 9b1). These assays clearly prove that Syt1 acts as a vesicle-docking factor and its binding partner is t-SNARE in vitro.

Since ALEX measures docking in solution, subsequent reactions after docking can be monitored. For example, the reaction of VAMP-2 to t-SNARE, after the Syt1-mediated docking, is monitored by adding soluble VAMP-2 to the **T–S** mixture (Fig. 9a3). The result shows that the docked population is dramatically reduced (Fig. 9b3), which implies that soluble VAMP-2 replaces Syt1 from t-SNARE to form ternary SNARE complex (Fig. 9a3).

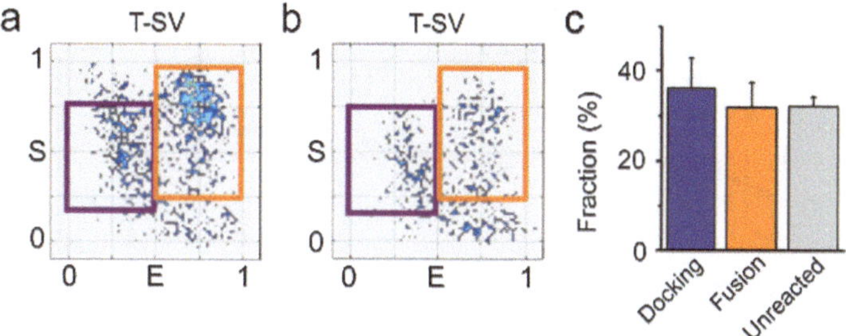

Fig. 10 Quantitative analysis of the subpopulations of docked and fused vesicles. (**a**, **b**) The ***E–S*** graphs of **T–SV** mixture obtained by (**a**) all photon burst searching method and by (**b**) acceptor photon burst searching method. The fractions of docked and fused vesicles are calculated by counting the number of bursts in the graphically selected areas in the ***E–S*** graph: $0.15<S<0.75$ and $0<E<0.5$ for docked vesicles (*purple box*), $0.25<S<1$ and $0.5<E<1$ for fused vesicles (*orange box*), and the rest for unreacted v-vesicles. (**c**) The docking and fusion populations obtained from (**b**) (Reprinted after modify, with permission, from [53])

When 10 μM Ca^{2+} is added to the **T–S** mixture with soluble VAMP-2 (Fig. 9a4), the docked population has reappeared in the ***E–S*** graph (Fig. 9b4). This shows that Ca^{2+} enhances the rebinding of Syt1, which is displaced by soluble VAMP-2, to the SNARE complex. This demonstrates that ALEX is useful for observing the vesicle–vesicle interactions, which could be applied further for other membrane proteins' interactions.

3.4 Quantitative Analysis of the Subpopulations of Docked and Fused Vesicles

Another strong advantage of ALEX is that the subpopulations of docked and fused vesicles are obtained in a quantitative manner. Typically the sum of all photon counts, $I_D^D + I_D^A + I_A^A$, is used for selecting bursts in diffusion-based single-molecule FRET because this method searches all fluorescent species in solution (Fig. 10a) [47]. However, since docked and fused vesicles contain both donor and acceptor dyes, these reacted vesicles are selected more efficiently than unreacted donor- and acceptor-only vesicles. Thus, for a quantitative analysis of subpopulations, the unbiased photon intensity, i.e., the acceptor emission by acceptor direct excitation (I_A^A), is used for selecting bursts (Fig. 10b) [44]. This acceptor burst search provides a quantitative measurement of subpopulations because unreacted v-vesicle, docked, and fused vesicles contain similar amounts of acceptor dyes. After searching bursts, the subpopulations of the bursts are analyzed by selecting molecules graphically using the ***E–S*** graph: $0.15 < S < 0.75$ and $0.0 < E < 0.5$ for docked vesicles (purple box), $0.25 < S < 1.0$ and $0.5 < E < 1.0$ for fused vesicles (orange box), and the rest for unreacted v-vesicles (Fig. 10c). This quantitative measurement ensures the measurement of docking and fusion kinetics as shown below.

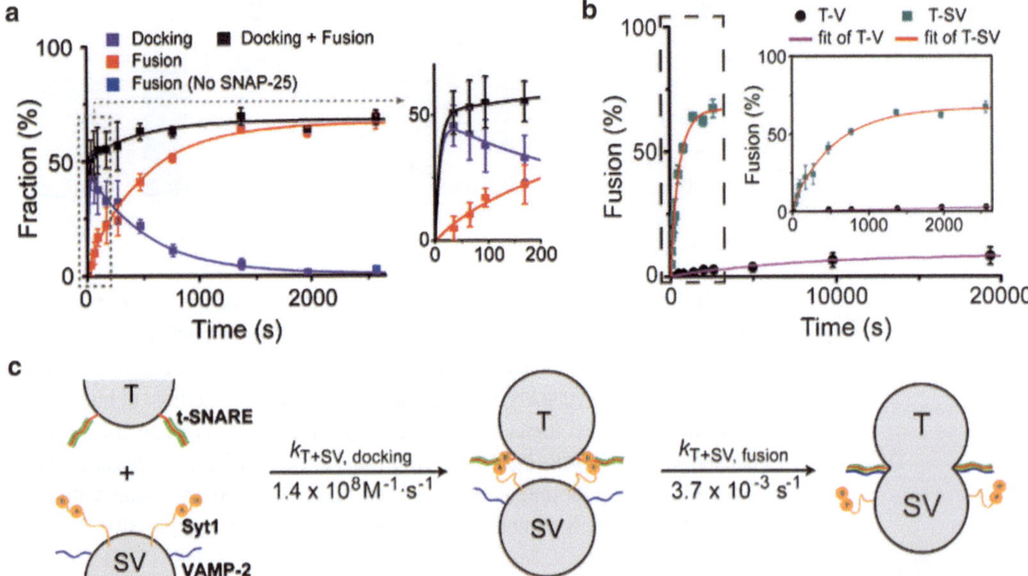

Fig. 11 Docking and fusion kinetics measurement by ALEX. (**a**) Kinetics of docked and fused vesicles of **T–SV** mixture. *Purple, red*, and *black squares* denote the fractions of docked vesicles, fused vesicles, and the sum of the docked and fused vesicles, respectively. (**b**) Comparison of the fusion kinetics of **T–SV** and **T–V** mixtures measured by ALEX. The data for **T–SV** fusion in (**b**) is the same as in (**a**) (*dark cyan square*). (**c**) Fusion kinetics obtained from (**a**) (Reprinted after modify, with permission, from [53])

3.5 Docking and Fusion Kinetics in Solution

Since ALEX records the detection time of each detected vesicle along with its fluorescent signal, the fusion kinetics of docked and fused vesicles can be measured in real time. Typically two vesicles are mixed to be 3 μM in a final lipid concentration for single-vesicle detection, and then ALEX measurement starts as soon as possible. Although currently it takes at least 20 s for starting a measurement after mixing, a microfluidic device will reduce the delay less than a second. After data acquisition, the time-trace data are divided into five time bins (20–50, 50–80, 80–120, 120–220, 220–320 s). For each time bin, the fractions of docked and fused vesicles are obtained. For longer time points than 5 min, the mixtures are preincubated in appropriate times, and then the docked and fused populations are measured.

From the ***E–S*** graph, the time-dependent variations of the three subpopulations, i.e., docked vesicles (purple square, Fig. 10b), fused vesicles (orange square, Fig. 10b), and unreacted v-vesicles, are obtained. Figure 11a presents the population change of the docked (purple line) and fused vesicles (red line) together with their sum (black line). The docking rate is obtained from the figure, since the increase of the sum (docked vesicles + fused vesicles) originates from vesicle docking. For the calculation of docking rate, (1) can be used:

$$\frac{d[D+F]}{dt} = k_{T+SV}^{docking}[\mathbf{T}][\mathbf{SV}] \quad (1)$$

where [D + F] denotes the concentration of the complex (docked vesicles + fused vesicles), k_{T+SV} is the docking rate, and [**T**] and [**SV**] are the concentrations of t-vesicle and v-vesicle, respectively. In Fig. 11a, the docking process is too fast that the docking population is saturated at 35 s, the first data point of the ALEX measurement. This fast docking process can be captured by combining ALEX with a microfluidic device in the near future. Thus, instead of getting an exact docking rate, the lower limit of a docking rate can be obtained from the initial increase of a complex using (1). In this measurement, 50 % of **SV** vesicles have been docked at 35 s when 100 pM vesicle concentrations (3 μM lipid concentration) of both **T** and **SV** are mixed together. As a result, the lower limit of the docking rate of **T–SV** is calculated to be $k_{T+SV,\ docking} = 1.4 \times 10^8\ M^{-1}\ s^{-1}$. As shown in Fig. 11a, fusion is the rate-limiting step for **T–SV**, i.e., the docking process is much faster than the fusion. However, when Syt1 is removed from v-vesicles (black-filled circles, Fig. 11b), the population of the docked vesicles is not observed. This indicates that the docking is the rate-limiting step for **T–V** mixture. Subsequently, the increase of the population of fused vesicles represents the docking rate. Therefore, from the increase of the fused fraction, the docking rate is estimated as $k_{T+V,\ docking} = 8.0 \times 10^4\ M^{-1}\ s^{-1}$, using (1), which is comparable with the previously reported docking rate for **T–V** [61, 62]. This result clearly shows that Syt1 increases docking rate by nearly 1,000-fold.

Since **T** and **SV** vesicles form docked vesicles immediately after mixing, the increase of fused population in Fig. 11a starts from a docked state. Thus, by fitting the increase of the fused population in Fig. 11a, the fusion rate of the docked vesicles can be obtained. The average fusion rate is $k_{T\text{-}SV,\ fusion} = 3.7 \pm 0.9 \times 10^{-3}\ s^{-1}$. Consequently, the time-dependent subpopulation analysis of ALEX provides the docking and fusion rates of vesicles in solution (Fig. 11c).

4 Summary and Perspectives

Here we describe a solution single-vesicle assay, ALEX, which has the capability to discriminate docked and fused vesicles and thus measure the kinetics of vesicle docking and fusion. ALEX is useful for studying docking factors and their sequential binding and unbinding reactions before and after lipid mixing. Since ALEX does not require the surface immobilization of vesicles, it bypasses the possible artifact due to the surface and obtains data in a high-throughput manner. Recently, content-mixing assays at the single-vesicle level have been developed using dye-labeled oligonucleotide and calcein [38, 63]. Although we introduce only two-color ALEX here, three- and four-color ALEX have been achieved before [29, 64]. Thus, multicolor ALEX can be easily used for vesicle fusion, which may report lipid mixing together with content mixing or SNARE complex formation. The multicolor ALEX approach will be extremely useful for studying the Ca^{2+}-triggering mechanism, dilation of fusion pore, and for understanding the correlation between SNARE complex formation and membrane fusion.

Acknowledgements

This work was supported by the National Research Foundation of Korea funded by the Ministry of Education, Science and Technology (MEST) (grant no. 2011-0016059) to N.K.L. and a National Institutes of Health grant (R01 GM051290) to Y.-K.S.

References

1. Fasshauer D, Sutton RB, Brunger AT, Jahn R (1998) Conserved structural features of the synaptic fusion complex: snare proteins reclassified as q- and r-snares. Proc Natl Acad Sci USA 95:15781–15786
2. Sutton RB, Fasshauer D, Jahn R, Brunger AT (1998) Crystal structure of a snare complex involved in synaptic exocytosis at 2.4 angstrom resolution. Nature 395:347–353
3. Sudhof TC, Rothman JE (2009) Membrane fusion: grappling with snare and sm proteins. Science 323:474–477
4. Brunger AT, Weninger K, Bowen M, Chu S (2009) Single-molecule studies of the neuronal snare fusion machinery. Annu Rev Biochem 78:903–928
5. Jahn R, Fasshauer D (2012) Molecular machines governing exocytosis of synaptic vesicles. Nature 490:201–207
6. Rizo J, Sudhof TC (2012) The membrane fusion enigma: snares, secl/muncl8 proteins, and their accomplices-guilty as charged? Annu Rev Cell Dev Biol 28:279–308
7. Jahn R, Scheller RH (2006) Snares—engines for membrane fusion. Nat Rev Mol Cell Biol 7:631–643
8. Lu XB, Zhang YH, Shin YK (2008) Supramolecular snare assembly precedes hemifusion in snare-mediated membrane fusion. Nat Struct Mol Biol 15:700–706
9. Weber T, Zemelman BV, McNew JA et al (1998) Snarepins: minimal machinery for membrane fusion. Cell 92:759–772
10. Tong JS, Borbat PP, Freed JH, Shin YK (2009) A scissors mechanism for stimulation of snare-mediated lipid mixing by cholesterol. Proc Natl Acad Sci USA 106:5141–5146
11. Tucker WC, Weber T, Chapman ER (2004) Reconstitution of Ca2+-regulated membrane fusion by synaptotagmin and snares. Science 304:435–438
12. Schaub JR, Lu XB, Doneske B, Shin YK, Mcnew JA (2006) Hemifusion arrest by complexin is relieved by Ca2+-synaptotagmin i. Nat Struct Mol Biol 13:748–750
13. Yang Y, Shin JY, Oh JM et al (2010) Dissection of snare-driven membrane fusion and neuroexocytosis by wedging small hydrophobic molecules into the snare zipper. Proc Natl Acad Sci USA 107:22145–22150
14. Shen JS, Tareste DC, Paumet F, Rothman JE, Melia TJ (2007) Selective activation of cognate snarepins by secl/muncl8 proteins. Cell 128:183–195
15. Struck DK, Hoekstra D, Pagano RE (1981) Use of resonance energy transfer to monitor membrane fusion. Biochemistry 20:4093–4099
16. McIntyre JC, Sleight RG (1991) Fluorescence assay for phospholipid membrane asymmetry. Biochemistry 30:11819–11827

17. Meers P, Ali S, Erukulla R, Janoff AS (2000) Novel inner monolayer fusion assays reveal differential monolayer mixing associated with cation-dependent membrane fusion. Biochim Biophys Acta 1467:227–243
18. van den Bogaart G, Holt MG, Bunt G, Riedel D, Wouters FS, Jahn R (2010) One snare complex is sufficient for membrane fusion. Nat Struct Mol Biol 17:358–364
19. van den Bogaart G, Thutupalli S, Risselada JH et al (2011) Synaptotagmin-1 may be a distance regulator acting upstream of snare nucleation. Nat Struct Mol Biol 18:805–812
20. Wilschut J, Papahadjopoulos D (1979) Ca2+-induced fusion of phospholipid vesicles monitored by mixing of aqueous contents. Nature 281:690–692
21. Smolarsky M, Teitelbaum D, Sela M, Gitler C (1977) A simple fluorescent method to determine complement-mediated liposome immune lysis. J Immunol Methods 15: 255–265
22. Nickel W, Weber T, McNew JA, Parlati F, Sollner TH, Rothman JE (1999) Content mixing and membrane integrity during membrane fusion driven by pairing of isolated v-snares and t-snares. Proc Natl Acad Sci USA 96:12571–12576
23. Duzgunes N (2003) Fluorescence assays for liposome fusion. Methods Enzymol 372:260–274
24. Duzgunes N, Wilschut J (1993) Fusion assays monitoring intermixing of aqueous contents. Methods Enzymol 220:3–14
25. Chang JY, Kim SA, Lu XB, Su ZL, Kim SK, Shin YK (2009) Fusion step-specific influence of cholesterol on snare-mediated membrane fusion. Biophys J 96:1839–1846
26. Myong S, Rasnik I, Joo C, Lohman TM, Ha T (2005) Repetitive shuttling of a motor protein on DNA. Nature 437:1321–1325
27. Weninger K, Bowen ME, Choi UB, Chu S, Brunger AT (2008) Accessory proteins stabilize the acceptor complex for synaptobrevin, the 1:1 syntaxin/snap-25 complex. Structure 16:308–320
28. Chung HS, McHale K, Louis JM, Eaton WA (2012) Single-molecule fluorescence experiments determine protein folding transition path times. Science 335:981–984
29. Choi UB, Strop P, Vrljic M, Chu S, Brunger AT, Weninger KR (2010) Single-molecule fret-derived model of the synaptotagmin 1-snare fusion complex. Nat Struct Mol Biol 17:318–324
30. Roy R, Hohng S, Ha T (2008) A practical guide to single-molecule fret. Nat Methods 5:507–516
31. Deniz AA, Dahan M, Grunwell JR et al (1999) Single-pair fluorescence resonance energy transfer on freely diffusing molecules: observation of Forster distance dependence and subpopulations. Proc Natl Acad Sci USA 96:3670–3675
32. Margittai M, Widengren J, Schweinberger E et al (2003) Single-molecule fluorescence resonance energy transfer reveals a dynamic equilibrium between closed and open conformations of syntaxin 1. Proc Natl Acad Sci USA 100:15516–15521
33. Torella JP, Holden SJ, Santoso Y, Hohlbein J, Kapanidis AN (2011) Identifying molecular dynamics in single-molecule fret experiments with burst variance analysis. Biophys J 100:1568–1577
34. Domanska MK, Kiessling V, Stein A, Fasshauer D, Tamm LK (2009) Single vesicle millisecond fusion kinetics reveals number of snare complexes optimal for fast snare-mediated membrane fusion. J Biol Chem 284:32158–32166
35. Fix M, Melia TJ, Jaiswal JK et al (2004) Imaging single membrane fusion events mediated by snare proteins. Proc Natl Acad Sci USA 101:7311–7316
36. Karatekin E, Di Giovanni J, Iborra C et al (2010) A fast, single-vesicle fusion assay mimics physiological snare requirements. Proc Natl Acad Sci USA 107:3517–3521
37. Liu T, Wang T, Chapman ER, Weisshaar JC (2008) Productive hemifusion intermediates in fast vesicle fusion driven by neuronal snares. Biophys J 94:1303–1314
38. Wang TT, Smith EA, Chapman ER, Weisshaar JC (2009) Lipid mixing and content release in single-vesicle, snare-driven fusion assay with 1-5 ms resolution. Biophys J 96:4122–4131
39. Yoon TY, Okumus B, Zhang F, Shin YK, Ha T (2006) Multiple intermediates in snare-induced membrane fusion. Proc Natl Acad Sci USA 103:19731–19736
40. Lee HK, Yang Y, Su ZL et al (2010) Dynamic ca^{2+}-dependent stimulation of vesicle fusion by membrane-anchored synaptotagmin 1. Science 328:760–763
41. Yoon TY, Lu X, Diao JJ, Lee SM, Ha T, Shin YK (2008) Complexin and ca2+ stimulate snare-mediated membrane fusion. Nat Struct Mol Biol 15:707–713
42. Kapanidis AN, Lee NK, Laurence TA, Doose S, Margeat E, Weiss S (2004) Fluorescence-aided molecule sorting: analysis of structure and interactions by alternating-laser excitation of single molecules. Proc Natl Acad Sci USA 101:8936–8941
43. Kim C, Kim JY, Kim SH, Lee BI, Lee NK (2012) Direct characterization of protein

oligomers and their quaternary structures by single-molecule fret. Chem Commun 48:1138–1140

44. Kapanidis AN, Margeat E, Ho SO, Kortkhonjia E, Weiss S, Ebright RH (2006) Initial transcription by rna polymerase proceeds through a DNA-scrunching mechanism. Science 314: 1144–1147
45. Kapanidis AN, Margeat E, Laurence TA et al (2005) Retention of transcription initiation factor sigma(70) in transcription elongation: single-molecule analysis. Mol Cell 20: 347–356
46. Santoso Y, Joyce CM, Potapova O et al (2010) Conformational transitions in DNA polymerase i revealed by single-molecule fret. Proc Natl Acad Sci USA 107:715–720
47. Lee NK, Kapanidis AN, Wang Y et al (2005) Accurate fret measurements within single diffusing biomolecules using alternating-laser excitation. Biophys J 88:2939–2953
48. Lee NK, Kapanidis AN, Koh HR et al (2007) Three-color alternating-laser excitation of single molecules: monitoring multiple interactions and distances. Biophys J 92:303–312
49. Lee NK, Koh HR, Han KY, Kim SK (2007) Folding of 8-17 deoxyribozyme studied by three-color alternating-laser excitation of single molecules. J Am Chem Soc 129: 15526–15534
50. Uphoff S, Holden SJ, Le Reste L et al (2010) Monitoring multiple distances within a single molecule using switchable fret. Nat Methods 7:831–836
51. Cordes T, Santoso Y, Tomescu AI et al (2010) Sensing DNA opening in transcription using quenchable Forster resonance energy transfer. Biochemistry 49:9171–9180
52. Le Reste L, Hohlbein J, Gryte K, Kapanidis AN (2012) Characterization of dark quencher chromophores as nonfluorescent acceptors for single-molecule fret. Biophys J 102:2658–2668
53. Kim JY, Choi BK, Choi MG et al (2012) Solution single-vesicle assay reveals pip2-mediated sequential actions of synaptotagmin-1 on snares. EMBO J 31:2144–2155
54. Kweon DH, Kim CS, Shin YK (2003) Insertion of the membrane-proximal region of the neuronal snare coiled coil into the membrane. J Biol Chem 278:12367–12373
55. Arac D, Chen XC, Khant HA et al (2006) Close membrane–membrane proximity induced by Ca2+-dependent multivalent binding of synaptotagmin-1 to phospholipids. Nat Struct Mol Biol 13:209–217
56. Hui EF, Gaffaney JD, Wang Z, Johnson CP, Evans CS, Chapman ER (2011) Mechanism and function of synaptotagmin-mediated membrane apposition. Nat Struct Mol Biol 18:813–821
57. Brose N, Petrenko AG, Sudhof TC, Jahn R (1992) Synaptotagmin: a calcium sensor on the synaptic vesicle surface. Science 256:1021–1025
58. Reist NE, Buchanan J, Li J, DiAntonio A, Buxton EM, Schwarz TL (1998) Morphologically docked synaptic vesicles are reduced in synaptotagmin mutants drosophila. J Neurosci 18:7662–7673
59. de Wit H, Walter AM, Milosevic I et al (2009) Synaptotagmin-1 docks secretory vesicles to syntaxin-1/snap-25 acceptor complexes. Cell 138:935–946
60. Liu HS, Dean C, Arthur CP, Dong M, Chapman ER (2009) Autapses and networks of hippocampal neurons exhibit distinct synaptic transmission phenotypes in the absence of synaptotagmin i. J Neurosci 29:7395–7403
61. Cypionka A, Stein A, Hernandez JM, Hippchen H, Jahn R, Walla PJ (2009) Discrimination between docking and fusion of liposomes reconstituted with neuronal snare-proteins using fcs. Proc Natl Acad Sci USA 106:18575–18580
62. Smith EA, Weisshaar JC (2011) Docking, not fusion, as the rate-limiting step in a snare-driven vesicle fusion assay. Biophys J 100:2141–2150
63. Diao JJ, Su ZL, Ishitsuka Y et al (2010) A single-vesicle content mixing assay for snare-mediated membrane fusion. Nat Commun 1
64. Yim SW, Kim T, Laurence TA et al (2012) Four-color alternating-laser excitation single-molecule fluorescence spectroscopy for next-generation biodetection assays. Clin Chem 58:707–716
65. Santoso Y, Hwang LC, Le Reste L, Kapanidis AN (2008) Red light, green light: probing single molecules using alternating-laser excitation. Biochem Soc Trans 36:738–744

Chapter 2

Imaging the Stages of Exocytosis in Epithelial Type II Pneumocytes

Thomas Haller and Paul Dietl

Abstract

Exocytosis proceeds through distinct stages. Based on morphological and functional criteria, they can be classified as pre- and hemifusion, fusion, and postfusion. During the prefusion stage, plasma and vesicle membranes approach each other. During the hemifusion stage, the outer membrane of the vesicle merges with the inner leaflet of the plasma membrane. During the fusion stage, both leaflets of plasma and vesicle membranes are fully merged, and the lumen of the vesicle opens to the extracellular space via an aqueous, small, and pore-like channel. During the postfusion stage, the fusion pore may further expand, and the vesicle content can be entirely expelled. The possibility to capture these events with imaging techniques depends on the speed they occur and the vesicle size in combination with the optical method used (temporal and spatial resolution). Type II pneumocytes of the lung have large (>1 μm) vesicles, termed lamellar bodies (LBs), which are useful for studies of exocytosis with live cell imaging techniques. Fluorescent dyes can be targeted to distinct compartments, and a certain exocytotic stage can be studied by measuring stage-specific diffusion and/or fluorescence enhancement of the selected probes. This paper overviews optical techniques and fluorescent dyes suitable for the investigation of the stages of LB exocytosis.

Key words Imaging, Epithelial, Fluorescent dye

1 Introduction

Pulmonary alveolar type II (AT II) cells synthesize, store, and release surfactant. This material, a complex of lipids and proteins, is essential to lower surface tension at the respiratory air–liquid interface—a decisive function in lung biomechanics. Without surfactant, our lungs would be prone to collapse, and the work of breathing (precisely speaking, the work of inspiration) would be prohibitively high. Thus, surfactant deficiency, or disturbances of the pulmonary surfactant system, can have health- and life-threatening consequences. The most prominent example with a clear-cut etiology is IRDS, the respiratory syndrome of preterm babies suffering from respiratory insufficiency due to an immaturity of their surfactant system [1].

Peter Thorn (ed.), *Exocytosis Methods*, Neuromethods, vol. 83,
DOI 10.1007/978-1-62703-676-4_2, © Springer Science+Business Media New York 2014

Surfactant is stored as compact structural assemblies within extraordinarily large vesicles (lamellar bodies, LBs). These vesicles reach diameters of several micrometers which is in the range of entire synaptic terminals. Probably due to these dimensions, the exocytotic machinery is astonishingly slow, and time scales of many seconds up to minutes are observed in a typical stimulus-secretion response. The regulation of LB exocytosis is complex, involving hormones as well as local chemical and physical factors (reviewed in [2]). Substantial evidence indicates that Ca^{2+} is a major second messenger to trigger intracellular fusion events and to integrate a number of different stimuli into a common cellular response [3]. In addition, many recent data indicate that not only the fusion process itself but also the postfusion phase is an important determinant of the secretory time scale and the amount of released substances [4–9]. In light of all these aspects—a unique but easily amenable exocytotic system, however sharing all the features with other Ca^{2+}-regulated systems—the AT II cells appear as a promising model in exocytosis research.

2 Stages of Exocytosis and Experimental Approaches for Their Visualization

A basic schematic presentation of the stages we can distinguish using live cell imaging techniques and which are further described in this article is shown in Fig. 1.

In our definition, the prefusion stage will comprise all maturation processes of vesicles that occur prior to any biophysical interaction of the two membranes—the limiting membrane of the vesicle and the plasma membrane of the cell. Thus, this stage will represent all pre-exocytotic vesicles regardless of their location within the cell (close to or still remote from the site of fusion, still in transport, or already functionally attached to the plasma membrane) and regardless of their endowment with ligands and molecules that are required to initiate, to facilitate, or to sustain the final fusion process. Our definition of this stage is reductionist and does not further discriminate between anchored, primed, and docked vesicles to give only examples of the plethora of biochemical,

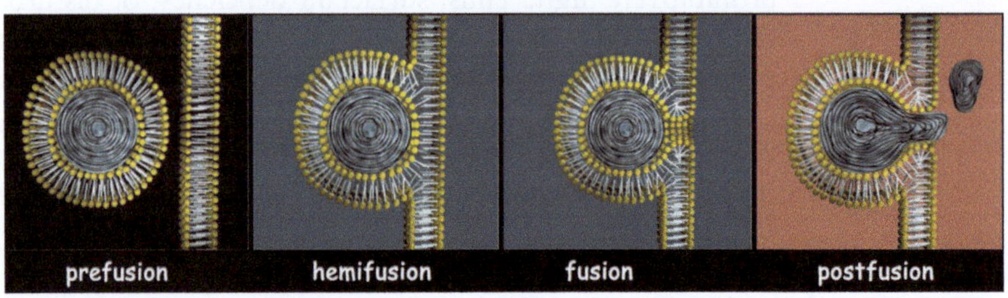

Fig. 1 Stages of exocytosis. See text for details

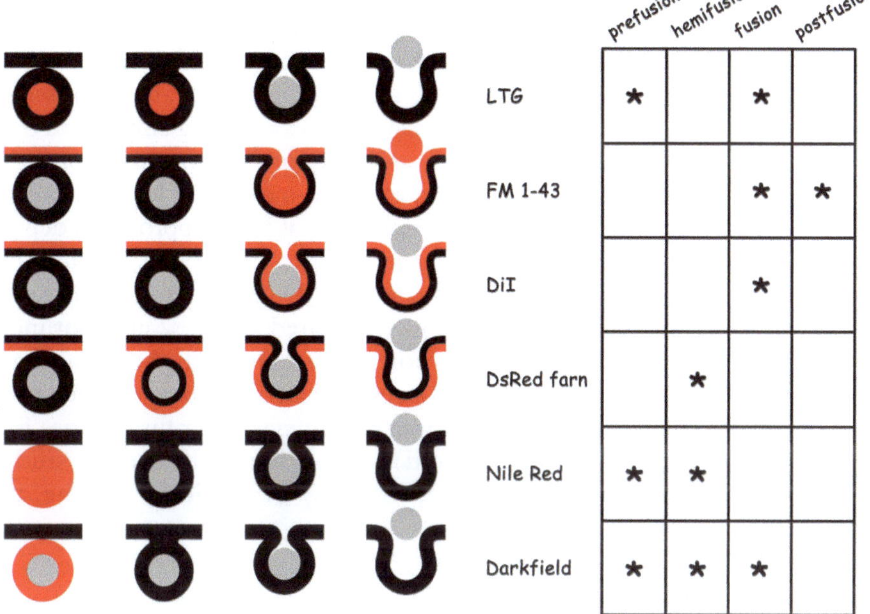

	prefusion	hemifusion	fusion	postfusion
LTG	*		*	
FM 1-43			*	*
DiI			*	
DsRed farn		*		
Nile Red	*	*		
Darkfield	*	*	*	

Fig. 2 Scheme of plasma and LB membranes, including their inner and outer leaflets, in the presence of fluorescence indicators in the prefusion, hemifusion, fusion, and postfusion stages (from *left* to *right*) of exocytosis. *Red color* indicates the postulated site of origin of fluorescence signals. *Gray circle* denotes surfactant. Note that the *upper* three dyes report fusion pore formation, whereas the *lower* ones (including dark field) report lipid merger. The *right table* illustrates preferential stage-specific use of these dyes. Modified from [19]

biophysical, and structural conditions that have been found to induce fusion competence. Basically, any dye that diffuses or is actively taken up into exocytotic vesicles or that accumulates within the limiting membrane prior to (hemi)fusion can be used to visualize LBs at this stage. Figure 2 presents a synopsis of all dyes/ microscopic techniques outlined in this article and delineates the ones that can be used in the prefusion stage.

The hemifusion stage (Fig. 1) denotes membrane merger and is based on the proximity model of membrane fusion, as outlined in Section 3.4 below. Apart from dark-field microscopy, a unique and new application of this method (see below), any dye that is able to report merger of the facing leaflets of plasma and vesicle membrane can in principle be used to visualize this phase (Fig. 2). In general, lipid mixing should precede content mixing if this phase is of measurable duration.

Fusion pore formation (Fig. 1), the hallmark of exocytosis, is most accurately assessed electrophysiologically (by cell membrane capacitance steps), but it is also accessible to microscopic measurements using approaches summarized in Fig. 2. Pioneering studies in other cell types revealed that early fusion pores can open and close again (fusion pore flickering), without detectable fluorophore discharge, suggesting that the early fusion pore is a narrow aqueous channel with similar characteristics than that of ion channels [10–12]. Evidently, these early stages of nascent fusion pores (lasting a few milliseconds) are inaccessible to fluorescence dyes. Later stages, however, can be assessed using either dyes that report content mixing (such as LTG) or lipid mixing along the outer plasma membrane/inner vesicle membrane leaflets (DiI) or both (FM 1-43). The detection limit of these approaches depends very much on the fluorescence signal-to-noise ratios, as outlined below.

The postfusion phase (Fig. 1) denotes fusion pore expansion and content release. As delineated in Fig. 2 and outlined below (*FM 1-43*), dye diffusion kinetics through the expanding fusion pore can be used to assess dynamic fusion pore behavior. Dyes that remain bound to surfactant even after fusion pore formation (FM 1-43 and Nile red) can in addition be used to study surfactant transformation processes, in particular at the air–liquid interface [13].

3 Specific Dyes/Methods Used to Study LB Exocytosis

3.1 LysoTracker® Green

Prefused vesicles can be visualized with the popular [14] LysoTracker and LysoSensor dyes which come in blue, green, yellow, and red emission properties (Invitrogen—Molecular Probes). These dyes belong to a larger group of fluorophores that are termed acidotropic (or lysosomotropic) since they preferentially accumulate in acidic cell compartments like lysosomes, late endosomes, phagosomes, *trans*-Golgi, or synaptic vesicles [15]. Nonetheless, the resulting emission intensities and spectral properties of the Tracker dyes are essentially pH independent, in contrast to the Sensor dyes that produce pH-dependent absorbance and emission shifts. The mechanism of cellular uptake is by nonionic diffusion of the uncharged molecule, whereas the accumulation and retention in acidic organelles is believed to require protonation in the presence of a low pH of the weakly basic amines connected to the fluorogenic multi-pyrrole ring structures [14]. Each protonated molecule acquired a net charge that retards back diffusion into the cytosol. Thus, LTG accumulation is proportional to the H^+ concentration in LBs [16], and since this is an energy (ATP)-dependent process, it is at the same time a vitality parameter for these cells [17]. Accordingly, all fluorophores of the LysoTracker group tested in AT II cells yielded a stable staining of prefused LBs up to days that may only vanish due to repeated or intense light exposures or in severely compromised cells. Staining by LTG is furthermore dose and time dependent. Due to that, care has to be taken not to overcharge these organelles during the loading conditions as a perturbation of intravesicular pH and osmotic swelling have to be expected

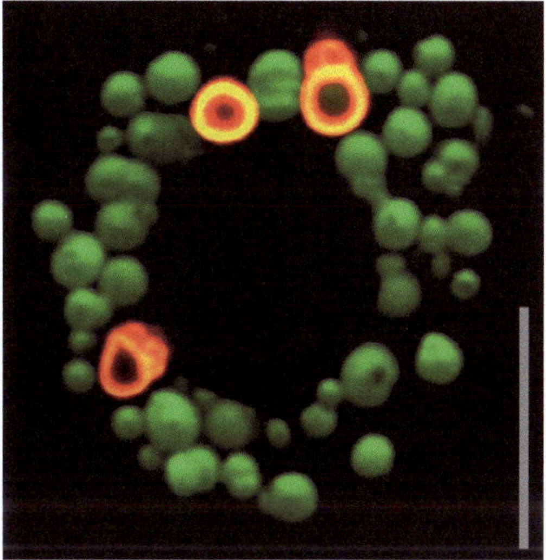

Fig. 3 Simultaneous use of two dyes reporting fusion pore formation in a single alveolar type II cell. Loss in LysoTracker green fluorescence (*green*) is matched by increase in FM 1-43 fluorescence (*red*). Bar = 10 μm. (With permission from [20])

(manufacturer's instructions). Usually, a 100 nM dye solution applied for 30 min followed by its washout is entirely sufficient to obtain satisfying imaging results with bright vesicular fluorescence and low background intensities (Fig. 3).

In the AT II cells, the LysoTracker dyes selectively partition into LBs, which has been demonstrated by co-localization with proteins specifically expressed on the limiting LB membrane such as LAMP3 [18] or by specific and unspecific lipid-staining fluorophores [19–21]. Furthermore, localization and appearance of LTG-stained structures largely correspond to LBs when compared with contrast-enhancing imaging techniques or dark-field microscopy [22, 23]. This indicates that the majority of acidic organelles in AT II cells, at least in terms of the entire enclosed volume, are LBs, whereas lysosomes and related structures only constitute a small and negligible fraction. The combination of unusually large secretory organelles in AT II cells and their specific targeting by the LysoTracker dyes is what made the use of them to excellent tools for optical tracking and live cell imaging of distinct stages during LB exocytosis.

Naturally, for that purpose, it is essential that LysoTracker dyes are stage specific. As we will discuss below, fluorescence of LysoTracker green disappears when the fusion pore has opened but remains stable during hemifusion. Thus, LTG fluorescence is a reliable marker of the prefusion stage but is not suited to discriminate between the pre- and hemifusion states.

LTG fluorescence intensity from single LBs declines entirely with a time constant of several seconds [24] after cell activation and when these vesicles to fuse with the cell membrane and establish fusion pores. This suggests that the dye is trapped within LBs in aqueous form rather than being inserted into lipid layers as has been suggested by the manufacturer (Invitrogen—Molecular Probes). Signal loss during opening of the fusion pore is then the consequence of a free outward diffusion through a water-filled fusion pore of adequate size. In line with this argument is the fact that we have not seen LTG redistribution from the vesicle lumen into the plasma membrane during hemifusion, a situation where the lumen of the vesicle is closed but the lipid membranes are already merged.

3.2 FM® 1-43 (N-(3-Triethylammoniumpropyl)-4-(4-(dibutylamino)styryl)pyridinium dibromide)

FM 1-43 is used as a general membrane probe but has been originally developed by Betz and colleagues to identify actively firing neurons and the recycling of synaptic vesicles (reviewed in [25]). In pulmonary physiology FM 1-43 is used, particularly by our group, to visualize surfactant exocytosis, to catch the instance of fusion pore formation by AT II cells, and to further trace the fate of the extruded surfactant on its route through the fusion pore into the extracellular space and its final adsorption at an air–liquid interface [13, 26].

FM 1-43 has a high partition coefficient for lipids, but in contrast to pure lipophilic probes like Nile red, it specifically adsorbs to hydrophilic/hydrophobic phase boundaries, so, for example, at air–liquid interfaces, in the outer leaflet of cell membranes when FM 1-43 is applied from outside, or in surfactant membranes when it has access through the fusion pore (Fig. 3). Permeation of FM 1-43 across membranes is largely obstructed by two permanently positively charged quaternary amino groups that are connected, via a fluorogenic group, with a lipophilic hydrocarbon tail. Thus, the amphiphilic nature makes this molecule to a tracer of the cell membrane that does not easily enter the cell interior [27]. In vital AT II cells, for example, fluorescence remains restricted to the cell surface in a diffuse but stable manner for usually up to hours [28]. A further important feature is the dramatic increase in FM 1-43 fluorescence quantum yield by more than two orders of magnitude (~350 [29]): It is almost nonfluorescent in water but becomes highly fluorescent after reversible insertion of the hydrocarbon tail into lipid membranes or other hydrophobic environments. This leads to a fast and localized brightening of fused LBs exposing their content—surfactant—to FM 1-43, with an intensity that exceeds by far the still present but faint signal from the plasma membrane (Fig. 2). Combined fluorescence and patch clamp measurements revealed that the onset in FM 1-43 fluorescence increase coincided with the stepwise change in cell capacitance, making it one of the fastest indicators of membrane insertion via exocytotic membrane fusion [20]. The delay between capacitance changes and fluorescence increase was below the temporal resolution applied in these investigations (200 ms).

Due to these features, FM 1-43 is stage specific to define the instance of pore formation. We want to emphasize, however, that the actual signal change does not primarily arise from the increase in membrane surface area which is the case in other exocytotic systems and which yields similar information than capacitance measurements [27, 30], but arises predominantly from the contents of LBs which act as a kind of signal amplifier. This is understandable considering that surfactant is assembled in tight lipid packages with an enormous interfacial area. Apart from a few other systems in which the plasma membrane together with granule contents is stained by FM 1-43 (e.g., in prolactin-secreting pituitary lactotrophs [29, 31]), this is very unique in exocytosis research and probably remains a specificity of the AT II cells. Even more, this approach allows at the same time analysis of the postfusion events in surfactant exocytosis. For example, repeated FRAP experiments with fused LBs were used to calculate the growth of the fusion pores [4]. This was possible because initial, narrow fusion pores keep the "FM 1-43 sensor" (= surfactant) within the vesicles and because staining kinetics of surfactant by this dye is dependent on the area enclosed by that pore, which increases only slowly and in dependence of intracellular Ca^{2+} [4, 24]. Only after the pore has expanded to a sufficient extent, is the bulky surfactant complex is able to pass, probably supported by an active squeeze out induced by vesicle coating with actin [21].

The same features of FM 1-43 that enable measurement of fusion pore formation preclude its ude for the measurement of hemifusion: The amphiphatic property causes an orientated insertion of FM 1-43 in the cis monolayer (the external leaflet facing the medium) and prevents its flip-flopping that would otherwise lead to its even distribution within the entire cell membrane [29].

3.3 DiI (1,1′-Dioctadecyl-3,3,3′,3′-tetramethylindocarbocyanine perchlorate)

DiI is one of the several related amphiphilic membrane stains (absorption/emission maxima = 549/565 nm) that have been frequently used to trace plasma membrane borders and lipid diffusion. DiI has a structural similarity to phospholipids and can selectively partition into different lipid phases (gel or fluid) depending on the matching between its alkyl chain lengths with that of the lipids. Therefore, DiI has been used to study membrane organization and dynamics [32–34]. The fluorescence lifetime of DiI depends on the accessibility to water and on the viscosity of the local microenvironment, and this is why it was used to detect lipid rafts in cells and phase separation in model membranes [33]. DiI was also employed to characterize tension-induced changes in lipid packing [35].

Due to its partitioning to the outer leaflet of the plasma membrane, DiI can be used to monitor LB fusion pore formation via its diffusion into the inner leaflet of the fused LB [19]. In that sense, it acts similar than FM 1-43, however with some significant differences:

DiI staining of the plasma membrane is slower and less reversible than with FM 1-43. Following its washout, DiI remains within the plasma membrane for a considerable time (>20 min) that is sufficient to perform most applications in live cell experiments. Hence, in the absence of the dye in the bath, surfactant is not readily stained when the fusion pore has opened. DiI staining of the fused LB therefore rather appears as a ring of fluorescence (i.e., staining of the limiting membrane) than a fluorescent sphere (i.e., staining of LB contents).

Owing to the immediate amplification of quantum yield when FM 1-43 interacts with surfactant lipids, its superior signal-to-noise ratio makes FM 1-43 to our opinion preferential over DiI to assess the instance of fusion pore formation.

3.4 DsRed-Farn

DsRed or drFP583, as originally named [36], is a 28 kDa GFP analogue from a *Discosoma* species with excitation and emission maxima at 558 and 583 nm, respectively, the longest yet reported for a wild-type spontaneously fluorescent protein, with excellent resistance to pH extremes [37]. The vector of this protein containing a farnesylation site (pDsRed-Monomer-F) is commercially available and can be expressed by adenoviral expression in type II pneumocytes as described recently [19]. Due to its farnesylation site, DsRed-Farn selectively incorporates into the inner leaflet of the plasma membrane by posttranslational protein farnesylation through membrane-bound enzymes [38].

In the context of stage-specific detection of exocytosis, the theory is that a molecule, which is anchored to the inner leaflet of the plasma membrane with a freely diffusing lipidic moiety, should report "membrane merger" by diffusion into the outer leaflet of the exocytotic vesicle. According to the hemifusion stalk-lipidic pore model (reviewed in detail by [39]), this stage is defined as "hemifusion," the initiation of the exocytotic process that terminates with full membrane merger and fusion pore formation. It should be kept in mind, however, that despite ample experimental evidence in favor of it, this is still one ("proximity model") of the two conflicting models ("protein pore model"), reviewed in detail by Sorensen [40]. Furthermore, lipid flow across proteinaceous domains which determine the edge of the nascent fusion pore may be limited as compared to free diffusion within the plasma membrane [41, 42].

Analysis of DsRed-Farn diffusion from the plasma membrane into the limiting LB membrane indeed confirmed long-lasting (several seconds) hemifusion states before fusion pore formation was demonstrated by dark-field microscopy (see below), adding another piece of evidence in favor of the proximity model [19]. However, these measurements also revealed a certain degree of unreliability, denoting a considerable scatter of onset of DsRed diffusion, in part extending to the postfusion stage. This reflected most probably not only variations in diffusibility due to biological heterogeneity, as mentioned above, but also methodological limitations (photobleaching, resolution limitation).

The use of DsRed-Farn should not be restricted to the measurement of hemifusion of LBs, for which we would rather recommend dark-field microscopy anyway (see below). The dye is also a reasonable tool to monitor trafficking of lipidic plasma membrane domains (such as via endocytosis and fusion with intracellular organelles), because once trapped by farnesylation, the dye will follow the "fate" of the inner leaflet of the plasma membrane, wherever it goes. In this way, it was shown that type II cell plasma membrane recycles to LBs, forming discrete domains [43].

3.5 Nile Red (9-Diethylamino-5-benzo[α]phenoxazinone)

Nile red is used to stain and quantitate lipids, typically neutral lipid droplets in hepatocytes by fluorescence microscopy or cytometry (reviewed in [44]). Like FM 1-43 it has a high partition coefficient for organic solvents (~200) and a high photochemical stability, it undergoes a dramatic enhancement in fluorescence intensity in nonpolar solvents with a large Stokes shift, and its spectral properties are similarly environment sensitive than those of FM 1-43, although these properties have not been investigated in detail in both cases (reviewed in [45]). In contrast to FM 1-43, its partitioning into lipids seems to be somewhat less reversible, and by its entirely apolar nature, Nile red easily crosses cell membranes and permeates into the cell interior where it reacts with intracellular membranes and organelles.

Nile red has been used to identify AT II cells in various lung cell preparations, particularly in AT I and AT II co-culture systems [46], but its use as a marker of LBs is of rather marginal significance due to a reported lack in specificity and some problems in reproducibility [47, 48]. In our investigation, Nile red could be well used to stain LBs and to monitor hemifusion in a similar way (and with similar limitations) as described for DsRed-Farn (see above). Naturally, in contrast to DsRed-Farn, Nile red diffusion along the hemifusion stalk leads to a destaining—instead of staining—of hemifused LBs [19]. Owing to its presumed diffusion along the hemifusion stalk, the onset of Nile red destaining was found to precede LTG destaining (fluid-phase marker; see above) by 1,36 s on average [19].

3.6 Dark-Field Microscopy

Dark-field microscopy, well known to (older) clinicians for the diagnosis of syphilitic lesions (*Treponema pallidum*), is a technique with superior resolution characteristics over bright-field microscopy, in a similar way than tiny dust particles become visible when illuminated from a flashlight in the dark. Basically, it is a contrast enhancement optical method enabling "scattered light" to pass the objective while shielding the incident light. LBs, together with their limiting membranes, are structures that contain a high

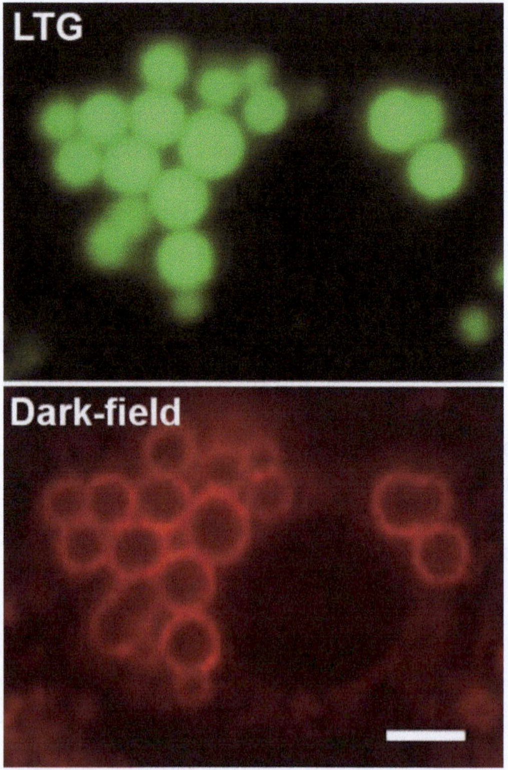

Fig. 4 LysoTracker green fluorescence (*above*) and dark-field illumination (*below*) of part of a single type II cell. *Green spheres* (LysoTracker green) and *red rings* (dark field) represent LBs. Scale bar = 4 μm

amount of lipids, representing a site where refraction, diffraction, and possible reflection of light may occur. Although the precise quantum optical nature of these phenomena lacks any detailed description (probably no one has yet seriously thought about it), our observations consistently reveal bright rings in dark-field microscopy which outshine by far any other organelles in type II pneumocytes as shown in Fig. 4 [19]. These rings co-localize with LTG (Fig. 4) and LAMP3-GFP fluorescence, a lysosomal-associated membrane protein selectively expressed on the limiting LB membrane [18, 19].

Interestingly, when LBs fuse with the plasma membrane, a sharp, biphasic drop of dark-field light intensity occurs at the entire limiting membrane of the fusing LB (Fig. 5). Coincidence analysis revealed that the first, slow phase of light intensity decay roughly matched with the diffusion of dyes reporting membrane merger (DsRed-Farn and Nile red), whereas the second, fast phase of decay initiated with the onset of diffusion of dyes reporting fusion pore formation (FM 1-43, LTG, or DiI). For a detailed description

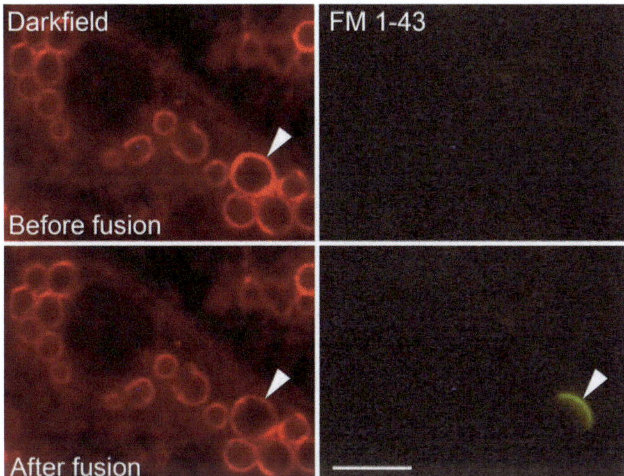

Fig. 5 Dark-field illumination (*left*) and FM 1-43 fluorescence (*right*) of several LBs within a single cell. Note the decrease of dark-field intensity of the limiting membrane of the fused LB (*arrow*). FM 1-43 fluorescence of this LB (*arrow*) indicates fusion. Scale bar = 8 μm. From ref. [20] with permission

we refer to ref. [19]. With the caveat of conceptual and methodological constraints in the definition of "prefusion," "hemifusion," and "fusion" as outlined above (*DsRed-Farn*), the far most plausible explanation of these phenomena was that the first, slow phase of light decay reported the hemifusion stage, whereas the onset of the second, fast phase reported the formation of a fusion pore.

We have been asked frequently why, at the stage of hemifusion, dark-field light intensity decay is observed to occur evenly, over the entire limiting membrane, and not exclusively at the site of hemifusion (i.e., where the membranes merge). Here again, plausibility on the basis of experimental evidence gives a reasonable explanation: The limiting membrane of secretory vesicles is under tension [49], causing considerable membrane flow from the plasma membrane to the vesicle membrane when these membranes fuse [11]. The behavior of light at this surface may in turn critically depend on the density of lipid packing. It is hence well conceivable that lipid flow from the inner leaflet of the plasma membrane to the outer leaflet of the LB membrane during hemifusion causes a change of the optical properties of the entire LB membrane and not only of the hemifusion stalk alone.

3.7 Fusion Assay

The properties of LTG as described above are principally suited to analyze fusion events in single cells by microscopy. However, we also sought to develop assay systems for a more quantitative but still time-resolved analysis of LB fusions, particularly one with a high sample throughput for toxicological and pharmacological purposes. Such assay systems using live mammalian cells are still relatively rare except those utilizing permeabilized cells, reconstituted cell-free systems, or invertebrate models like isolated sea urchin eggs [50].

One of these assays uses LTG and is based on the presence of a quenching molecule, a highly light-absorbing dye, in the supernatants of adherent AT II cells grown on 96-well plates [51]. The quencher, for which we used the nontoxic Brilliant Black, dramatically increases the specificity of the LTG signals because fluorescence excitation and light collection by the plate reader are restricted to a finite plane—that of the cells at the plate bottom. This configuration then creates a kind of pseudo-confocality in the instrument and minimizes contribution of scattered and back-reflected light and non-vesicular fluorescence due to LTG release from fused vesicles. As a result, the measured signal is *per se* background corrected and correlates with the pool of prefused vesicles. Conversely, this pool and the ensuing signals decline rapidly in actively exocytosing cells. The measurement of fluorescence decay in comparison to that of quiescent cells (untreated controls) is then a direct and proportional readout of the rate and amount of vesicle fusions. In contrast to end-point measurements of released compounds (e.g., phospholipids in case of the surfactant-secreting AT II cells), the fusion assay revealed a dramatic faster onset in the rate of vesicle fusion upon stimulation, which is due to the different mechanisms how fusion and postfusion events are regulated in these cells.

4 Synopsis of Methods and Results

The prefusion stage of LB exocytosis is remarkably slow, lasting between seconds and many minutes depending on the mode of cell stimulation (purinergic agonists, cell stretch, etc.) as reviewed in detail elsewhere [52, 53]. It should be noted that LBs are quite well visible using bright-field microscopy alone, allowing the tracking of intracellular vesicle movement in the absence of any fluorescence probe. Since dark-field microscopy yields a sharp image of the limiting LB membrane before fusion and enables the assessment of hemifusion and fusion pore formation too, this noninvasive method is ideally suitable to assess all stages except for surfactant release (postfusion). The hemifusion stage, which is (with all limitations discussed above) most precisely resolved with dark-field microscopy, has a lifetime of up to 10 s [19]. Dark-field microscopy can be combined with essentially all fluorescence measurements, making this tool versatile for single-cell multiparameter analysis. The combination of patch clamp with fluorescence microscopy (FM 1-43, LTG [20]) and the combination of dark-field microscopy with fluorescence microscopy (LTG, Nile red, DsRed-Farn, DiI [19]) allow the conclusion that the instance of fusion pore formation can be reliably assessed using appropriate dyes (Fig. 2), with a detection limit that is only delayed between a few milliseconds but less than a second following the formation of a nascent pore. The postfusion stage, i.e., the time between pore formation and surfactant release, was found to be variable and long lasting (up to hours) and dependent on actin coating and active coat compression [18, 21, 24]. FM 1-43, when present in the bath in essentially infinite amounts (and hence hardly subject to any bleaching), is the dye of choice to study postfusion events in surfactant secretion, from the dynamics of fusion pore expansion [4] to surfactant conversion at the air–liquid interface [13].

References

1. Griese M (1999) Pulmonary surfactant in health and human lung diseases: state of the art. Eur Respir J 13:1455–1476
2. Andreeva AV, Kutuzov MA, Voyno-Yasenetskaya TA (2007) Regulation of surfactant secretion in alveolar type II cells. Am J Physiol Lung Cell Mol Physiol 293: L259–L271
3. Frick M, Eschertzhuber S, Haller T, Mair N, Dietl P (2001) Secretion in alveolar type II cells at the interface of constitutive and regulated exocytosis. Am J Respir Cell Mol Biol 25:306–315
4. Haller T, Dietl P, Pfaller K, Frick M, Mair N, Paulmichl M, Hess MW, Furst J, Maly K (2001) Fusion pore expansion is a slow, discontinuous, and Ca^{2+}-dependent process regulating secretion from alveolar type II cells. J Cell Biol 155:279–289
5. Miklavc P, Frick M, Wittekindt OH, Haller T, Dietl P (2010) Fusion-activated Ca^{2+} entry: an "active zone" of elevated Ca^{2+} during the postfusion stage of lamellar body exocytosis in rat type II pneumocytes. PLoS One 5:e10982
6. Singer W, Frick M, Haller T, Bernet S, Ritsch-Marte M, Dietl P (2003) Mechanical forces impeding exocytotic surfactant release revealed by optical tweezers. Biophys J 84:1344–1351
7. Haller T, Pfaller K, Dietl P (2001) The conception of fusion pores as rate-limiting structures for surfactant secretion. Comp Biochem Physiol A Mol Integr Physiol 129:227–231
8. Dietl P, Haller T, Mair N, Frick M (2001) Mechanisms of surfactant exocytosis in alveolar type II cells in vitro and in vivo. News Physiol Sci 16:239–243
9. Dietl P, Haller T (2000) Persistent fusion pores but transient fusion in alveolar type II cells. Cell Biol Int 24:803–807
10. Fernandez JM, Neher E, Gomperts BD (1984) Capacitance measurements reveal stepwise fusion events in degranulating mast cells. Nature 312:453–455
11. Monck JR, Alvarez de Toledo G, Fernandez JM (1990) Tension in secretory granule membranes causes extensive membrane transfer through the exocytotic fusion pore. Proc Natl Acad Sci U S A 87:7804–7808
12. Neher E, Marty A (1982) Discrete changes of cell membrane capacitance observed under conditions of enhanced secretion in bovine adrenal chromaffin cells. Proc Natl Acad Sci U S A 79:6712–6716
13. Haller T, Dietl P, Stockner H, Frick M, Mair N, Tinhofer I, Ritsch A, Enhorning G, Putz G (2004) Tracing surfactant transformation from cellular release to insertion into an air-liquid interface. Am J Physiol Lung Cell Mol Physiol 286:L1009–L1015
14. Freundt EC, Czapiga M, Lenardo MJ (2007) Photoconversion of Lysotracker red to a green fluorescent molecule. Cell Res 17:956–958
15. Abreu BJ, Guimaraes M, Uliana LC, Vigh J, von Gersdorff H, Prado MA, Guatimosim C (2008) Protein kinase C modulates synaptic vesicle acidification in a ribbon type nerve terminal in the retina. Neurochem Int 53: 155–164
16. Anderson RG, Orci L (1988) A view of acidic intracellular compartments. J Cell Biol 106:539–543
17. Chander A, Johnson RG, Reicherter J, Fisher AB (1986) Lung lamellar bodies maintain an acidic internal pH. J Biol Chem 261: 6126–6131
18. Miklavc P, Wittekindt OH, Felder E, Dietl P (2009) Ca^{2+}-dependent actin coating of lamellar bodies after exocytotic fusion: a prerequisite

for content release or kiss-and-run. Ann N Y Acad Sci 1152:43–52

19. Miklavc P, Albrecht S, Wittekindt OH, Schullian P, Haller T, Dietl P (2009) Existence of exocytotic hemifusion intermediates with a lifetime of up to seconds in type II pneumocytes. Biochem J 424:7–14
20. Mair N, Haller T, Dietl P (1999) Exocytosis in alveolar type II cells revealed by cell capacitance and fluorescence measurements. Am J Physiol 276:L376–L382
21. Miklavc P, Hecht E, Hobi N, Wittekindt OH, Dietl P, Kranz C, Frick M (2012) Actin coating and compression of fused secretory vesicles are essential for surfactant secretion—a role for Rho, formins and myosin II. J Cell Sci 125:2765–2774
22. Wemhoner A, Jennings P, Haller T, Rudiger M, Simbruner G (2011) Effect of exogenous surfactants on viability and DNA synthesis in A549, immortalized mouse type II and isolated rat alveolar type II cells. BMC Pulm Med 11:11
23. Wemhoner A, Hackspiel I, Hobi N, Ravasio A, Haller T, Rudiger M (2010) Effects of perfluorocarbons on surfactant exocytosis and membrane properties in isolated alveolar type II cells. Respir Res 11:52
24. Miklavc P, Mair N, Wittekindt OH, Haller T, Dietl P, Felder E, Timmler M, Frick M (2011) Fusion-activated Ca^{2+} entry via vesicular P2X4 receptors promotes fusion pore opening and exocytotic content release in pneumocytes. Proc Natl Acad Sci U S A 108:14503–14508
25. Betz WJ, Mao F, Smith CB (1996) Imaging exocytosis and endocytosis. Curr Opin Neurobiol 6:365–371
26. Possmayer F, Hall SB, Haller T, Petersen NO, Zuo YY, Bernardino de la Serna J, Postle AD, Veldhuizen RA, Orgeig S (2010) Recent advances in alveolar biology: some new looks at the alveolar interface. Respir Physiol Neurobiol 173(Suppl):S55–S64
27. Betz WJ, Bewick GS (1992) Optical analysis of synaptic vesicle recycling at the frog neuromuscular junction. Science 255:200–203
28. Haller T, Ortmayr J, Friedrich F, Volkl H, Dietl P (1998) Dynamics of surfactant release in alveolar type II cells. Proc Natl Acad Sci U S A 95:1579–1584
29. Brumback AC, Lieber JL, Angleson JK, Betz WJ (2004) Using FM1-43 to study neuropeptide granule dynamics and exocytosis. Methods 33:287–294
30. Smith CB, Betz WJ (1996) Simultaneous independent measurement of endocytosis and exocytosis. Nature 380:531–534
31. Cochilla AJ, Angleson JK, Betz WJ (2000) Differential regulation of granule-to-granule and granule-to-plasma membrane fusion during secretion from rat pituitary lactotrophs. J Cell Biol 150:839–848
32. Klausner RD, Wolf DE (1980) Selectivity of fluorescent lipid analogues for lipid domains. Biochemistry 19:6199–6203
33. Packard BS, Wolf DE (1985) Fluorescence lifetimes of carbocyanine lipid analogues in phospholipid bilayers. Biochemistry 24:5176–5181
34. Kahya N, Scherfeld D, Bacia K, Schwille P (2004) Lipid domain formation and dynamics in giant unilamellar vesicles explored by fluorescence correlation spectroscopy. J Struct Biol 147:77–89
35. Muddana HS, Gullapalli RR, Manias E, Butler PJ (2011) Atomistic simulation of lipid and DiI dynamics in membrane bilayers under tension. Phys Chem Chem Phys 13:1368–1378
36. Matz MV, Fradkov AF, Labas YA, Savitsky AP, Zaraisky AG, Markelov ML, Lukyanov SA (1999) Fluorescent proteins from nonbioluminescent Anthozoa species. Nat Biotechnol 17:969–973
37. Baird GS, Zacharias DA, Tsien RY (2000) Biochemistry, mutagenesis, and oligomerization of DsRed, a red fluorescent protein from coral. Proc Natl Acad Sci USA 97: 11984–11989
38. Fujiyama A, Tsunasawa S, Tamanoi F, Sakiyama F (1991) S-farnesylation and methyl esterification of C-terminal domain of yeast RAS2 protein prior to fatty acid acylation. J Biol Chem 266:17926–17931
39. Chernomordik LV, Kozlov MM (2008) Mechanics of membrane fusion. Nat Struct Mol Biol 15:675–683
40. Sorensen JB (2009) Conflicting views on the membrane fusion machinery and the fusion pore. Annu Rev Cell Dev Biol 25:513–537
41. Zimmerberg J, Chernomordik LV (1999) Membrane fusion. Adv Drug Deliv Rev 38:197–205
42. Chernomordik LV, Kozlov MM (2008) Mechanics of membrane fusion. Nat Struct Mol Biol 15:675–683
43. Albrecht S, Usmani SM, Dietl P, Wittekindt OH (2010) Plasma membrane trafficking in alveolar type II cells. Cell Physiol Biochem 25:81–90
44. Greenspan P, Mayer EP, Fowler SD (1985) Nile red: a selective fluorescent stain for intracellular lipid droplets. J Cell Biol 100:965–973
45. Demchenko AP, Mely Y, Duportail G, Klymchenko AS (2009) Monitoring biophysical

properties of lipid membranes by environment-sensitive fluorescent probes. Biophys J 96:3461–3470

46. Mishra A, Chintagari NR, Guo Y, Weng T, Su L, Liu L (2011) Purinergic P2X7 receptor regulates lung surfactant secretion in a paracrine manner. J Cell Sci 124:657–668
47. Chen J, Chen Z, Narasaraju T, Jin N, Liu L (2004) Isolation of highly pure alveolar epithelial type I and type II cells from rat lungs. Lab Invest 84:727–735
48. Liu L, Wang M, Fisher AB, Zimmerman UJ (1996) Involvement of annexin II in exocytosis of lamellar bodies from alveolar epithelial type II cells. Am J Physiol 270: L668–L676
49. Chizmadzhev YA, Kuzmin PI, Kumenko DA, Zimmerberg J, Cohen FS (2000) Dynamics of fusion pores connecting membranes of different tensions. Biophys J 78:2241–2256
50. Avery J, Jahn R, Edwardson JM (1999) Reconstitution of regulated exocytosis in cell-free systems: a critical appraisal. Annu Rev Physiol 61:777–807
51. Wemhoner A, Frick M, Dietl P, Jennings P, Haller T (2006) A fluorescent microplate assay for exocytosis in alveolar type II cells. J Biomol Screen 11:286–295
52. Dietl P, Haller T (2005) Exocytosis of lung surfactant: from the secretory vesicle to the air–liquid interface. Annu Rev Physiol 67:595–621
53. Dietl P, Haller T, Frick M (2012) Spatio-temporal aspects, pathways and actions of Ca^{2+} in surfactant secreting pulmonary alveolar type II pneumocytes. Cell Calcium 52:296–302

Chapter 3

Carbon-Fiber Amperometry in the Study of Exocytosis

Michael D. Duffield, Ravinarayan Raghupathi, and Damien J. Keating

Abstract

Understanding how signaling molecules are released from cells is essential for furthering our knowledge of the basic biological mechanisms controlling many significant biological pathways. These molecules, including neurotransmitters, hormones, growth factors, and peptides, are released from cells via a process called exocytosis. Our laboratory has utilized a noninvasive method of measuring the release of oxidizable molecules from cells, known as carbon-fiber amperometry. In this chapter we will describe how we undertake such measurements, how the resulting data is analyzed, and what the outcomes mean in terms of physiology. We provide examples of our work measuring catecholamine release in single chromaffin cells as well as serotonin release from intact sections of colon.

Key words Exocytosis, Amperometry, Chromaffin cells, Enterochromaffin cells, Voltammetry

1 Introduction

1.1 Why Study Exocytosis?

Exocytosis is the process whereby intracellular vesicles fuse with the external plasma membrane of the cell and a pore is formed between these membranes to allow the release of vesicle contents from the cell. This process is exquisitely regulated and is vital to cell survival, homeostasis, and a wide variety of specialized cell functions. Two major types of exocytosis are said to occur within cells—constitutive and regulated (for a review, see [1]). Constitutive exocytosis, as its name suggests, occurs in all cells and on an ongoing basis and is vital for a variety of "benign" cellular housekeeping functions, including plasma membrane recycling and insertion of both plasma membrane and membrane-localized proteins. Regulated exocytosis, in contrast, occurs in response to specific signals—such as an action potential reaching a nerve terminal—causing an increase in intracellular calcium levels and subsequent activation of the exocytotic machinery [2].

While both kinds of exocytosis are obviously essential to normal cell function, it is regulated exocytosis which is more often thought of when discussing exocytosis. This is the form of

Peter Thorn (ed.), *Exocytosis Methods*, Neuromethods, vol. 83,
DOI 10.1007/978-1-62703-676-4_3, © Springer Science+Business Media New York 2014

exocytosis which is responsible for the release of neurotransmitters, hormones, and other intracellular communication molecules and as such has an important role in both health and a variety of disease states. As might be expected for such an important cellular process, regulated exocytosis has been extensively studied using a variety of techniques and in a number of model systems—both in vivo and in vitro—and while we now have some understanding of the proteins and cellular processes involved in the control of exocytosis, there remains much to be elucidated about this process. Amperometry is a technique which has its roots outside of biology and neuroscience but which has been used since the early 1970s to study exocytosis of neurotransmitters and hormones [3]. In brief, it utilizes the fact that in the presence of a high-voltage electrode, some chemicals are able to undergo oxidation, resulting in a measureable flow of current, in order to measure exocytotic release of these chemicals. The following chapter will discuss amperometry and the valuable information this technique can provide in the study of exocytosis.

1.2 Amperometric Theory

A number of the molecules which undergo exocytosis are capable of being oxidized—that is, they are able to undergo a chemical reaction involving the donation of electrons. Such molecules include the important hormones and neurotransmitters noradrenaline (norepinephrine), adrenaline (epinephrine), dopamine, and serotonin. The technique of amperometry utilizes this property to create a detector for such molecules, consisting of a polarizable electrode in solution, to which a voltage greater than the redox potential of the species in question is applied [4]. Under resting conditions (when no oxidizable molecules are present at the electrode surface), there will be no current flow. In the presence of oxidizable chemicals, however, molecules diffusing close to the surface of such an electrode will be oxidized, transferring electrons to the electrode and resulting in a flow of current which can be recorded and measured using appropriate circuitry (in practice such circuitry operates very similarly to that of a patch-clamp amplifier in voltage-clamp mode). This chapter will primarily discuss constant-voltage amperometry—in which, as the name applies, the voltage applied to the electrode is held constant. Although there are other variations on the technique of amperometry, the constant-voltage technique provides the finest temporal and spatial resolution and is therefore arguably the most useful for investigation of the exocytotic process.

Holding the voltage used at the amperometry electrode sufficiently above the redox potential of the chemical species in question will ensure that the oxidation reaction occurs very quickly and that the rate-limiting step of the process will be the diffusion of the molecule to the electrode. Thus, by placing the electrode in very close proximity to the source of these molecules—such as a cell undergoing exocytosis—it is possible to record very high time resolution events. This time resolution is sufficient that in many cases it is possible to utilize amperometry to detect the release of oxidizable molecules from individual vesicles [5].

Not only is it possible to detect the presence and release of these chemical species, but if the number of electrons transferred during each oxidation reaction is known, then it is also possible to use Faraday's law to directly quantify the number of molecules undergoing oxidation at the electrode surface. This enables us to directly estimate the number of molecules released during exocytosis and, under the right experimental conditions, down to the number of molecules released during fusion of a single exocytotic vesicle.

1.3 Amperometry in the Study of Exocytosis

Amperometry was originally used in the biological sciences to measure secretion in whole-tissue samples, in particular looking at neurotransmitter release in the brain (e.g., see [3, 6]). Such studies initially utilized electrodes made from carbon paste packed into Teflon tubes. Later amperometry electrodes were fabricated using carbon fibers, and these have remained the most popular form of electrode for use in these studies. While providing significant information on the release of neurotransmitters within the brain in the in vivo situation, the full power of amperometry to investigate exocytosis was not appreciated until these techniques were extended to the single-cell level [7]. In the remainder of this chapter, we will discuss in finer detail the methodologies for undertaking amperometric measurements from both whole-tissue and single cells, using our own experience with measurement of epinephrine release from single chromaffin cells [8, 9] and of serotonin release in whole-gut preparations [10, 11] as examples.

In whole-tissue samples amperometric measurements are generally limited to the levels of oxidizable molecules present in the experimental preparation—this is due to the fact that in a whole-tissue preparation, there are usually numerous cells simultaneously releasing the molecule of interest, limiting the spatial resolution of the technique, while the distance of the probe from most individual cells limits the time resolution of detection of release. Nevertheless, while the achievable time/space resolution is insufficient to detect individual release events, it is still more than adequate to demonstrate changes in molecule release from whole-tissue samples in response to experimental manipulation of the tissue.

At the single-cell level, the time and spatial resolution of constant-voltage amperometry is able to provide valuable information on exocytosis down to the level of individual release events—on amperometric recordings these are observed as current "spikes" as individual release "quanta" of molecules from exocytosis of a single vesicle are oxidized. The height and area of these spikes are a measure of the number of molecules released per event, while the width of spikes is an indicator of the speed of release, and the frequency is a direct measurement of the rate of vesicle fusion. In addition to the current "spikes," amperometric recordings can also display "foot signals"—smaller current signals occurring prior to, or even without, an accompanying spike. These signals are believed to represent leak of oxidizable molecules through a forming "fusion pore" [12, 13], with the profile of these signals reflecting parameters of fusion pore formation.

1.4 Comparison of Amperometry and Other Techniques

Amperometry provides a tremendous complement to the range of techniques utilized in the investigation of exocytosis. As with all techniques, it has a number of advantages and disadvantages which should to be taken into account when determining whether this is the optimal technique for answering exocytosis-related hypotheses.

Firstly, amperometry is a noninvasive technique which does not disrupt the normal function of the cells being used. Another major advantage of amperometry over other techniques utilized in the investigation of exocytosis is that it provides direct detection of the exocytosed molecules. This is a major advantage over methodologies such as membrane-bound dyes or other protein-bound fluorescent markers, or membrane capacitance measurements, which may provide measures of the amount of vesicle fusion occurring, but not necessarily the amount of release from these vesicles. Of course, this is offset to a large extent by the primary limitation of amperometry, which is clearly its ability to detect only oxidizable molecules—and while a range of different hormone and neurotransmitter molecules are oxidizable, this obviously does affect the usefulness of amperometry in some applications. Some research groups have been able to alleviate this limitation to some extent by, for example, artificially loading vesicles with oxidizable molecules or using specialized electrode coatings—nonetheless, that ability to only detect oxidizable chemical species is a base limitation of the amperometry technique.

Constant-voltage amperometry is able to provide very fine time resolution of exocytotic events, with spatial resolution limited only by the size of the electrode and the specific sample preparation. The technique, however, provides only limited information regarding the specific chemical species being oxidized. The use of amperometry alone is therefore insufficient to demonstrate the presence of specific oxidizable chemical species (although the oxidation potential utilized at the electrode may be suggestive of certain molecules). Other amperometric techniques, such as cyclic voltammetry—which uses both oxidation and reduction cycles—are able to provide more information in the form of electrochemical "fingerprints" but at the expense of the time resolution of recording. In practice other non-electrochemical techniques are generally required to confirm the presence of the species of interest in a sample.

On a related note, the usefulness of amperometry as a technique for measuring exocytosis is often tied to its ability to detect specific molecules. This usually relies on there being no other oxidizable molecules present—or, at least, the presence of other oxidizable molecules in only low concentrations—or, alternatively, that the oxidizable molecules present can be differentiated based on their redox potential. Obviously whether or not this proves a limitation to the use of amperometry is dependent on the specific samples and experimental model in use—but is necessary to consider when designing studies utilizing amperometry.

In summary then, the technique of amperometry—specifically constant-voltage amperometry—is a valuable and powerful technique in the study of exocytosis, although, as with all techniques, it is often best utilized in conjunction with other techniques to overcome the advantages and disadvantages intrinsic to any methodology. In the remainder of this chapter, we will describe the use of constant-voltage amperometry in two different experimental methodologies—in the measurement of adrenaline exocytosis from single chromaffin cells and in the release of serotonin from whole-tissue gastrointestinal preparations.

2 Materials and Methods

2.1 Technique 1: Adrenaline Release from Single Chromaffin Cells

2.1.1 Chromaffin Cells and Exocytosis

The adrenal chromaffin cell (so called because it can be visualized by staining with chromium salts) is a widely used cellular model for the study of calcium-dependent neurotransmitter release. Chromaffin cells share the same lineage and biochemical characteristics as sympathetic neurons, and the observation over 40 years ago that acetylcholine induced catecholamine release from these cells led to the term "stimulus–secretion coupling" being coined [14]. Chromaffin cells synthesize and secrete the catecholamines adrenaline (epinephrine) and noradrenaline (norepinephrine) which are stored in and released from large dense-core vesicles known as chromaffin granules [14, 15]. Chromaffin cells can be stimulated to release adrenaline using established secretagogues, such as high external K^+. Adrenaline is readily oxidized in solution and is therefore an ideal candidate for amperometric detection. While a large number of studies have utilized bovine adrenal chromaffin cells [15], rat and mouse models have also been employed [16]. Our laboratory routinely isolates murine adrenal chromaffin cells to measure adrenaline release by amperometry as detailed below. The advantages of the mouse model are that animals are easy to maintain, chromaffin cells can be easily and rapidly cultured and can be readily differentiated from cortical cells due to their smooth and round appearance [17], and a large number of amperometric events can be measured from a single cell (e.g., [8]). In addition, the use of a mouse model allows investigations of cells from a range of knockout and transgenic mouse models.

2.1.2 Procedures

Primary Culture of Mouse Adrenal Chromaffin Cells

Most primary cultures involve isolating the tissue or organ of interest, digesting it with an enzyme or enzymes to disrupt connective tissue, filtering out the undigested material, and plating the isolated cells on plastic or glass that has been pretreated to assist with cell adhesion. In our laboratory, the following protocol is used to culture adrenal chromaffin cells from mice:

1. Mice are humanely killed and the adrenal glands removed. It is ideal to start with around six to eight adrenal glands per culture.
2. The adrenal medulla is dissected out from each gland in ice-cold Locke's buffer (145 mM NaCl, 5.6 mM KCl, 3.6 mM $NaHCO_3$, 5.6 mM glucose, 5.0 mM HEPES, pH 7.4) (see Note 1).
3. The medullae are then incubated for 30 min in 5 ml collagenase type A (Roche, Germany, 3 mg/ml in Locke's buffer) in a shaking water bath set at 37 °C. During this time the suspension is triturated carefully with a pipette at 15 min, 25 min, and 30 min, respectively (see Note 2).
4. The enzyme solution is diluted in ice-cold Locke's buffer, and the cells are pelleted at 1,000 × *g* for 10 min at 4 °C.
5. The cell pellet is resuspended in DMEM (Dulbecco's modified Eagle's medium supplemented with 10 % (v/v) heat-inactivated fetal calf serum, 100 units/ml penicillin, and 100 mg/ml streptomycin (Invitrogen, Carlsbad, CA, USA) and filtered through a 40 μm nylon mesh (Sigma-Aldrich, USA).
6. The filtrate is centrifuged as above to pellet cells which are resuspended in supplemented medium and plated onto pretreated 35 mm^2 plastic tissue culture dishes (Iwaki, Asahi Glass Co., Japan) (see Note 3).
7. The cells are incubated at 37 °C with 5 % CO_2 for 3–4 days prior to amperometry (see Note 4).

Notes

1. The aim is to maximize the ratio of medullary/cortical cells harvested, although this step is clearly a trade-off between a higher proportion of medullary cells and an increased time of preparation (which will result in decreased cell health). Improved medullary dissection times come with practice of the procedure.
2. It is more important to have healthy cells rather than concentrate on high yields. The trituration step is extremely important

and must be performed with great care so as not to damage cells. We use a 1 ml pipette and triturate only ten times between shaking (three times in total) and are careful to not introduce bubbles during this process to avoid shear stress on cells. We feel it better have less tissue broken down rather than risk the health of those cells which are ultimately cultured.

3. For amperometric recordings, it is advisable to have cells clustered in the center of the plate (this maximizes the number of cells which are accessible to the amperometry electrode). To achieve this we recommend placing 200 μl of the cell suspension in step 6 above onto the center of each dish, incubating this for an hour or two at 37 °C/5 % CO_2 to ensure cell attachment and then topping up the medium to desired levels (normally 2 ml/dish).
4. At least 2 days incubation is required for cells to regain function following the digestions and isolation protocol.

Amperometric Measurement of Adrenaline Release

During amperometry chromaffin cells are constantly perfused with a standard bath solution (140 mM NaCl, 5 mM KCl, 2 mM $CaCl_2$, 1 mM $MgCl_2$, 5 mM D-glucose, 10 mM HEPES, pH 7.4). The temperature of the bath solution is maintained at 37 °C using a temperature controller (TC-344B; Warner Instrument Corporation, USA) through which the solution flows, and the outlet of the perfusion system is placed within 500 μm of the cell being recorded.

For amperometric measurements a carbon-fiber electrode (5 μm diameter, ProCFE, Dagan Corporation, USA) (see Note 1) is backfilled with mercury and mounted on an electronic micromanipulator (Sutter MP-285; Sutter Instruments, CA, USA) and placed in close apposition to a chromaffin cell, with a voltage of +800 mV applied to the electrode under voltage-clamp conditions. Adrenaline release is stimulated by changing the perfusing solution to one containing a 70 mM K^+ for 60 s (this solution is identical in composition to the standard bath solution but with 70 mM K^+ replacing an equimolar amount of NaCl). Current "spikes" due to oxidation of released adrenaline are recorded under baseline and stimulated conditions (see Note 2), using an EPC-7 amplifier (List Medical, Darmstadt, Germany), an ITC-18 A–D interface (InstruTECH Corporation, NY, USA), and Pulse software (HEKA Elektronik, Germany), sampled at 10 kHz, using a 1 kHz low-pass hardware filter. Essentially any patch-clamp setup can be used to measure these currents. An example amperometric trace recorded from these cells is shown in Fig. 1.

Notes

1. A carbon-fiber electrode should typically not be used for more than five recordings in order to avoid probe desensitization.

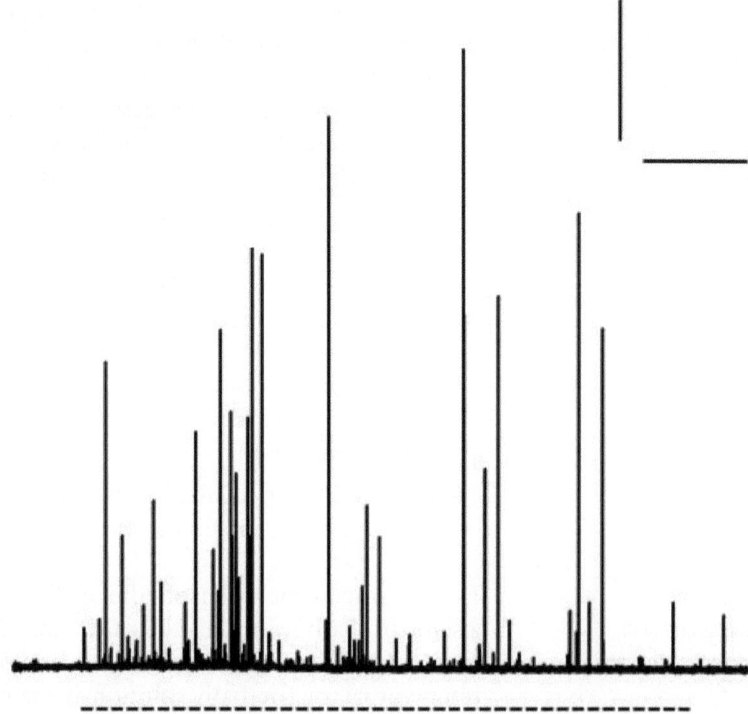

Fig. 1 A typical amperometric trace of stimulus-induced catecholamine secretion from a chromaffin cell. Typical responses of chromaffin cells to stimulation with a 70 mm K^+ solution (indicated by *dashed line* under traces). *Spikes* indicate the release of catecholamines from a single vesicle in mouse chromaffin cells. Scale bar represents 10 s and 100 pA

2. In order to remove the effects of variance between different probes, comparisons of spike kinetics should preferentially be made by recording from one cell per experimental group with the same electrode.

Data Analysis

In our laboratory the Pulse (HEKA Elektronik, Germany) data files are converted to Axon Binary Files (ABF Utility, version 2.1, Synaptosoft, USA) and secretory spikes analyzed (Mini Analysis, version 6.0.1, Synaptosoft, USA) for the period of stimulation (generally 60 s) (see Note 1).

Each spike (Fig. 2) corresponds to a single release event, and these can be measured at a sub-millisecond resolution. Among the various parameters that can be analyzed from a single spike are the amplitude, rise time, half-width, and decay time (Fig. 3). Often, a spike may display what is termed a pre-spike foot (PSF) signal that corresponds to the formation of the fusion pore at the plasma membrane (Fig. 2b) or a stand-alone foot signal (SAF) (Fig. 2c), which is believed to represent kiss-and-run exocytosis, where the

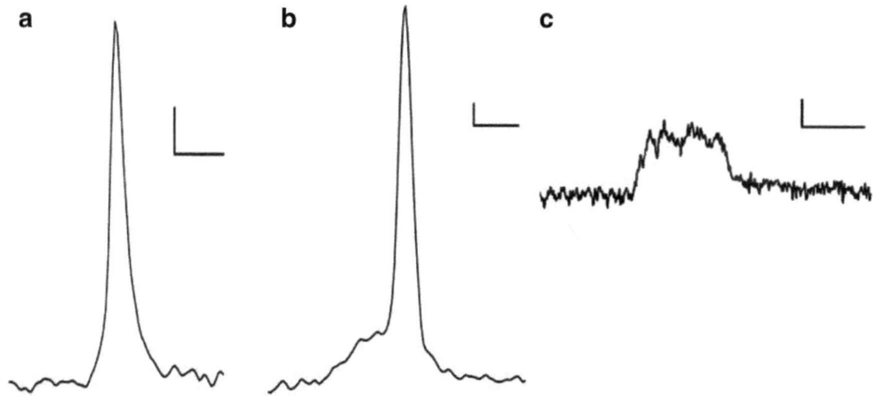

Fig. 2 Different types of individual amperometric spikes as observed using carbon-fiber amperometry. (**a**) A full fusion event without a pre-spike foot signal, (**b**) a full fusion event with a pre-spike foot signal, and (**c**) a kiss-and-run event represented as a stand-alone foot signal are shown. Scale bars represent 2 ms and 10 pA

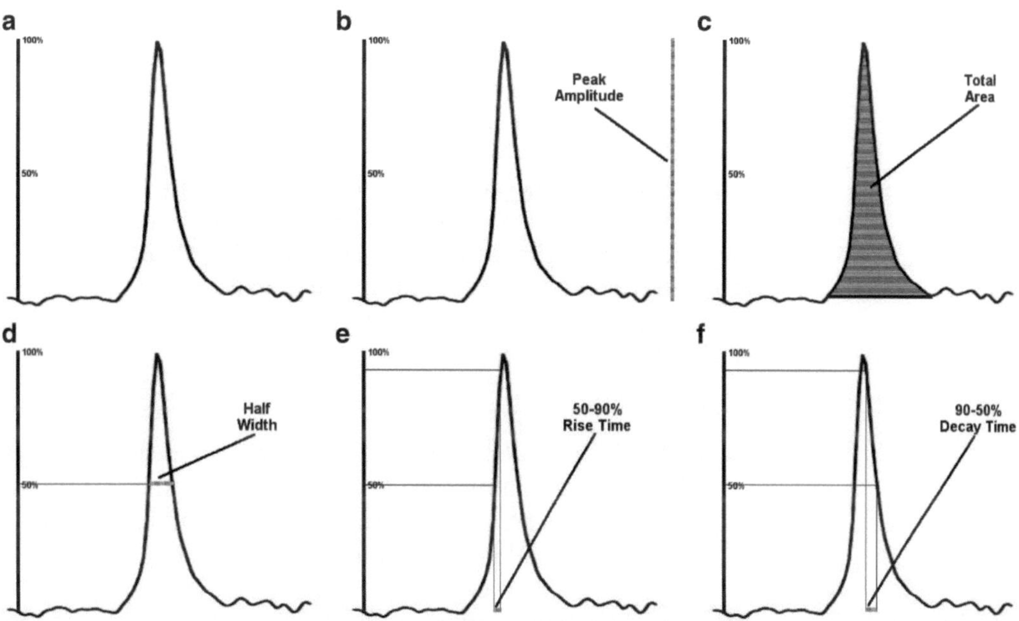

Fig. 3 Analysis of amperometric spikes. A number of variables are routinely extracted from amperometric spikes (**a**), including (**b**) peak amplitude, (**c**) total spike area, (**d**) width (at 50 % height), (**e**) rise time (50–90 % height), and (**f**) decay time (90–50 % height)

fusion pore closes before it has fully stabilized resulting in no current spike being seen after the initial foot signal.

In our experiments we select amperometric spikes for analysis if spike amplitude exceeds 10 pA or at least 2.5 times the root-mean-squared (RMS) noise of the baseline. In our laboratory only cells with less than 10 or more than 200 events within the 60 s stimulation period are excluded from analysis. For kinetic analysis of spikes and PSF signals, only those events that are not overlapping are included. Only PSF signals longer than 1 ms and above 2.5 times the root-mean-squared noise of the baseline are analyzed as foot signals. Rise time of each spike is calculated from the 50–90 % rising phase in order to avoid skewing caused by PSF signals (Fig. 3d).

In analyzing spike kinetics, all spikes that meet our threshold criteria are included in calculating the median values of each spike parameter for each cell. The averages of these median values are then used to compare each parameter between cell populations [18]. This is done to avoid errors associated with pooling large numbers of spikes from cells where there is a large cell to cell variability. Analysis of PSF signal kinetics is normally performed using pooled data from all recorded cells, as many recordings may contain a low number of foot signals, which precludes obtaining an adequate median value for each cell. PSF signal onset is defined when the signal exceeds the peak-to-peak noise of a 5 ms time segment, while the end of the PSF is defined as the inflection point between the PSF signal and the spike. PSF signal lifetime, τ, is taken as the intervening time interval. The total event area is taken as the integral from the PSF onset to the time where the spike current falls to 2.5 times the RMS noise of the baseline.

Spike kinetic data is non-parametrically distributed and can be evaluated for statistical significance using the Mann–Whitney *U* test. Alternatively, cube-rooting all values in each data set to obtain a parametrically distribution can be done prior to using a one-way ANOVA.

Note

1. It should be noted that this is the analysis pipeline employed within our laboratory and for the particular equipment setup we use—there are many alternative software and analysis packages which could be utilized to the same ends.

2.2 Technique 2: Serotonin Release in the Gastrointestinal Tract

2.2.1 Serotonin and Enterochromaffin Cells

Serotonin (5-hydroxytryptamine, 5-HT) is a hormone and neurotransmitter with a broad range of physiological functions. These include well-described roles in the central nervous system such as the regulation of sleep, metabolism, body temperature, appetite, and mood. In the periphery 5-HT is important for a range of homeostatic mechanisms and plays a significant role in multiple GI functions and in several GI disorders. Approximately 95 % [19] of total body 5-HT is synthesized and released from enterochromaffin (EC) cells, specialized endocrine cells located in the epithelial layer lining the gastrointestinal (GI) tract.

EC cells are the major site of peripheral 5-HT synthesis and secretion. EC cells are located among the epithelial cells lining the gut lumen, and, as such, much of the research on EC cell function has focused on their role in the gastrointestinal (GI) tract. These cells are endoderm-derived enteroendocrine cells sharing a common stem cell origin with other epithelial cells [20]. Our current knowledge regarding the regulation of 5-HT secretion from EC cells has largely been gained from studies on intact gastrointestinal tissue. Such studies include in vitro measurements of 5-HT release from segments of dog, pig, guinea pig, ferret, rat, mouse, and human small intestine [21–24]. These intact tissue samples contain a mixture of endocrine, epithelial, muscular, and neuronal cell types, and potential signaling effects from these other cells make it difficult to interpret whether experimental interventions directly or indirectly affect EC cell function. 5-HT release in such whole-tissue samples may also be secondary to muscle contraction [25]. Nonetheless, such studies have provided the majority of mechanistic insight into 5-HT release from EC cells and indicate that secretion is triggered by Ca^{2+} entry through voltage-gated Ca^{2+} channels, with release stimulated by muscarinic, nicotinic, 5-HT_3, and β-adrenergic receptor activation [21–24]. 5-HT release, both in vivo [26] and in vitro [27], is highly correlated with motor complexes.

Several groups have successfully measured 5-HT release from the GI tract using carbon-fiber amperometry. These include measurements from the colon [10, 11] and ileum [28] of various rodent species. In this section we will explain how we use carbon-fiber amperometry to measure 5-HT release from the colon in either an open sheet [11] or an intact tube [10] conformation. While we will outline how we have done this in the colon in mice, the protocol is applicable to almost any section of the GI tract and in any species.

2.2.2 Procedures

Preparation of Tissue

1. C57BL/6 mice (20–90 days old) of either sex are euthanized humanely by inhalation of anesthetic (Nembutal) followed by cervical dislocation.
2. The entire colon is removed and placed in room temperature Krebs solution (118 mM NaCl, 4.7 mM KCl, 1.0 mM $NaHPO_4 \cdot 2H_2O$, 25 mM $NaHCO_3$, 1.2 mM $MgCl \cdot 6H_2O$, 11 mM D-glucose, 2.5 mM $CaCl_2 \cdot 2H_2O$) constantly bubbled with carbogen gas (95 % O_2, 5 % CO_2).
3. When working with an open sheet conformation, a midline incision is made along the mesenteric border and the entire colon pinned mucosal side uppermost in a Sylgard-lined petri dish containing oxygenated Krebs solution.

Amperometric Measurements of Serotonin Release from the Gastric Mucosal Layer

Amperometric recordings from the mouse colon are made using essentially the same technique and equipment we have previously described in this chapter for amperometric recordings on single chromaffin cells. The major deviation from this protocol is that we adjust the electrode holding potential to ~+400 mV to selectively detect the oxidation current that is attributed to release of 5-HT from enterochromaffin cells in these preparations [28].

In an open sheet conformation, the whole colon was placed mucosa uppermost in a Sylgard-lined organ bath and continuously perfused with oxygenated Krebs solution heated to 35–37 °C using an automatic temperature controller (TC-344B; Warner Instrument Corporation). In order to avoid contact with the mucosa or tissue surface during colonic contractions (see Note 1), the carbon-fiber electrode is placed 100 μm above the tissue surface, and +375 to 400 mV applied to the electrode under voltage-clamp conditions.

We have also measured 5-HT release in intact tube preparations in order to retain the circumferential integrity of the colon and preserve peristalsis. The whole colon is placed in a Sylgard-lined organ bath and continuously perfused with oxygenated Krebs solution that is temperature controlled at 35–37 °C as described. The carbon-fiber electrode is placed within 100 μm of the epithelial layer at the opening of the anal end of the intact distal colon tube (see Note 1). At this electrode location, we are able to measure 5-HT ejected from the anal end of the colon during peristalsis.

Note

1. This ensures no 5-HT release or signal artifacts are evoked by the electrode compressing the tissue.

Calibration

In order to measure the concentration of 5-HT being released from the mucosa layer, each electrode should be calibrated both before and after each experiment. It is important to first equilibrate the electrode in normal bath solution prior to beginning this procedure as each electrode loses sensitivity rapidly after first being exposed to physiological buffer. To calibrate an electrode we place the carbon fiber into increasing concentrations of exogenous 5-HT and measure the increasing oxidation currents at the required oxidizing potential (~400 mV). Once we obtain a stable reading at various 5-HT concentrations, we can create a calibration curve of current vs. 5-HT concentration (Fig. 4). As probe sensitivity reduces during an experiment, we undertake this calibration both before and after each experiment and use the average of these readings for our final calibration.

Measurements of Contractile Force in Colonic Preparations

To correlate direct release of 5-HT from enterochromaffin cells with cyclical colonic migrating motor complexes (CMMCs), it is necessary to record the mechanical activity of the circular muscle in at least two independent sites along the whole isolated colon. To do this, we use two independent Grass (FT-03C) isometric force transducers (Grass, Quincy, MA, USA) connected via fine suture thread to two spring stainless steel claws anchored to the colon (Fig. 5). Mechanical recordings are made under isotonic conditions,

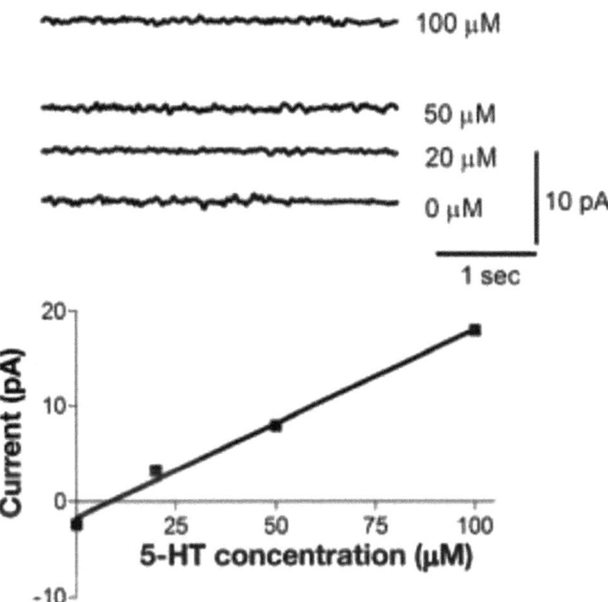

Fig. 4 Calibration curve of current vs. 5-HT concentration. *Upper traces* show amperometric oxidation currents recorded with electrode in solution of indicated 5-HT concentration. *Lower curve* shows calibration curve generated from these recordings to enable calculation of 5-HT concentrations from experimental data

whereby the circular muscle layer can shorten during contraction (see Fig. 5). The isometric force transducers are connected to two custom-made preamplifiers (Biomedical Engineering, Flinders University, SA, Australia) and then to a PowerLab (model 4/30; AD Instruments, Bella Vista, NSW, Australia).

In a tube preparation in order to correlate direct release of 5-HT from EC cells with the initiation of peristalsis, it is necessary to record the propulsive force generated by the colon on an inserted fecal pellet. To do this we use an isometric force transducer connected via fine suture thread to a natural fecal pellet covered in epoxy resin (Fig. 6). When peristalsis is triggered, the string attached to the pellet prevents the pellet from being expelled, and the tension transducer can accurately record the propulsive force that is exerted by the colon on the pellet.

Analysis

We are able to measure multiple aspects of 5-HT release in either open sheet or tube preparations from our amperometric recordings. From our calibration curves we can obtain the peak, steady state, and basal 5-HT levels in these preparations. We also measure the number of 5-HT peaks over an experimental time course to obtain the frequency of peristalsis- or contraction-evoked 5-HT release. These are typically correlated with coincident contractions measured simultaneously (Figs. 5 and 6).

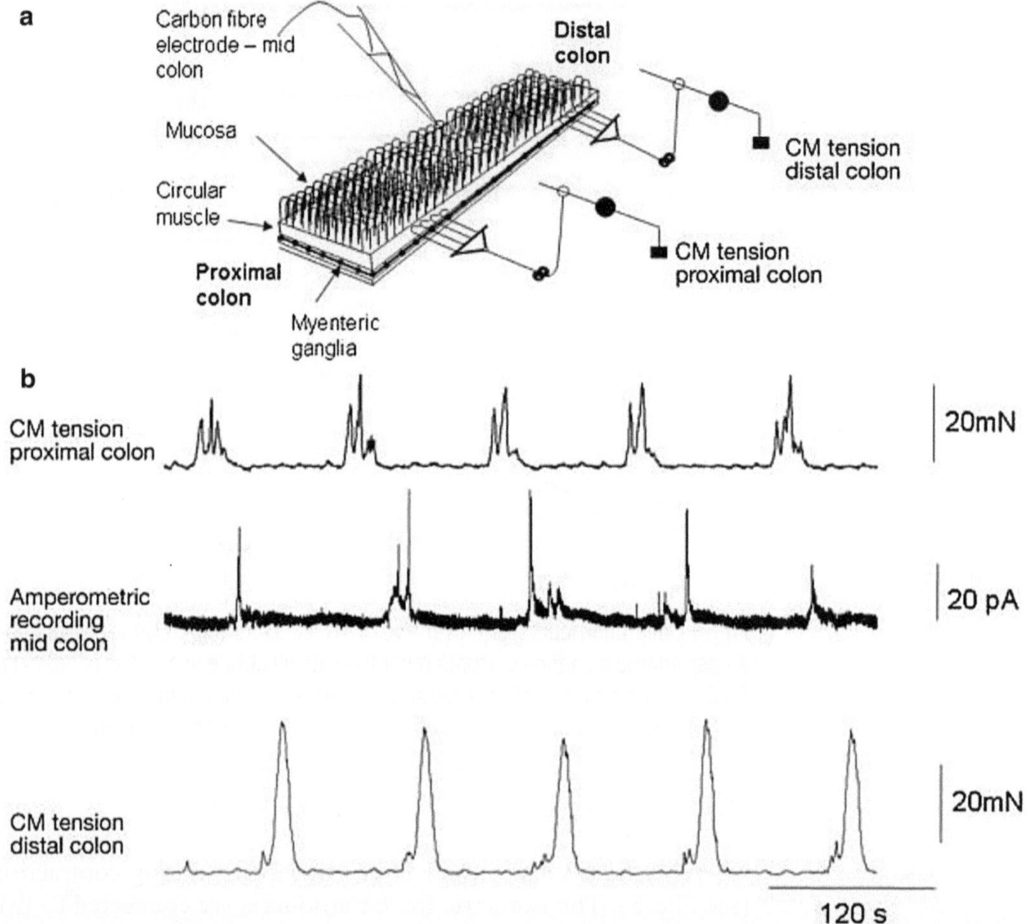

Fig. 5 Amperometric recordings reveal cyclic release of 5-HT from enterochromaffin cells during spontaneous colonic migrating motor complexes (CMMCs). (**a**) Diagrammatic representation of the recording setup for simultaneous mechanical recordings and amperometric recordings. (**b**) Spontaneous anally propagating CMMCs are shown where the CMMC contraction starts in the proximal colon and propagates to the distal colon. As the CMMC propagates past the midcolon, the carbon-fiber electrode detects a release of 5-HT

3 Conclusions

This chapter provides a detailed outline of how we prepare tissue/cells for our experiments, how we measure the release of oxidizable molecules from cells and tissues, and how we analyze this data. Amperometry provides a noninvasive approach to understanding the pathways and proteins that control multiple stages of the exocytosis process. This technique allows measurement of the number of vesicles undergoing exocytosis over a specified time period, the amount of neurotransmitter released from each individual vesicle,

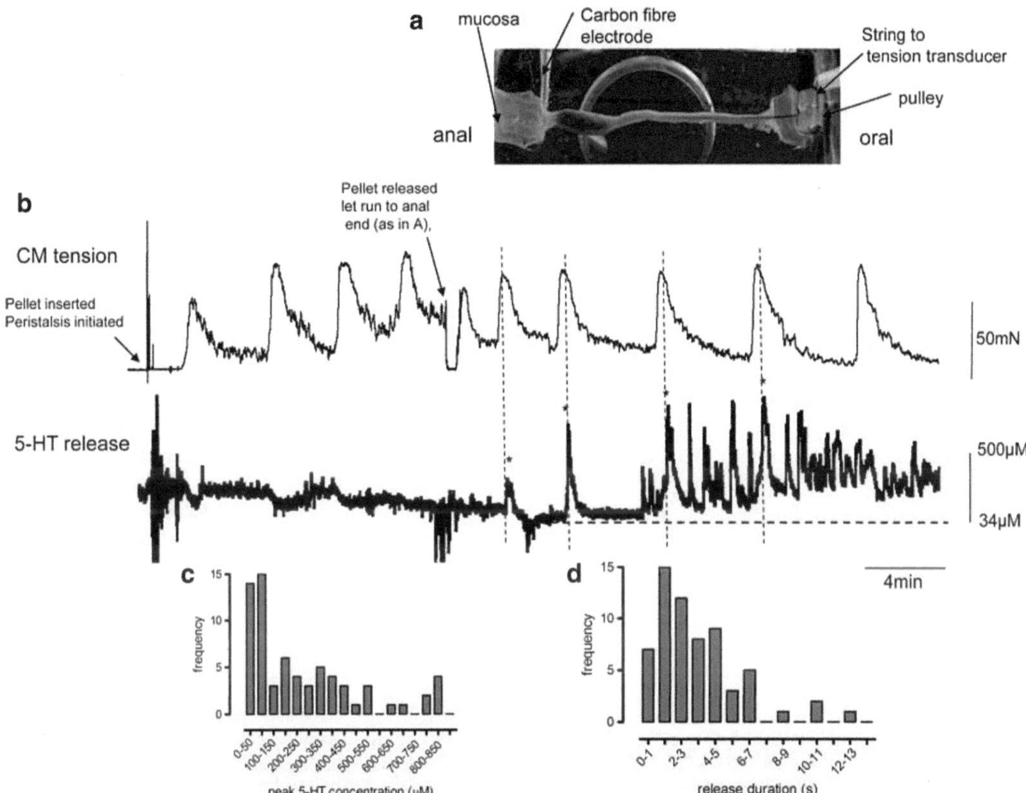

Fig. 6 The measurement of 5-hydroxytryptamine (5-HT) release during peristalsis using amperometry in a colonic tube preparation. In an intact colon, real-time amperometric recordings of 5-HT release were made from a population of enterochromaffin (EC) cells at the same time as cyclical peristaltic waves were initiated in response to a fixed pellet. (**a**) Photomicrograph showing the preparation used, showing the location of the carbon-fiber electrode and fixed pellet. Amperometric recordings were made ~2–3 mm anal to the fixed pellet. (**b**) A simultaneous recording of 5-HT release at the same time as cyclical peristaltic waves in response to a fixed pellet. *Left*, insertion of fecal pellet and initiation of cyclical peristaltic waves. Transient 5-HT release events occurred, which were uncorrelated with peristaltic contractions. The release of the pellet (*arrow*) induced propagation of the pellet to the anal end of the colon as shown in **a**. At this site, there was increased amplitude of 5-HT release events, some of which occurred at the same time as peristalsis and other release events that had no temporal correlation with peristalsis (**b**, *right*). *CM* circular muscle. (**c**) A frequency histogram of the concentrations of 5-HT release events. (**d**) A frequency histogram of the duration of 5-HT release transients

and changes related to the initial fusion pore formation and stability. The technique is also applicable to measurement of the release of oxidizable molecules directly from tissue, such as 5-HT release in sections of gut. Combining this technique with pharmacological and genetic manipulations can provide highly informative data on this important biological process.

References

1. Burgoyne RD, Morgan A (2003) Secretory granule exocytosis. Physiol Rev 83:581–632
2. Sudhof TC (2004) The synaptic vesicle cycle. Annu Rev Neurosci 27:509–547
3. Kissinger PT, Hart JB, Adams RN (1973) Voltammetry in brain tissue—a new neurophysiological measurement. Brain Res 55:209–213
4. Chow R, Rüden L (2009) Electrochemical detection of secretion from single cells. In: Sakmann B, Neher E (eds) Single-channel recording. Springer, New York, pp 245–275
5. Wightman RM, Jankowski JA, Kennedy RT et al (1991) Temporally resolved catecholamine spikes correspond to single vesicle release from individual chromaffin cells. Proc Natl Acad Sci U S A 88:10754–10758
6. Stamford JA (1986) In vivo voltammetry: some methodological considerations. J Neurosci Methods 17:1–29
7. Leszczyszyn DJ, Jankowski JA, Viveros OH, Diliberto EJ, Near JA, Wightman RM (1990) Nicotinic receptor-mediated catecholamine secretion from individual chromaffin cells. Chemical evidence for exocytosis. J Biol Chem 265:14736–14737
8. Keating DJ, Dubach D, Zanin MP et al (2008) Dscrl/rcanl regulates vesicle exocytosis and fusion pore kinetics: implications for Down syndrome and Alzheimer's disease. Hum Mol Genet 17:1020–1030
9. Zanin MP, Phillips L, Mackenzie KD, Keating DJ (2011) Aging differentially affects multiple aspects of vesicle fusion kinetics. PLoS One 6:e27820
10. Spencer NJ, Nicholas SJ, Robinson L et al (2011) Mechanisms underlying distension-evoked peristalsis in guinea pig distal colon: is there a role for enterochromaffin cells? Am J Physiol 301:G519–G527
11. Keating DJ, Spencer NJ (2010) Release of 5-hydroxytryptamine from the mucosa is not required for the generation or propagation of colonic migrating motor complexes. Gastroenterology 138:659–670, e652
12. Chow RH, von Ruden L, Neher E (1992) Delay in vesicle fusion revealed by electrochemical monitoring of single secretory events in adrenal chromaffin cells. Nature 356:60–63
13. Zhou Z, Misler S, Chow RH (1996) Rapid fluctuations in transmitter release from single vesicles in bovine adrenal chromaffin cells. Biophys J 70:1543–1552
14. Douglas WW (1968) Stimulus-secretion coupling: the concept and clues from chromaffin and other cells. Br J Pharmacol 34:451–474
15. Burgoyne RD (1991) Control of exocytosis in adrenal chromaffin cells. Biochim Biophys Acta 1071:174–202
16. Domínguez N, Rodríguez M, Machado JD, Borges R (2012) Preparation and culture of adrenal chromaffin cells # T neurotrophic factors. Methods Mol Biol 846:223–234
17. Moser T, Neher E (1997) Rapid exocytosis in single chromaffin cells recorded from mouse adrenal slices. J Neurosci 17:2314–2323
18. Colliver TL, Hess EJ, Ewing AG (2001) Amperometric analysis of exocytosis at chromaffin cells from genetically distinct mice. Neurosci Methods 105:95–103
19. Erspamer V (1954) Pharmacology of indolealkylamines. Pharmacol Rev 6:425–487
20. Thompson M, Fleming KA, Evans DJ, Fundele R, Surani MA, Wright NA (1990) Gastric endocrine cells share a clonal origin with other gut cell lineages. Development 110:477–481
21. Racke K, Schworer H (1991) Regulation of serotonin release from the intestinal mucosa. Pharmacol Res 23:13–25
22. Racke K, Reimann A, Schworer H, Kilbinger H (1996) Regulation of 5-ht release from enterochromaffin cells. Behav Brain Res 73:83–87
23. Minami M, Tamakai H, Ogawa T et al (1995) Chemical modulation of 5-ht3 and 5-ht4 receptors affects the release of 5-hydroxytryptamine from the ferret and rat intestine. Res Commun Mol Pathol Pharmacol 89:131–142
24. Hirafuji M, Ogawa T, Kato K et al (2001) Noradrenaline stimulates 5-hydroxytryptamine release from mouse ileal tissues via alpha(2)-adrenoceptors. Eur J Pharmacol 432:149–152
25. Bertrand PP (2006) Real-time measurement of serotonin release and motility in guinea pig ileum. J Physiol 577:689–704
26. Tanaka T, Mizumoto A, Mochiki E, Haga N, Suzuki H, Itoh Z (2004) Relationship between intraduodenal 5-hydroxytryptamine release and interdigestive contractions in dogs. J Smooth Muscle Res 40:75–84
27. Bertrand PP, Bertrand RL (2010) Serotonin release and uptake in the gastrointestinal tract. Auton Neurosci 153:47–57
28. Bertrand PP (2004) Real-time detection of serotonin release from enterochromaffin cells of the guinea-pig ileum. Neurogastroenterol Motil 16:511–514

Chapter 4

Imaging of Insulin Exocytosis from Pancreatic Beta Cells

Mica Ohara-Imaizumi, Kyota Aoyagi, and Shinya Nagamatsu

Abstract

The pancreatic beta cells are highly sensitive to the glucose concentration, and the minute to minute regulation of insulin secretion by glucose occurs at the level of exocytosis of insulin, transcriptional rate of the insulin gene, translation of the mRNA, and processing of the proinsulin to mature insulin (Mahato et al., Biochem J 174:517–526, 1978; Hedeskov, Physiol Rev 60:442–509, 1980; Welsh et al., J Biol Chem 260:13590–13594, 1985; Welsh et al., Biochem J 235:459–467, 1986; Newgard and McGarry, Annu Rev Biochem 64:689–719,1995). Thus, glucose is the most important physiological regulator of insulin gene transcription, biosynthesis, and secretion. The insulin biosynthetic rate is not always proportional to the secretion rate. Insulin secretion occurs by the fusion of insulin granules and plasma membrane (Nagamatsu, Pancreatic beta cells in health and disease 177–194, 2008). Insulin granules exist as large-dense core structures that are discernible by electron microscopy as an electron-dense interior surrounded by a clear region in a 300–350-nm intracellular membrane-delineated compartment (Greider et al., J Cell Biol 41:162–166, 1969; Lange, J Ultrastruct Res 46:301–307, 1974).

It is generally accepted that SNARE proteins play a crucial role in insulin exocytotic process (Wollheim et al, Diabetes Rev 4:276–297, 1996). Several reports revealed that t-SNARE protein syntaxin 1(syt1) and SNAP-25 are plasma membrane localized, whereas the v-SNARE protein VAMP2 is associated with insulin secretory granules (Regazzi et al., EMBO J 14:2723–2730, 1995; Sadoul et al., J Cell Biol 128:1019–1028, 1995; Boyd et al., J Biol Chem 270:18216–18218, 1995; Nagamatsu et al., J Biol Chem 271:1160–1165, 1996). Furthermore, we previously reported that alteration of SNARE proteins expression is closely associated with type 2 diabetes (Nagamatsu et al., Diabetes 48:2367–2373, 1999; Ohara-Imaizumi et al., Diabetologia 47:2200–2207, 2004). In order to explore the mechanism of insulin exocytosis, methods for gene transfer and imaging system are powerful tools. Here, we describe about the methods for studying the insulin exocytosis from the point of view of imaging techniques and gene transfer system.

Key words Exocytosis, Insulin, Electroporation, Fluorescence, TIRF

1 A New Electroporation System Gene Transfer System in Pancreatic Beta Cells

1.1 Introduction

Gene transfection is a powerful tool to study the functional role of proteins in the glucose-induced insulin secretion. Due to their capacity to mediate highly efficient gene transfer in nondividing cells, virus-mediated gene transfer has emerged as the first choice for engineering beta cells. Together with Lentiviruses [16] and adeno-associated viruses [17], adenoviruses are the most

Peter Thorn (ed.), *Exocytosis Methods*, Neuromethods, vol. 83,
DOI 10.1007/978-1-62703-676-4_4,

commonly used viral vectors in current beta-cell research [18]. Because of their high transfection efficacy, the amount of secreted insulin from recombinant virus-infected islets is widely used to evaluate the effect of transgene on the insulin secretion from infected beta cells. However, adenoviruses can infect not all but only a subpopulation of beta cells in pancreatic islets. Thus, the amount of secreted insulin from recombinant virus-infected islets should consist of that from infected and uninfected beta cells in infected islets, which results in the underestimation of the effect of transgene expression.

To circumvent of this problem, human growth hormone (hGH) is widely used as a transfection reporter [19]. In many cell types, exogenously expressed hGH is shown to be stored in secretory vesicles and secreted into extracellular medium in response to stimulation; thus, the measurement of secreted hGH can be used to evaluate the secretion only from transfected cells. Indeed, in beta-cell-derived cell lines, INS-1 and Min6 cells, co-transfection of hGH with a gene of interest by lipofection and evaluation of secreted hGH to examine the effect of transgene on secretion is commonly used [20, 21]. However, it is technically difficult for adenoviruses to transfer two independent transgene simultaneously. Nonviral methods are commonly used for co-transfection in several cell types, but their transfection efficacies in pancreatic islets are very low [22–24] because pancreatic islets are terminally differentiated cell clusters that are difficult to introduce exogenous plasmids.

Here, we described a new transfection system based on electroporation, which enables us to co-transfect hGH and a gene of interest and thereby assess the effect of transgene on the glucose-induced secretion. Our method uses standard plasmids to introduce transgene, which will cut down the time needed to produce recombinant adenoviruses, thus accelerates projects to elucidate the beta-cell functions.

1.2 Materials and Methods

1.2.1 Overview of Islet Electroporation

Freshly isolated pancreatic islets are suspended in Hank's balanced salt solution (HBSS) (Invitrogen). To prevent islets from adhering to manipulating pipette and electrode chamber, 0.5 % BSA is supplemented in the HBSS. We confirmed that 0.5 % BSA did not affect transfection efficacy and islet viability. Islets are transferred into an open electrode chamber (CUY520P5, Nepa gene, Chiba, Japan) filled with 0.5 % BSA-containing HBSS supplemented with plasmid DNA at a concentration of 1.0–1.5 mg/ml. To increase the transfection efficacy, aggregated islets should be dispersed by pipetting. After application of electric pulses, islets are handpicked using pipette and suspended into culture medium. After two times wash with culture medium, electroporated islets are placed in 35- or 60-mm culture dish (Iwaki, Tokyo, Japan) and cultured for 2 days.

1.2.2 DNA Concentration

High plasmid DNA concentration will result in high transfection efficacy. On the other hand, high DNA concentration will generate big sticky materials mainly made from DNA on the surface of the anode, which will entangle islets and result in severe difficulty in islet recovery from the electrode chamber after electroporation. Therefore, DNA solution at a concentration of 1.0–1.5 mg/ml is usually used for our electroporation. Plasmid DNA is prepared using Qiagen Mega or Giga prep kit (Qiagen) and dissolved into TE at a concentration of >8 mg/ml for stock. Quality of plasmid DNA severely affects the transfection efficacy. The plasmid DNA prepared by CsCl centrifugation followed by phenol/chloroform extraction reduces the transfection efficacy in our procedure.

1.2.3 Number of Islets for Electroporation

The number of islets subjected to electroporation does not affect the transfection efficacy if they are not aggregated. However, if too much islets are used for one electroporation, a part of islets will be lost by sticking the DNA-containing materials on the surface of the anode. Therefore, to avoid the deprivation of islets, 100–150 islets are usually used for one electroporation.

1.2.4 Electric Pulse Setting

Nepa21 pulse generator (Nepa gene) is used to apply electric pulses. This pulse generator enables to apply electric pulses with two distinct settings. For islet electroporation, short, high-voltage pulses (poring pulses) are applied to break the plasma membrane and form small pores, and then plasmid DNA is introduced into islet cells by long, low-voltage pulses (transfer pulses). The settings of transfer pulses have little impact on transfection efficacy and islet viability. We observed that the voltage (50–80 V) and pulse length (50–70 ms) did not affect the transfection efficacy. On the other hand, the poring pulse setting dramatically affects the transfection efficacy and islet viability. In order to optimize the poring pulse setting, we examined the relationship between the voltages of poring pulses and the amount of exogenously expressed hGH in transfected islets. In the experiments, various voltage pulses (6 pulses, decay constant (10 %), pulse interval (50 ms)) were applied as poring pulses and the amount of hGH expressed in electroporated islets was measured by ELISA (Roche) 2 days after electroporation. As shown in Fig. 1, the amount of expressed hGH showed bell-shaped relationship against the 1 ms poring pulses. On the other hand, the amount of expressed hGH was inversely correlated with the voltages of 2.5 ms poring pulses. In these experiments, we found that application of poring pulses at 200 V for 2.5 ms resulted in the highest expression of hGH, but we also found that isolated islets were seriously damaged by 2.5 ms pulses. As for the 1 ms pulses, 220 V gave higher expression of hGH than 200 V, but electroporated islets with 200 V pulses looked better than those with 220 V. From these results, we apply six pulses (200 V, 1 ms duration, 50 ms interval, decay constant (10 %)) as

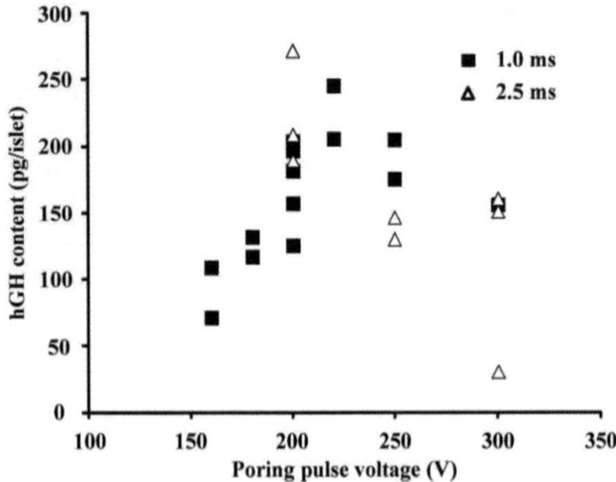

Fig. 1 Relationship between poring pulse setting and the amount of exogenously expressed hGH. Size-matched pancreatic islets were subjected to 1.0-ms (*filled rectangle*) or 2.5-ms (*open triangle*) poring pulse with indicated voltages, and the amounts of hGH were determined by ELISA 2 days after electroporation

poring pulses followed by five pulses (50 V, 50 ms duration, 50 ms interval, decay constant (40 %)) and another five pulses in the opposite direction of the electric field as transfer pulses.

1.2.5 Western Blotting Analysis of Exogenously Expressed Gene

To examine whether the exogenously expressed gene can be expressed in electroporated islets, we performed western blotting. Freshly isolated islets were electroporated with a plasmid encoding V5-tagged human Arf6 (Arf6-V5) or a mock vector and cultured for 2 days. Seventy-five electroporated islets were washed with KRB containing 2.2 mM glucose and recovered in 50 μl of sample buffer (2 % (w/v) SDS, 10 % (w/v) sucrose, 0.05 % (w/v) Bromophenol blue, 62.5 mM Tris–HCl, pH 7.4) supplemented with 20 μM DTT. Sodium dodecyl sulfate (SDS)-polyacrylamide gel electrophoresis (PAGE) was performed on 11 % polyacrylamide gel equipped with 5 % stacking gel. After separation of 20 μl of the sample by SDS-PAGE, the proteins were transferred to polyvinylidene difluoride (PVDF) membranes following standard procedures with a semidry transblotting apparatus. The membrane was blocked in 5 % (w/v) nonfat milk in TBST (150 mM NaCl, 0.05 % (w/v) Tween 20, 25 mM Tris–HCl, pH 7.5) for 30 min at room temperature followed by overnight incubation with anti-V5 antibody (Invitrogen) diluted in Can Get Signal solution (TOYOBO, Tokyo, Japan). After washing in TBST, the membrane was incubated for 1 h at room temperature with horse radish peroxidase-labeled anti-mouse IgG (DAKO) in Can Get Signal solution. After washing, the immunoreactive bands were visualized by SuperSignal (PIERCE) and a luminescence image analyzer with

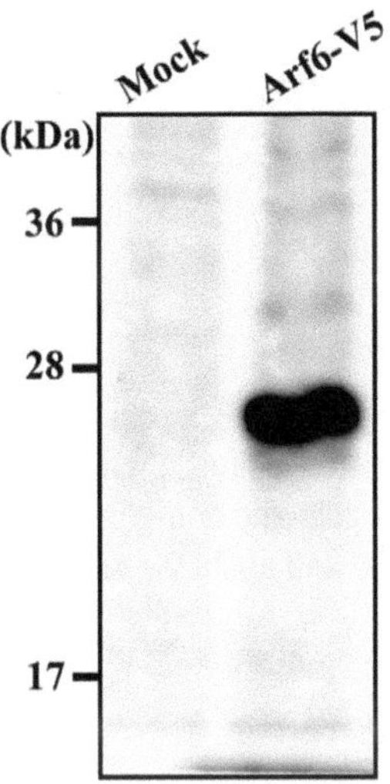

Fig. 2 Western blotting analysis in transfected islets. A mock vector or the vector encoding V5-tagged human Arf6 transfected islets were subjected to western blotting using anti-V5 antibody

an electronically cooled charge-coupled device camera system (LAS-4000, GE Healthcare). As shown in Fig. 2, a specific signal can be detected in Arf6-V5-transfected islets.

1.2.6 Identification of Transfected Islet Cell Types

Pancreatic islets consist of four different types of endocrine cells: insulin (β)-, glucagon (α)-, somatostatin (δ)-, and pancreatic polypeptide (PP)-producing cells. In mouse islets, alpha and beta cells are composed of 18 and 77 % of total islet cell population, respectively [25]. Thus, exogenous plasmid DNA will be introduced into these four types of cells by electroporation of islets and identification of cell type of transfected cells should be required. To this end, we examined which types of cells were transfected by our electroporation.

Electroporated islets with a plasmid encoding CMV promoter-driven EGFP (pEGFP-N1, Clontech) were dissociated into single cells by incubation in Ca^{2+}-free KRB containing 1 mM EGTA and cultured on fibronectin-coated (Koken Co., Tokyo, Japan) coverslips in RPMI1640 medium supplemented with 10 % FBS (Gibco BRL), 200 units/ml penicillin, and 200 μg/ml streptomycin at 37 °C in an atmosphere of 5 % CO_2.

Two days after electroporation, cells were fixed for 15 min at room temperature in 4 % paraformaldehyde (PFA) and washed three times in PBS for 5 min each. They were then incubated with the anti-insulin (Sigma) or anti-glucagon (Sigma) antibodies at 4 °C overnight, respectively. After washing three times with PBS for 5 min each, they were incubated with alexa546-tagged

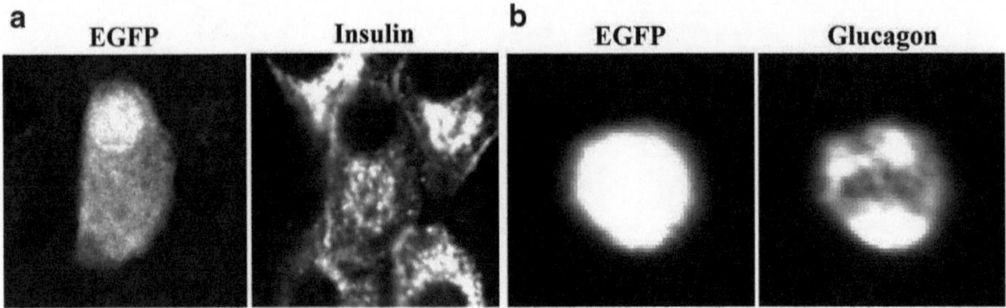

Fig. 3 Immunostaining of transfected islet cells. The CMV promoter-driven EGFP transfected islet cells were immunostained with anti-insulin and anti-glucagon antibody. About 50 % of EGFP positive cells were stained with anti-insulin antibody (**a**) and about 30 % of EGFP positive cells were immunostained with anti-glucagon antibody (**b**)

anti-mouse IgG (Invitrogen) at room temperature for 1 h, washed three times with PBS, and then mounted with DABCO. Immunostained islet cells were observed using confocal laser microscopy, FV1000 with a 40× 1.35 NA UApo objective (Olympus, Tokyo, Japan).

As shown in Fig. 3, we observed GFP-positive cells labeled with anti-insulin and anti-glucagon antibodies, respectively. By counting the number of GFP-positive cells, we found that about 50 and 30 % of transfected cells were insulin-positive β-cells and glucagon-positive α-cells, respectively.

1.2.7 Secretion Assay from Transfected β-Cells in Islets

In order to demonstrate that electroporated islets cultured for 2 days can maintain their function, we next examined the secretagogue-induced secretion in transfected islet cells. Because exogenously expressed human growth hormone was shown to be stored into insulin granule and to be secreted into extracellular medium by exocytosis in pancreatic beta-cell lines [20, 21], freshly isolated islets were electroporated with a plasmid encoding CMV promoter-driven hGH. Two days after electroporation, 15 size-matched islets were preincubated for 30 min in 500 μl of KRB containing 2.2 mM glucose and then transferred into 250 μl of KRB containing 2.2 or 16 mM glucose. After 30 min incubation, islets were removed by pipette and supernatant was recovered as the secreted sample. The picked islets were transferred into 1 ml of KRB containing 1 % Nonidet P-40 and 100 mM KCl, then sonicated on ice to recover the total cellular contents of insulin and hGH. The secreted and total cellular content of insulin and hGH were measured by an insulin ELISA kit (Morinaga, Japan) and a hGH ELISA kit (Roche). As shown in Fig. 4, 16 mM glucose, a physiological secretagogue for insulin from pancreatic beta cells, stimulated hGH secretion, indicating that β-cells in electroporated islets can maintain their function after 2 days cultivation.

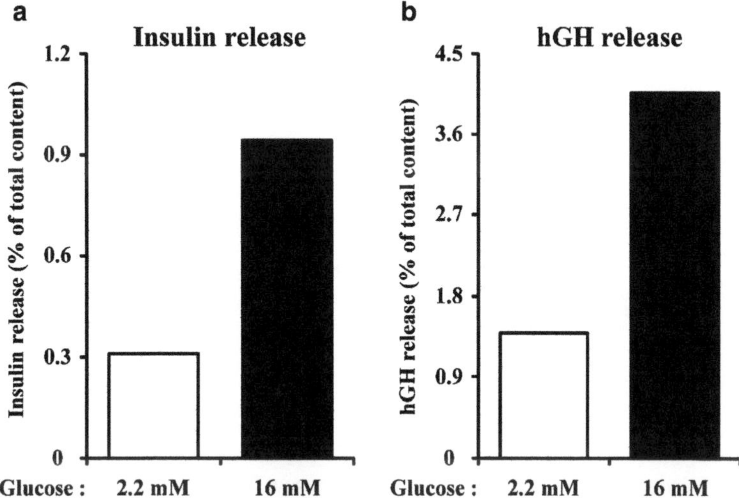

Fig. 4 An example of insulin/hGH secretion assay. Fifteen size-matched islets transfected with hGH were stimulated with 2.2 or 16 mM glucose for 30 min. The amount of insulin and hGH secreted from the same batch of islets was measured by ELISA. The amount of secreted insulin (**a**) and hGH (**b**) is expressed as the percentage of total cellular contents

Pancreatic beta cells show a characteristic biphasic insulin secretion consisting of a rapidly developing and transient first phase followed by a sustained second phase [26, 27]. To further examine whether transfected β-cells in electroporated islets can respond to the glucose stimulation, we next performed the perifusion assay to demonstrate biphasic hGH secretion in transfected beta cells. Freshly isolated islets were electroporated with the vector encoding the CMV promoter-driven hGH and cultured for 2 days. The size-matched 200 islet s were housed in a small chamber and perifused with KRB containing 2.8 mM glucose for 30 min at a flow rate of 0.3 ml/min. Insulin/hGH secretion was stimulated by 22 mM glucose for 30 min. After the experiments, islets were solubilized by 1 % Nonidet P-40 to recover the total cellular content of insulin and hGH. As show in Fig. 5, hGH secretion showed a biphasic pattern, which looks very similar to that of insulin secretion from the same batch of islets.

2 TIRF Imaging in Pancreatic Beta Cells

2.1 Introduction

TIRF microscopy is a technique that specifically illuminates fluorophores within a closely restricted layer just adjacent to the interface at which total internal reflection occurs [28, 29]. When angled light strikes the interface and the light undergoes total internal reflection, some of the light penetrates the

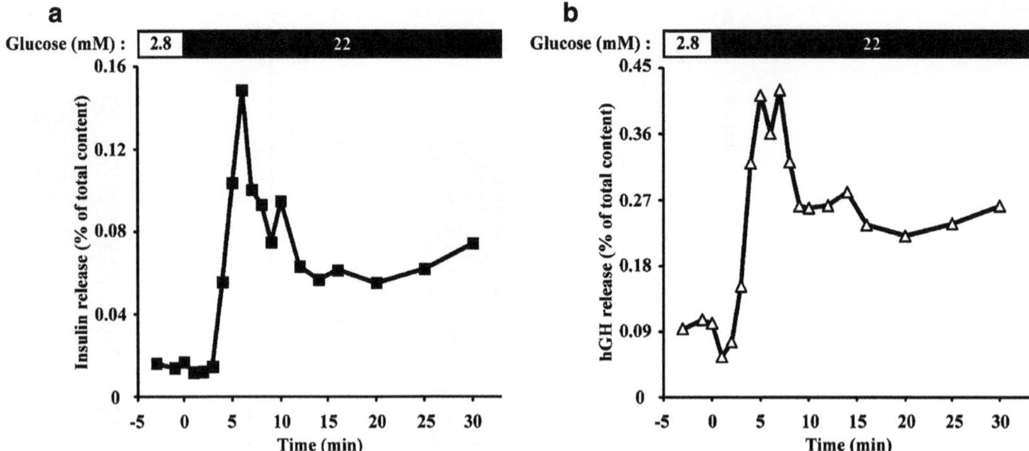

Fig. 5 An example of insulin/hGH perifusion assay. Two hundred of size-matched islets transfected with hGH were perifused with 2.8 or 22 mM glucose. The amount of insulin and hGH secreted from the same batch of islets was measured by ELISA. Both insulin (**a**) and hGH (**b**) showed the characteristic biphasic secretion pattern with almost the same time course. The amount of secreted insulin and hGH is expressed as the percentage of total cellular contents

interface as an electromagnetic field called "evanescent wave." Because the evanescent wave vanishes exponentially with distance from the interface, the penetration depth of the evanescent field is usually less than 100 nm (Fig. 6). Therefore, when cells are grown on a coverslip, a very thin layer adjacent to the coverslip can be specifically illuminated by the evanescent wave formed by the total internal reflection occurring at the interface between the coverslip and extracellular medium. In pioneering work by Daniel Axelrod, this technique was first applied to the study of live cells, primarily to investigate cell surface adhesions [30]. Then, several studies of protein dynamics, topography of cell–substrate contacts, endocytosis, and exocytosis using this technique were reported [31–36].

Because the exocytotic process occurs in the vicinity of the plasma membrane, TIRF microscopy is a powerful tool to analyze the exocytotic process of individual insulin granules in living pancreatic beta cells. To observe the insulin secretory process under TIRF microscopy, insulin granules should be labeled with fluorescent probes. Previously, a weak base fluorescent dye, acridine orange, was used for labeling granules because the inside of most vesicles is acidic [37]. However, acridine orange accumulates not only in insulin granules but also other secretory vesicles such as GABA-containing small synaptic-like vesicles. Thus, the expression of GFP-tagged insulin would be best to specifically analyze the motion of insulin granules. Therefore, we made a recombinant adenovirus encoding human preproinsulin tagged with GFP, which is localized in insulin secretory granules [38, 39], allowing us to

a

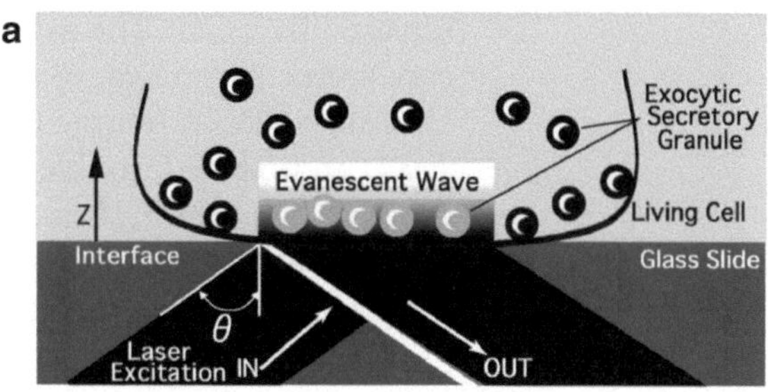

b

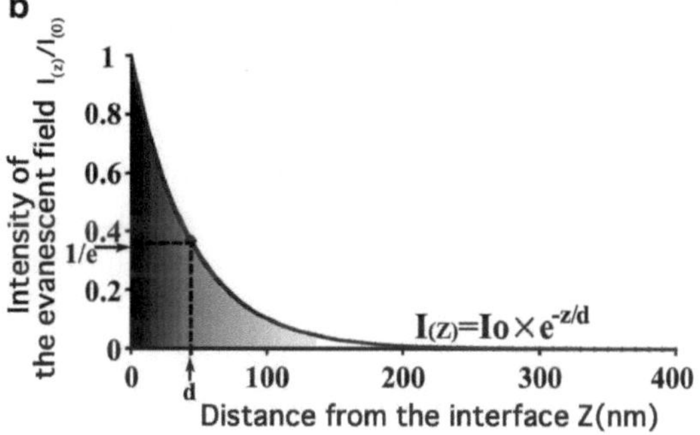

Fig. 6 TIRF excitation using an evanescent field. (**a**) Schematic drawing of the evanescent wave excitation. When a light beam propagating through a transparent medium of high refractive index n_1 (e.g., coverglass) encounters an interface with a medium of lower refractive index n_2 (e.g., adhering cells), total internal reflection occurs at incidence angles (θ), which are greater than critical angle. Although being totally reflected, some of the incident light beam evokes an evanescent field that penetrates into the medium of lower refractive index. (**b**) The intensity of the evanescent field exponentially decays with increasing distance from the interface: $I_{(z)} = I_{(0)}e^{-Z/d}$, where $I_{(z)}$ represents the intensity at a perpendicular distance z from the interface and $I_{(0)}$ is the intensity at the interface. The theoretical penetration depth (d: the distance where the intensity has decreased to $I_{(0)}/e$) depends on the incidence angle (θ), wavelength (λ), as well as the refractive index of the glass (n_1) and medium (n_2): $d = (\lambda/4\pi)(n_1^2\sin^2\theta - n_2^2)^{-1/2}$. The penetration depth, which usually ranges 30–100 nm, decreases with increasing θ. For example, when a laser light (λ = 488 nm) propagates through the high-refractive-index glass ($n_1 = 1.8$) at the incidence angle (θ = 65°) onto the cells ($n_2 = 1.37$), d equals 44 nm calculated from the formula (see refs. [28, 29])

analyze the real-time motion of insulin granules (**Note 1**). We describe the methodology of TIRF microscopy in insulin exocytosis of pancreatic beta cells.

2.2 Materials

2.2.1 Probes Used for Labeling of Insulin Granules

1. Plasmids Used for Labeling of Insulin Granules in MIN6 Cells Human pre-proinsulin cDNA pchi 1–19 (provided by Professor G. I. Bell, University of Chicago, Chicago, IL, USA) lacking a TGA stop codon was amplified by PCR using forward primer, 5-*GAATTC*CGGGGGTCCTTCTGCCATG-3 (*Eco*RI site is italicized), and reverse primer, 5-*GGATCC*-CAGTTGCAGTAGTTCTCCAGC-3, where TGA was

replaced by TGG. The product was subcloned into a pGEMT easy vector (Promega, Madison, WI, USA). The approx. 0.3-kb insulin cDNA fragment lacking stop codon was cleaved by *Eco*RI and *Bam*HI double digestion, which was subcloned into the *Eco*RI and *Bam*HI site of multiple cloning sites of pEGFP-N1 (Clontech/Takara Bio, Otsu, Japan) [38].

2. Recombinant Adenovirus Used for Labeling of Insulin Granules in Primary Cultured Beta Cells
   For the recombinant adenovirus production of insulin–GFP, a cDNA fragment containing pre-proinsulin and GFP was cut by *Eco*RI and *Not*I restriction enzymes, which was ligated into the pAdex1CA cosmid vector, and Adex1CA insulin–GFP was prepared and amplified by the standard protocol (Takara Shuzo Co., Kyoto, Japan) as described previously [40]. The reverse primer used here encodes the C-terminus of pre-pro-insulin, ending LENYCNWD. The fusion with the N-terminus of eGFP includes a short linker (provided by the pEGFP-N1 plasmid) such that the sequence reads LENYCNWDPPVATM (the linker sequence is underlined and M is the first residue of eGFP) [39].

2.2.2 Cell Culture

1. MIN6 cells
   - MIN6 cells (a gift from Prof. J.-I. Miyazaki (Osaka University, Osaka, Japan)) [41].
   - DMEM (Dulbecco's modified Eagle's MEM, high glucose) supplemented with 10 % fetal bovine serum (FBS), 50 units/ml penicillin, 50 μg/ml streptomycin, 28.4 μM 2-mercaptoethanol (GIBCO, Carlsbad, CA, USA).
   - Trypsin–EDTA: 0.25 % trypsin with 1 mM EDTA (ethylenediaminetetraacetic acid) solution (GIBCO).
   - PBS (phosphate buffered saline, pH 7.4, without $CaCl_2$, without $MgCl_2$).
   - Fibronectin (KOKEN, Tokyo, Japan).
   - Effectene transfection reagent (Qiagen, Hilden, Germany) (**Note 2**).
   - Coverglass: High-refractive-index ($n_d = 1.788$) coverglass (Olympus). Coverglass were treated with fibronectin in PBS for 15 min.
2. Primary Cultured pancreatic beta cells
   - C57BL/6 mice age 10–14 weeks (**Note 3**).
   - RPMI1640 medium (11 mM glucose, GIBCO/Invitrogen) supplemented with 10 % fetal bovine serum (FBS), 200 units/ml penicillin, 200 μg/ml streptomycin.

- Hank's balanced salt solution (HBSS, GIBCO/ Invitrogen).
- Ca^{2+}-free KRB (Krebs-Ringer buffer; containing 1 mM EDTA).
- Liberase RI (Roche, Mannheim, Germany).
- Fibronectin (KOKEN, Tokyo, Japan).
- High-refractive-index (n_d=1.788) coverglass (Olympus) treated with fibronectin in PBS for 15 min.

2.2.3 Buffers and Reagents for Live Cell TIRF Imaging

1. Low-glucose KRB: KRB (110 mM NaCl, 4.4 mM KCl, 1.45 mM KH_2PO_4, 1.2 mM $MgSO_4$, 2.3 mM calcium gluconate, 4.8 mM $NaHCO_3$, 2.2 mM glucose, 10 mM HEPES, 0.3 % BSA; pH 7.4) was prepared.
2. High-glucose KRB: KRB (glucose concentration was changed to 22 mM) was prepared.
3. High KCl–KRB (NaCl was reduced to maintain the isotonicity of the solution) was prepared.
4. Immersion oil: Diiodomethane sulfur immersion oil (n_d = 1.78, Cargille Laboratories, NJ) was used to make contact between the objective lens and the coverslip (**Note 4**).

2.2.4 Buffers and Reagents for Fixed Cell TIRF Imaging

1. Paraformaldehyde buffer (0.1 M phosphate buffer, pH 7.4 with 4 % paraformaldehyde (Wako Pure Chemical Industries, Osaka, Japan)) was prepared.
2. Low-glucose KRB (Sect. 2.2.3) was prepared.
3. Blocking buffer (PBS with 3 % BSA (Free HG, Wako) and 0.1 % Triton X-100 (Surfact-Amps, Thermo, IL, USA)) was prepared.
4. Mounting media (PBS), 50 % Glycerol (Sigma-Aldrich), 2.5 % DABCO (1,4-diazabicyclo[2.2.2]octane, Sigma-Aldrich) was prepared.

2.2.5 Culture and Transfection to MIN6 Cells

1. MIN6 cells were seeded in a T-75 plastic flask, incubated in an atmosphere of 5 % CO_2 at 37 °C, and subcultured by regular trypsinization (wash with PBS and treatment with Trypsin–EDTA) and reseeding. For TIRF observation, cells were seeded on fibronectin-coated high-refractive-index coverglass.
2. MIN6 cells were transfected with insulin–GFP vector using Effectene transfection reagent according to the manufacturer's protocol. All experiments were performed between 2 and 3 days after transfection.

2.2.6 Culture and Adenoviral Infection to Primary Beta Cells

1. Preparation of beta cells: Pancreatic islets of Langerhans were isolated as described previously [42]. In brief, islets were isolated by collagenase (Liberase) digestion of C57BL/6 mouse pancreas and selected by handpicking. Isolated islets were

dissociated into single cells by incubation in Ca^{2+}-free KRB and cultured on fibronectin-coated high-refractive-index glass coverglass in RPMI1640 at 37 °C, in an atmosphere of 95 % air/5 % CO_2.

2. Infection: For the infection of the pancreatic beta cells with the recombinant adenovirus, 1-day cultured single cells were incubated with RPMI1640 medium (5 % FBS) and the required adenovirus (Adex1CA insulin–GFP: 30 multiplicity of infection per cell) for 1 h at 37 °C, after which RPMI1640 medium with 10 % FBS was added. All experiments were performed between 1 and 2 days after infection.

2.3 Methods

2.3.1 TIRF Microscopy System

Two types TIRF microscopy have been developed: prism-type TIRF microscopy and objective-type (through-the-lens-type) TIRF microscopy. The prism-type TIRF microscopy is mostly homemade setup. This setup is easily accomplished, since it requires only the microscope, prism, and laser, all components that are readily available. The drawback for this setup is the requirement that the specimen be positioned between the prism and the microscope objective, making it difficult to perform manipulations, to inject media into the specimen space, and to carry out physiological measurements. Recently, objective-type TIRF microscopy generates the evanescent field with the objective lens that is used for viewing the cell. The objective-type microscopes have been commercially available from Olympus, Nikon, and Zeiss Co. This setup requires that the laser be introduced through the microscope and greatly benefits from an objective lens with a numerical aperture (NA) >1.4. The specimen must be located on the top surface of the coverglass. The preparation, facing away from the objective, is accessible and permits the use of other instrumentation such as micromanipulators. We describe the objective-type TIRF microscopy system used in our laboratory (Fig. 7). This system is based on an Olympus objective-type TIRF microscope with minor modifications.

1. Light from an Ar laser (488 nm: Melles Griot, Carlsbad, CA) or a He/Ne laser (543 nm: Melles Griot) was introduced to an inverted microscope (IX70, Olympus Co., Tokyo, Japan) through a single-mode fiber and two illumination lenses; the light was focused at the back focal plane of a high-aperture objective lens (Apo 100× OHR; numerical aperture (NA) 1.65, Olympus) (**Note 5**). The focal point was moved off-axis to the most peripheral position in the objective lens by simply shifting the position of the fiber using micrometer laser adjustment.
2. To observe GFP alone, we used a 488-nm laser line for excitation and a 515-nm long-pass filter for the barrier. To observe the fluorescent image of RFP or Cy3, we used a 543-nm laser

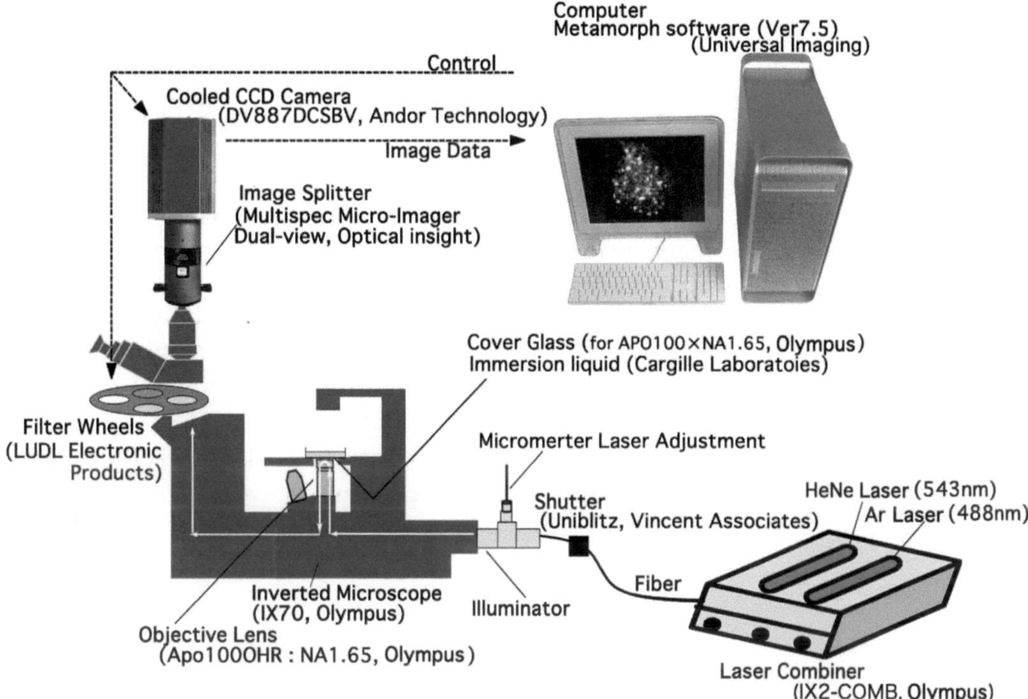

Fig. 7 Schematic representation of the TIRF setup described in the text

line and a long-pass 590-nm filter (Chroma, Brattleboro, VT). Two lasers can be introduced simultaneously with the laser combiner unit (Olympus). A filter wheel system (LUDL Electronic Products, Hawthorne, NY) is used with several interchangeable filters that can be rotated to bring the appropriate filter into the light path in a synchronized fashion. A computer-controlled shutter (Uniblitz, Vincent Associates, Rochester, NY) is placed between the laser combiner unit and illuminator to prevent photobleaching of cells between exposures.

3. Images were collected by a cooled charge-coupled device (CCD) camera (DV887DCSBV, ANDOR Technology, Belfast, Northern Ireland; operated with Metamorph version 7.5, Universal Imaging Corporation, Downingtown, PA) (**Note 6**). Images were acquired at 50–300-ms intervals.

4. To observe the fluorescence image of GFP and Cy3 simultaneously under TIRF illumination, we used the 488-nm laser line for excitation and an image splitter (Dual-View, MultiSpec Micro-Imager, Optical Insight, Santa Fe, NM) that divided the green and red components of the images with 565-nm dichroic mirror (Chroma) and passed the green component through a 530 ± 15-nm band-pass filter (Chroma) and the red

component through a 630 ± 25-nm band-pass filter or a 630-nm< long-pass filter (Chroma) (**Note 7**). Images were then projected side by side onto the CCD camera. The two images were brought into focus in the same plane by adding weak lenses to one channel, and they were brought into register by careful adjustment of the mirrors in the image splitter [43, 44]. Before each experimental session, we took an alignment image that showed the density by means of scattered 90-nm TetraSpeck fluorescent beads (Molecular Probes, Eugene, OR). They were visible in both the green and red channels and thus provided markers in the *x*–*y* plane. Beads in the two images were brought into superposition by shifting one image using Metamorph software.

5. Most analyses, including tracking (single projection of difference images) and area calculations, were performed using Metamorph software.

2.3.2 Live Cell TIRF Imaging

1. Turn TIRF system on 30 min before use.
2. Mount the insulin–GFP-transfected MIN6 cells or infected beta cells on the high-refractive-index coverslips in an open chamber and incubated in low-glucose KRB for 30 min at 37 °C.
3. Transfer coverslips to the thermostat-controlled stage (37 °C) of the TIRF microscope.
4. Find insulin–GFP expressing cells in epifluorescence first.
5. Incline of optic fiber of TIRF microscopy is switched from epifluorescent illumination mode to evanescent wave illumination mode using micrometer laser adjustment.
6. Precisely focus the image with z focus drive.
7. Lower laser intensity and use ND (neutral density) filters to minimize photodamage and increase CCD sensitivity/gain and image contrast.
8. Use Metamorph software to acquire images at 50–300-ms intervals.
9. Stimulate cells by switching buffers in a mounted chamber from low-glucose KRB to high-glucose KRB or high KCl–KRB. For bath application, stimulation with glucose and KCl was achieved by addition of 52 mM glucose–KRB or 100 mM KCl–KRB into the chamber (final = 50 mM KCl or 22 mM glucose).
10. Continue acquiring images for 20 min after high-glucose stimulation.
11. Save the acquired images to hard disk drive on a computer or to other physical media.

2.3.3 Analysis of Images

Most analyses, including tracking and area calculations, are performed using MetaMorph software. To analyze the data, fusion events are manually selected and the average fluorescence intensity of individual granules in a 1-μm × 1-μm square placed over the granule center is calculated. The number of fusion events is manually counted while looping 6,000 frame time lapses. Sequences are exported as single TIFF files or converted into AVI or QuickTime movies.

2.3.4 Fixed Cell TIRF Imaging

To look at docked insulin granules and endogenous protein in plasma membrane, we have used fixed cell TIRF imaging:

1. Culture primary beta cells or MIN6 cells on high-refractive-index glass coverglass.
2. Wash twice with low-glucose KRB.
3. Fix with 4 % paraformaldehyde for 15 min at room temperature.
4. Wash 3× 5 min with PBS.
5. When you image a fluorescent protein-expressing cells, skip to Section 2.3.4.11.
6. Block and permeabilize with blocking buffer (PBS with 3 % BSA and 0.1 % Triton X-100).
7. Incubate cells overnight at 4 °C with diluted primary antibodies.
8. Wash 3× 5 min with PBS.
9. Incubate cells for 1 h at room temperature in the dark with corresponding fluorescent-labeled secondary antibody conjugates.
10. Wash 3× 5 min with PBS.
11. To mount cells, mount media (PBS, 50 % Glycerol, 2.5 % DABCO) onto the cells on coverslip. Mounted slides should be stored at 4 °C protected from light.
12. Acquire images using TIRF microscopy.

2.3.5 Example of Insulin Exocytosis Imaging by TIRF

Beta cells release insulin on glucose stimulation with a biphasic pattern that consists of a rapidly initiated and transient first phase followed by a sustained second phase. TIRF imaging has allowed us to directly observe the dynamic motion of single insulin granules undergoing exocytosis during biphasic insulin release in living beta cells. We transfected mouse primary pancreatic beta cells with recombinant adenovirus encoding insulin–GFP in order to label the insulin secretory granules and then examined the dynamic motion of granules near the plasma membrane [45].

TIRF microscopy revealed that glucose stimulation induced interesting exocytotic responses that originated from two distinct types of insulin granules with completely different behaviors prior to fusion: previously docked granules that are visible before the onset of stimulation under TIRF microscopy and newcomer granules that cannot be detected by TIRF microscopy or are dimly visible before stimulation (Fig. 8a). It should be noted that in primary cultured pancreatic beta cells, newcomer granules fuse immediately after they reach the plasma membrane. The time from landing to fusion is less than 50 ms [39]. More interestingly, these two types of fusion events reveal different temporal patterns. Fusions originating from previously docked granules are detected preferentially during the first phase, whereas fusion events during the second phase mostly arose from newcomer granules (Fig. 8b). Thus, it is likely that there is a mechanistic difference between first and second phase release [39, 45].

We also examined the relationship between t-SNARE syntaxin 1A and biphasic insulin release by TIRF microscopy [45]. Syntaxin 1A was shown to play a crucial role in secretory granule docking and subsequent fusion steps in the exocytotic process. We examined the docking status of insulin granules in syntaxin 1A-knockout (syntaxin 1A$^{-/-}$) beta cells using fixed cell TIRF imaging with immunostaining for insulin (Fig. 8c). We interpret the individual fluorescent spots shown in the TIRF image in Fig. 8c to be equivalent to morphologically docked granules. We rarely observed morphologically docked granules in syntaxin 1A$^{-/-}$ cells. In live cell beta-cell TIRF imaging, syntaxin 1A$^{-/-}$ mice beta cells showed fewer previously docked granules with no fusion during the first phase, whereas fusion from newcomers, which are responsible for second phase release, was still preserved (Fig. 8d, e). Thus, we proposed a model for biphasic insulin release wherein docking and fusion of insulin granules is Synt1A-dependent during the first phase but Synt1A-independent during the second phase [45].

3 Notes

1. GFP-tagged neuropeptide Y, phogrin, and IAPP were also used to label insulin granules [46–48].
2. We have also used Lipofectamine 2000 (Life Technologies, Carlsbad, CA).
3. We have also used Wistar rats age 8–10.
4. In order to maintain the NA (1.65) and to achieve the high quality of image, a special coverglass of high-refractive-index glass ($n_d = 1.788$, Olympus) and special immersion medium of high refractive index (immersion liquid: $n_d = 1.78$, Cargille Laboratories) must be used.
5. We use a high-aperture objective lens (Olympus APO 100×/NA 1.65 apochromatic objective lens). Living cells typically have a refractive index about 1.37. To achieve total internal

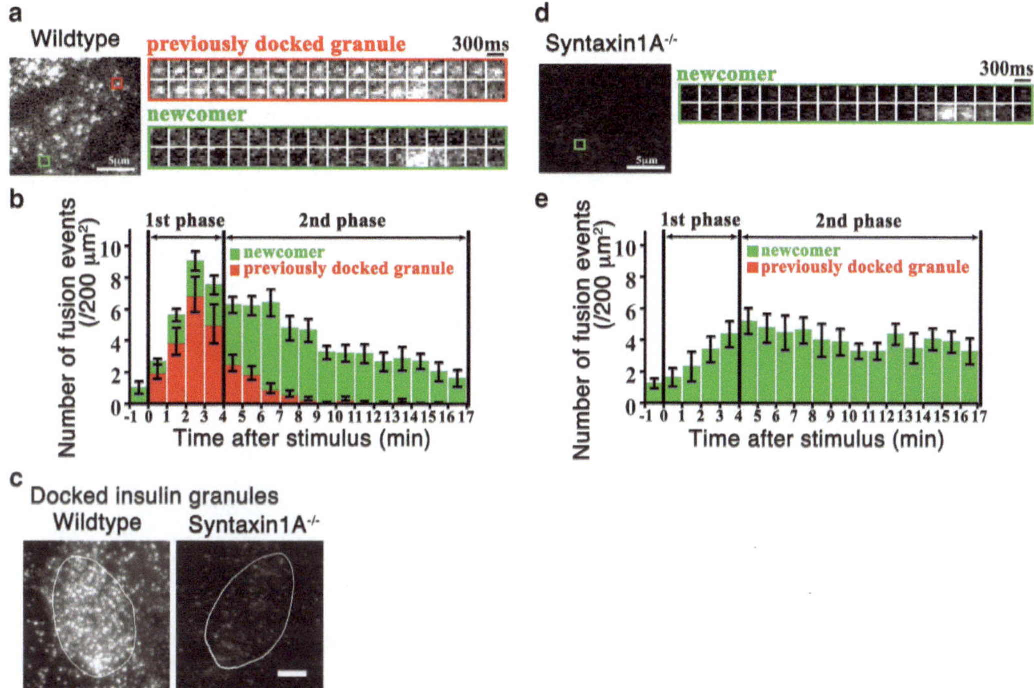

Fig. 8 TIRF images and analysis of single GFP-labeled insulin granule motion in mouse primary beta cells during glucose stimulation. (**a**) *Left panel*: The real-time motion of GFP-labeled insulin granules in WT beta cell was imaged to the plasma membrane. This is the TIRF image of a single cell harboring insulin–GFP, showing that many insulin granules docked to the plasma membrane. *Right panel*: Sequential images (1 μm × 1 μm per 300-ms interval) of a granule docking and fusing with the plasma membrane were presented during 22-mM glucose stimulation. (**b**) Histogram of the number of fusion events (per 200 μm^2) in WT β-cells at 60-s intervals after stimulation ($n = 10$ cells). The *black column* shows fusion from previously docked granules, and the *gray column* shows fusion from newcomers. During the first phase, fusion occurred mostly from previously docked granules. (**c**) TIRF images of docked insulin granules in WT or Synt1A$^{-/-}$ β-cells (fixed cell TIRF imaging). The surrounding *lines* represent the outline of cells that attached to the coverglass. Scale bar: 5 μm. Pancreatic β-cells were prepared from WT and Synt1A$^{-/-}$ mice, fixed, and immunostained for insulin. (**d**) TIRFM during glucose stimulation in Synt1A$^{-/-}$ β-cells and sequential images of a newcomer granule docking and fusing under glucose stimulation. (**e**) Histogram of the number of fusion events (per 200 μm^2) in the Synt1A$^{-/-}$ cells at 60-s intervals after stimulation (modified from our previous data [45])

reflection, such a specimen must be illuminated with NA greater than 1.37. Recently, 1.4–1.45 NA objectives using normal coverglass and immersion oil have been available from Olympus, Nikon, and Zeiss Co.

6. Underlying all direct imaging of living cells is the desire to preserve the living subject for as long as possible, through minimization of both phototoxic cell damage and also photobleaching of the incorporated fluorophores. Higher sensitive CCD (Electron Multiplying CCD) is suitable for TIRF microscopy system, which not only exhibits the sensitivity and speed to capture dynamic membrane events from within very thin small excitation volume but furthermore enables the laser excitation power to be attenuated.

7. The penetration depth is dependent on the wavelength (λ). To precisely analyze the colocalization of GFP and RFP/Cy3, we only use the 488-nm laser line for excitation. However, the fluorescence intensity of red component by 488-nm excitation is relatively low compared to that by 543-nm excitation. Recently, multicolor, multi-angle illuminator that allows simultaneous multi-channel laser TIRF illumination, has been available from Olympus (RFAEVAW, Cell TIRF). Users can control each laser's penetration depth to simultaneously capture TIRF for multi-channels.

References

1. Mahato RI, Henry J, Narang AS, Sabek O, Fraga Ashcroft SJ, Bunce J, Lowry M, Hansen SE, Hedeskov CJ (1978) The effect of sugars on (pro)insulin biosynthesis. Biochem J 174:517–526
2. Hedeskov CJ (1980) Mechanism of glucose-induced insulin secretion. Physiol Rev 60:442–509
3. Welsh M, Nielsen DA, MacKrell AJ, Steiner DF (1985) Control of insulin gene expression in pancreatic beta-cells and in an insulin-producing cell line, RIN-5F cells. II. Regulation of insulin mRNA stability. J Biol Chem 260:13590–13594
4. Welsh M, Scherberg N, Gilmore R, Steiner DF (1986) Translational control of insulin biosynthesis. Evidence for regulation of elongation, initiation and signal-recognition-particle-mediated translational arrest by glucose. Biochem J 235:459–467
5. Newgard CB, McGarry JD (1995) Metabolic coupling factors in pancreatic beta-cell signal transduction. Annu Rev Biochem 64:689–719
6. Nagamatsu S (2008) Mechanism of insulin exocytosis analysed by imaging techniques. In: Seino S, Bell GI (eds) Pancreatic beta cells in health and disease. Springer, pp 177–194
7. Greider MH, Howell SL, Lacy PE (1969) Isolation and properties of secretory granules from rat islets of Langerhans. II. Ultrastructure of the beta granule. J Cell Biol 41:162–166
8. Lange RH (1974) Crystalline islet B-granules in the grass snake (*Natrix natrix* (L.)): tilting experiments in the electron microscope. J Ultrastruct Res 46:301–307
9. Wollheim CB, Lang J, Regazzi R (1996) The exocytotic process of insulin secretion and its regulation by Ca^{2+} and G-proteins. Diabetes Rev 4:276– 297
10. Regazzi R, Wollheim CB, Lang J, Theler JM, Rossetto O, Montecucco C, Sadoul K, Weller U, Palmer M, Thorens B (1995) VAMP-2 and cellubrevin are expressed in pancreatic beta-cells and are essential for Ca(2+)-but not for GTP gamma S-induced insulin secretion. EMBO J 14:2723–2730
11. Sadoul K, Lang J, Montecucco C, Weller U, Regazzi R, Catsicas S, Wollheim CB, Halban PA (1995) SNAP-25 is expressed in islets of Langerhans and is involved in insulin release. J Cell Biol 128:1019–1028
12. Boyd RS, Duggan MJ, Shone CC, Foster KA (1995) The effect of botulinum neurotoxins on the release of insulin from the insulinoma cell lines HIT-15 and RINm5F. J Biol Chem 270:18216–18218
13. Nagamatsu S, Fujiwara T, Nakamichi Y, Watanabe T, Katahira H, Sawa H, Akagawa K (1996) Expression and functional role of syntaxin 1/HPC-1 in pancreatic beta cells. Syntaxin 1A, but not 1B, plays a negative role in regulatory insulin release pathway. J Biol Chem 271:1160–1165
14. Nagamatsu S, Nakamichi Y, Yamamura C, Matsushima S, Watanabe T, Ozawa S, Furukawa H, Ishida H (1999) Decreased expression of t-SNARE, syntaxin 1, and SNAP-25 in pancreatic beta-cells is involved in impaired insulin secretion from diabetic GK rat islets: restoration of decreased t-SNARE proteins improves impaired insulin secretion. Diabetes 48:2367–2373
15. Ohara-Imaizumi M, Nishiwaki C, Nakamichi Y, Kikuta T, Nagai S, Nagamatsu S (2004) Correlation of syntaxin-1 and SNAP-25 clusters with docking and fusion of insulin granules analysed by total internal reflection fluorescence microscopy. Diabetologia 47:2200–2207
16. Ju Q, Edelstein D, Brendel MD, Brandhorst D, Bretzel RG, Brownlee M (1998) Transduction of non-dividing adult human pancreatic beta cells by an integrating lentiviral vector. Diabetologia 41:736–739
17. Flotte T, Agarwal A, Wang J, Song S, Fenjves ES, Inverardi L, Chesnut K, Afione S, Loiler S,

Wasserfall C, Kapturczak M, Ellis T, Nick H, Atkinson M (2001) Efficient ex vivo transduction of pancreatic islet cells with recombinant adeno-associated virus vectors. Diabetes 50:515–520

18. Ceste ME, Afra R, Mullen Y, Drazan KE, Benhamou PY, Shaked A (1994) Adenoviral-mediated gene transfer to pancreatic islets does not alter islet function. Transplant Proc 6:1545–1551
19. Wick PF, Senter RA, Parsels LA, Uhler MD, Holz RW (1993) Transient transfection studies of secretion in bovine chromaffin cells and PC12 cells. Generation of kainate-sensitive chromaffin cells. J Biol Chem 268: 10983–10989
20. Coppola T, Perret-Menoud V, Lüthi S, Farnsworth CC, Glomset JA, Regazzi R (1999) Disruption of Rab3-calmodulin interaction, but not other effector interactions, prevents Rab3 inhibition of exocytosis. EMBO J 18:5885–5891
21. Iezzi M, Regazzi R, Wollheim CB (2000) The Rab3-interacting molecule RIM is expressed in pancreatic beta-cells and is implicated in insulin exocytosis. FEBS Lett 474:66–70
22. Mahato RI, Henry J, Narang AS, Sabek O, Fraga D, Kotb M, Gaber AO (2003) Cationic lipid and polymer-based gene delivery to human pancreatic islets. Mol Ther 7:89–100
23. Lakey JR, Young AT, Pardue D, Calvin S, Albertson TE, Jacobson L, Cavanagh TJ (2001) Nonviral transfection of intact pancreatic islets. Cell Transplant 10:697–708
24. Gainer AL, Korbutt GS, Rajotte RV, Warnock GL, Elliott JF (1996) Successful biolistic transformation of mouse pancreatic islets while preserving cellular function. Transplantation 61:1567–1571
25. Cabrera O, Berman DM, Kenyon NS, Ricordi C, Berggren PO, Caicedo A (2006) The unique cytoarchitecture of human pancreatic islets has implications for islet cell function. Proc Natl Acad Sci U S A 103:2334–2339
26. Straub SG, Sharp GW (2002) Glucose-stimulated signaling pathways in biphasic insulin secretion. Diabetes Metab Res Rev 18:451–463
27. Rorsman P, Renstrom E (2003) Insulin granule dynamics in pancreatic beta cells. Diabetologia 46:1029–1045
28. Axerlod D (2001) Total internal reflection fluorescence microscopy in cell biology. Traffic 2:764–774
29. Steyer JA, Almers W (2001) A real-time view of life within 100 nm of the plasma membrane. Nat Rev Mol Cell Biol 2:268–275
30. Axelrod D (1981) Cell-substrate contacts illuminated by total internal reflection fluorescence. J Cell Biol 89:141–145
31. Truskey GA, Burmeister JS, Grapa E, Reichert WM (1992) Total internal reflection fluorescence microscopy (TIRFM). II. Topographical mapping of relative cell/substratum separation distances. J Cell Sci 103:491–499
32. Vale RD, Funatsu T, Pierce DW, Romberg L, Harada Y, Yanagida T (1996) Direct observation of single kinesin molecules moving along microtubules. Nature 380:451–453
33. Sund SE, Axelrod D (2000) Actin dynamics at the living cell submembrane imaged by total internal reflection fluorescence photobleaching. Biophys J 79:1655–1669
34. Dickson RM, Norris DJ, Tzeng YL, Moerner WE (1996) Three-dimensional imaging of single molecules solvated in pores of poly(acrylamide) gels. Science 274:966–969
35. Oheim M, Loerke D, Stühmer W, Chow RH (1998) The last few milliseconds in the life of a secretory granule. Docking, dynamics and fusion visualized by total internal reflection fluorescence microscopy (TIRFM). Eur Biophys J 27:83–98
36. Sako Y, Minoghchi S, Yanagida T (2000) Single-molecule imaging of EGFR signalling on the surface of living cells. Nat Cell Biol 2:168–172
37. Tsuboi T, Zhao C, Terakawa S, Rutter GA (2000) Simultaneous evanescent wave imaging of insulin vesicle membrane and cargo during a single exocytotic event. Curr Biol 10:1307–1310
38. Ohara-Imaizumi M, Nakamichi Y, Tanaka T, Ishida H, Nagamatsu S (2002) Imaging exocytosis of single insulin secretory granules with evanescent wave microscopy: distinct behavior of granule motion in biphasic insulin release. J Biol Chem 277:3805–3808
39. Ohara-Imaizumi M, Nishiwaki C, Kikuta T, Nagai S, Nakamichi Y, Nagamatsu S (2004) TIRF imaging of docking and fusion of single insulin granule motion in primary rat pancreatic beta-cells: different behaviour of granule motion between normal and Goto-Kakizaki diabetic rat beta-cells. Biochem J 381:13–18
40. Nagamatsu S, Watanabe T, Nakamichi Y, Yamamura C, Tsuzuki K, Matsushima S (1999) α-Soluble N-ethylmaleimide-sensitive factor attachment protein is expressed in pancreatic β -cells and functions in insulin, but not γ -amino butyric acid secretion. J Biol Chem 274:8053–8060
41. Miyazaki J-I, Araki K, Yamato E, Ikegami H, Asano T, Shibasaki Y, Oka Y, Yamamura K-I (1990) Establishment of a pancreatic beta cell line that retains glucose-inducible insulin secretion: special reference to expression of

glucose transporter isoforms. Endocrinology 127:126–132

42. Lacy PE, Kostianovsky M (1967) Method for the isolation of intact islet of Langerhans from the rat pancreas. Diabetes 16:35–39
43. Ohara-Imaizumi M, Nishiwaki C, Kikuta T, Kumakura K, Nakamichi Y, Nagamatsu S (2004) Site of docking and fusion of insulin secretory granules in live MIN6 beta cells analyzed by TAT-conjugated anti-syntaxin 1 antibody and total internal reflection fluorescence microscopy. J Biol Chem 279:8403–8408
44. Ohara-Imaizumi M, Ohtsuka T, Matsushima S, Akimoto Y, Nishiwaki C, Nakamichi Y, Kikuta T, Nagai S, Kawakami H, Watanabe T, Nagamatsu S (2005) ELKS, a protein structurally related to the active zone-associated protein CAST, is expressed in pancreatic beta cells and functions in insulin exocytosis: interaction of ELKS with exocytotic microscopy. Mol Biol Cell 16:3289–3300
45. Ohara-Imaizumi M, Fujiwara T, Nakamichi Y, Okamura T, Akimoto Y et al (2007) Imaging analysis reveals mechanistic differences between first- and second-phase insulin exocytosis. J Cell Biol 177:695–705
46. Pouli AE, Emmanouilidou E, Zhao C, Wasmeier C, Hutton JC, Rutter GA (1998) Secretory-granule dynamics visualized in vivo with a phogrin-green fluorescent protein chimaera. Biochem J 333:193–199
47. Tsuboi T, Rutter GA (2003) Multiple forms of "kiss-and-run" exocytosis revealed by evanescent wave microscopy. Curr Biol 13:563–567
48. Michael DJ, Xiong W, Geng X, Drain P, Chow RH (2007) Human insulin vesicle dynamics during pulsatile secretion. Diabetes 56:1277–1288

Chapter 5

Functional, Quantitative, and Super-Resolution Imaging and Spectroscopic Approaches for Studying Exocytosis

Rory R. Duncan and Colin Rickman

Abstract

Regulated exocytosis is the process of secretion in specialized cells. Decades of intensive research have defined the molecules that drive the process of membrane fusion, as well as a plethora of accessory factors that shape and modulate the exocytotic function. As regulated exocytosis is generally a rapid event with a millisecond timescale, involving organelles on the nanometer scale, the field has long employed high spatial and temporal resolution techniques, including electrophysiology and, importantly, imaging approaches. Huge gains in the spatial and temporal resolution in biological microscopy have been delivered by recent hardware, engineering, and software breakthroughs; it is now possible to visualize and quantity the distributions, movements, and interactions of large cohorts of single protein molecules inside living cells. These technically demanding approaches have the potential to test long-standing hypotheses in the fields of membrane trafficking and exocytosis, building upon an extraordinary foundation of detailed biochemical and electrophysiological understanding.

Key words Fluorescence lifetime imaging microscopy, FLIM, Fluorescence correlation spectroscopy, FCS, Super-resolution, Photoactivation localization microscopy, PALM, Single-particle tracking, Stochastic optical reconstruction microscopy, STORM, Total internal reflection fluorescence microscopy, TIRFM

1 Introduction: Regulated Exocytosis

In neuronal and neuroendocrine cells, exocytosis is mediated by the plasma membrane proteins (tSNARE) syntaxin and SNAP-25 (synaptosome-associated protein 25 kDa) and the vesicular protein synaptobrevin (v-SNARE) [1–3]. The cytoplasmic regions of these three proteins interact to form a trimeric, four-helical complex [4–6]; this is proposed to overcome the thermodynamic barrier to membrane fusion, driving the final, Ca^{2+}-dependent exocytotic events [7–13]. Studies in vitro have determined that the tSNARE can form a heterodimer binary complex [14–22] and that this occurs before the v-SNARE enters the ternary, or core, complex [17, 23–25]. While the SNARE proteins alone can cause spontaneous

Peter Thorn (ed.), *Exocytosis Methods*, Neuromethods, vol. 83,
DOI 10.1007/978-1-62703-676-4_5, © Springer Science+Business Media New York 2014

fusion of liposomes in vitro, this process is far slower than that in cells [26]. Calcium sensitivity is contributed by the synaptotagmins, the putative calcium sensors of exocytosis, via a direct interaction with negatively charged lipids, syntaxin and the binary SNARE complex [11–13, 22, 27, 28]. In addition, the entire SNARE process is regulated by a conserved set of accessory proteins, which operate throughout the trafficking pathway.

The Sec1p/Munc18 (SM) protein family represents one such set of modulators, with mutations characterized by a severe disruption of general secretion or neurotransmitter release [29–31]. The mammalian SM protein munc18-1 was isolated through its interaction with the monomeric form of syntaxin 1, rendering the tSNARE unable to form the SDS-resistant ternary SNARE complex [31]. This closed form of syntaxin 1, in which its N-terminal Habc domain interacts with the SNARE helix, exhibits a high affinity for munc18-1. Syntaxin 1 can also adopt a structurally distinct conformation; in the open form, the SNARE helix does not interact with the N-terminal three helical regulatory domain (termed Habc) and open syntaxin has been thought not to interact with munc18-1 [32].

Munc13-1 was isolated as the mammalian homolog of the *C. elegans unc13* gene product, known to bind phospholipids and calcium by virtue of multiple C2 domains [33] and can act as diacylglycerol receptors [34]. Munc13-1 may interact with the N-terminal region of syntaxin [35]; genetic and electrophysiological data suggest that munc13 is a priming factor, affecting the kinetics of fast vesicle release but not the overall number of docked vesicles [34, 36, 37]. The molecular nature of vesicle "priming" remains undefined but likely involves protein interactions at specific exocytotic sites.

Regulated exocytosis does not proceed spontaneously but instead requires influx of calcium ions through voltage-gated calcium channels (VGCC). This triggers the "calcium sensor" synaptotagmin to transduce the calcium signal to membrane fusion of secretory vesicles [11–13, 22, 27, 28]. The final fusion event at the plasma membrane is driven by synaptotagmin-dependent mechanotransduction to the core SNARE complex comprising syntaxins, SNAP-25/23/29 (both target or tSNARE), and synaptobrevins (a vesicular, or v-SNARE) as well as SM proteins, such as munc18 [9, 38]. In both neurons and endocrine cells, the voltage dependent calcium influx is typically shaped by a classical negative feedback role of large conductance calcium- and voltage-activated potassium (BK) channels [39–45]. Central to this model is the requirement for secretory vesicles to be appropriately spatially localized with the secretory machinery and cognate VGCC and BK channels—it is currently thought that direct interactions between the exocytotic machinery and channel molecules may regulate this [46–49].

Decades of biochemical and electrophysiological research have thus defined many of the key proteins involved in regulated exocytosis. The membrane trafficking field has also employed imaging techniques extensively, often at the cutting edge, probably because such dynamic, spatiotemporally regulated events in the cell present an attractive and tractable process for imaging. It is now possible to extend imaging studies from the previously simple observational studies to functional imaging at the molecular level—i.e., not only how things appear but how they interact, move, function, and change over time, with spatial and temporal resolution sufficient to detect and quantify the behavior of single molecules. This brief review covers the recent technological advances made in super-resolution imaging and single-molecule spectroscopy and how these may impact on the membrane trafficking and exocytosis fields.

1.1 Imaging Exocytosis

Neuroendocrine and neuronal cells perform regulated exocytosis in order to secrete messenger cargo—peptide hormones and neurotransmitters, respectively—to the extracellular space. These soluble cargo molecules are packaged inside secretory vesicles either at the Golgi apparatus, in the case of neuroendocrine cells, or locally in the neuronal synapse. Vesicles vary in size according to cell type; for example, large dense-cored vesicles in neuroendocrine cells are several hundred nanometers in size, whereas synaptic vesicles in central neurons are much smaller at around 50 nm in diameter.

The different biology, spatial organization, and scale of neuroendocrine and neuronal cells present distinct challenges for the experimentalist aiming to visualize secretion. In neuronal preparations, FM-dye recycling has been a method of choice for two decades [50]. The FM dyes are amphipathic styryl dye molecules that increase their quantum efficiency (i.e., brightness) dramatically when they insert into the outer leaflet of cell membranes. This is advantageous as the dyes can be used to label plasma membrane—in neurons an additional benefit comes from the local membrane recycling that occurs during neurotransmission [50–53]. Stimulating the cells to secrete results in a round of exocytosis followed within a few seconds by compensatory endocytosis—if FM molecules are present in the synaptic outer membrane, then this endocytosis retrieves the fluorescent dye within recycling synaptic vesicle membranes. Perfusing the cellular bath to remove excess FM dye post-stimulation (termed "loading") results in the rapid de-partitioning of plasma membrane FM-dye molecules, whereas the synaptic vesicle dye remains. Subsequent rounds of stimulation result in the fusion of these labelled vesicles with the neuronal plasma membrane that can be assayed as a loss in fluorescence from these cells. This approach has been used in various incarnations to great effect over the years [50, 51, 54–56].

The lack of an intimately interconnected secretory vesicle cycle involving local vesicle reuse means that this assay is of more limited use in neuroendocrine cells. These cells have the advantage for the researcher of a larger surface area (compared to the synapse) as well as single large dense-cored vesicles (LDCVs) easily within the resolution of the light microscope. Imaging exocytosis in such preparations has employed total internal reflection fluorescence microscopy in recent years, providing high-contrast imaging with temporal resolution compatible with single-vesicle exocytosis. This technique involves coupling laser fluorescence excitation to the sample with an angle of incidence sufficient to induce total internal reflection at the interface between two refractive indices: in this case the excitation light penetrates a short distance into a sample with the lower refractive index. In practice, this is achieved by using an oil immersion lens with an aqueous sample—for example, cells in aqueous medium [57, 58]. The excitation evanescent wave that is established in the aqueous medium penetrates in a nonlinear way only a few hundred nanometers from the refractive index interface—in this case the glass coverslip that the cells are cultured on. The net effect of this is that only fluorescence within this distance is excited, suppressing out of focus haze and blur, resulting in an extremely thin optical section and high-contrast images. This is inherently a widefield approach, so data are acquired using a camera as opposed to scanning pixel by pixel; modern EM-CCD cameras are easily capable of acquiring images on the tens-of-milliseconds timescale at the low light levels experienced in live-cell imaging, so that visualizing exocytosis is now relatively straightforward. The first example of this used a cell permeant acidophilic dye, acridine orange, to label LDCVs in adrenal chromaffin cellsste [59]. Problems with the specificity of this dye when used to report exocytosis mean that it has fallen from favor [60]; the advent of encodable vesicle lumen probes based on GFP [61], often with enhanced pH sensitivity [62], means that this technique remains the approach of choice not only for imaging single-vesicle exocytosis but also for tracking vesicle behaviors pre- and post-exocytosis [58, 61, 63].

A caveat to this is that TIRFM data can only be acquired from cells (usually cultured in isolation) that adhere firmly to glass coverslips. An alternative imaging approach to visualize insulin granule fusion with the beta cell membrane inside semi-intact pancreatic islets was recently described—two-photon extracellular polar-tracer imaging-based quantification (TEPIQ) [64]. This technique uses an extracellular polar dye (sulforhodamine B) that cannot cross the cell plasma membrane so that intracellular secretory granules remain unlabelled. Upon stimulation the nascent fusing granules appear as transient bright fluorescent spots as the high concentration of dye molecules in the bath fills the granule through the fusion pore. This approach successfully visualized insulin granule secretion and enabled the quantification of the extent and the timings of exocytosis in response to physiological stimuli [8, 64].

The recent advent of new sCMOS cameras capable of acquiring thousands of frames per second at these light levels raises the exciting possibility of imaging exocytosis in "real time"—fast enough to capture data describing the fusion of single vesicles and perhaps even to test directly long-standing hypotheses based on amperometric data that fusion pore dynamics are modulated by SNARE accessory proteins.

1.1.1 Imaging Vesicle Dynamics

In addition to the use of TIRFM to image single-vesicle fusion events, the high-resolution tracking of single vesicles has been invaluable to describe cellular behavior leading up to exocytosis. Early work using this approach concluded that vesicles destined for fusion became "immobile" immediately before exocytosis, concluding that this was an indicator of protein–protein interactions preceding the final fusion event ("docking") [59, 65]. However, as acquisition speeds and spatial resolution increased, alongside better analytical approaches, it became clear that although vesicles do indeed behave as if "tethered" prior to fusion, they are not immobile. In fact, recent work demonstrated that the final motion before large dense-cored vesicle fusion is commonly a "jump" of a larger magnitude than the preceding "docked" activity [57]. The reason for the increased motion immediately pre-fusion remains unclear, although it is tempting to speculate that this behavior would allow vesicles to encounter a molecular environment necessary for exocytosis. The analysis of vesicle motion observed in TIRFM has also been used in an attempt to categorize vesicles into the functional "pools" theorized from capacitance and modelling studies—i.e., "reserve" or "readily releasable" poolso [57, 65]. These efforts clearly go further than simply stating that vesicles are "morphologically docked" although it remains to be tested whether the categorization of vesicles into "pools" based on their mobility holds for all cell types [66]. One key caveat for all vesicle tracking data presented thus far is that the rate of data acquisition is undoubtedly too slow to capture the fine detail of vesicle movements over time. This will certainly change as newer, faster camera technologies become available and perhaps will lead to new discoveries or the better testing of current hypotheses.

1.2 Imaging the Molecules Underlying Exocytosis

1.2.1 Imaging Lipids

A number of different approaches have been developed to image lipid membranes in cells—e.g., the FM dye approach described above, using Pleckstrin homology-domain GFP-fusion proteins to highlight phosphatidylinositol lipids [67], or cholesterol-binding antibiotics, such as filipin (which possesses weak fluorescent properties [68]). In addition, the use of the fluorescent dye "laurdan" (6-Dodecanoyl-2-dimethylaminonaphthalene)—a microenvironment-sensitive dye that intercalates between phospholipids in bilayers and changes its spectral properties according the packing of surrounding lipids [69]. It has been used to study bilayer properties in artificial and cellular membranes because of its sensitivity to the solvent relaxation effect [69–71]. For example, the maximum emission wavelength in phosphatidylcholine bilayers shifts approximately 45 nm from about 435 nm to about 480 nm when switched from a rigid membrane environment to one that is fluid (i.e., it senses and reports lipid phase or order). Apparently, the change represents differences in the number and/or mobility of water molecules present at the level of the phospholipid glycerol backbones: accordingly, laurdan is sensitive to membrane phase transitions and other alterations to membrane fluidity. Whether the structures reported by laurdan are lipid "rafts" (as recently reported [72]) remains unclear (as does the physical makeup of such rafts [73])—however, it is accepted that laurdan behaves in a predicable manner in vitro and in cellular membranes [69–71]. Imaging laurdan in plasma membranes has been limited because of an apparent requirement for 2-photon excitation in order to decrease photobleaching (as laurdan is quite photolabile [72]); however, we recently found that laurdan is compatible with total internal reflection fluorescence microscopy, permitting the visualization of lipid phase heterogeneities in living plasma membranes, in agreement with an earlier study. Other lipid microenvironment dyes have also been described, both for intensity-based and fluorescent lifetime imaging that may also be useful (e.g., [74]).

We recently used laurdan to quantify the effect of cholesterol sequestration on plasma membrane lipid order, finding a decrease in lipid order as cholesterol concentration decreased [16]. Also of interest was the observation that the apparent lipid order of large dense-cored vesicles in neuroendocrine cells was substantially lower than that of the plasma membrane, suggesting a different vesicular lipid composition or cholesterol concentration, compared to the cell surface [16].

One issue surrounding imaging lipid dynamics is that the heterogeneities in membranes are highly likely to reorganize on a timescale faster than image data can be acquired. To overcome this, imaging spectroscopies can be used to quantify the diffusional rates of lipid molecules, or dye reporters, in different lipid environments. Fluorescence correlation spectroscopy was thus used elegantly to determine that SNARE proteins, in vitro, absolutely preferred a lipid-disordered environment over lipid order—whether this is reiterated in a cellular context remains to be seen [75]. The use of such advanced spectroscopic approaches in cell and neurobiology is now becoming recognized as a powerful tool in the research armory [76, 77].

1.3 Imaging Proteins

There is a long history of imaging these proteins in the membrane trafficking and exocytosis fields, with the hypotheses surrounding protein localization, interactions, and dynamics broadly mirroring the improvements in spatial and temporal resolutions available as technologies advance. It is currently thought that the tSNARE reside in membrane "clusters"—when first observed, these were estimated to be 300–500 nm in diameter [78]. Importantly, these SNARE "clusters" have been used as exemplars for the way in which all membrane proteins are organized spatially at the cell surface [79]. As imaging technologies improved, however, it became clear that the size of these protein "clusters" was diffraction limited; i.e., they appeared at the limit of resolution of the microscope used for the imaging, whatever that was [79].

More recently, there have been very large gains made in both spatial and temporal resolutions available to life scientists with the advent of "super-resolution" imaging approaches [80–84]. As has been the case throughout the history of exocytosis research, the most advanced approaches have always been adopted early in order to gain further insight into membrane trafficking and fusion; super-resolution imaging is no exception to this rule. Broadly speaking, super-resolution imaging may be subdivided into two groups: hardware and computational based. The former group includes new microscope modalities that genuinely increase the spatial resolution of the microscope, "breaking the diffraction limit." Included in this is stimulated emission depletion microscopy, or STED. Originally developed over a number of years by Stefan Hell [85], this approach has been used in exocytosis research in a number of different studies [79, 82, 86, 87].

1.3.1 STED Imaging of Exocytotic Proteins

STED imaging was initially used in the study of the exocytosis to test a hypothesis about the fate of synaptic vesicles post-fusion: do their contents remain together, or do they diffuse away on the plasma membrane? This long-standing question proved difficult to address because the small size of synaptic vesicles (~40 nm) is below the resolution of diffraction-limited imaging (~250 nm). STED imaging allowed the visualization of synaptotagmin (a vesicular protein) during and after evoked exocytosis, finding that at least some of the vesicular content remained associated during synaptic vesicle recycling [82]. This important paper also established the idea that light microscopy could be used to visualize and quantify cellular behaviors involving structures on the nanometer scale, leading to another study that was able to push STED imaging to video rate. This approach allowed the tracking of synaptic vesicles in "real time" (28 frames per second) inside living synaptic boutons, finding that the vesicles exhibited highly restricted motions [88]. STED has not been used widely for live-cell studies, however, probably because of the relatively high levels of excitation energy that are required not being entirely compatible with living samples. This drawback appears to have been overcome with the advent of "gated STED" which employs pulsed excitation in combination with time-gated detection [89, 90]: this combination means that lower excitation energies can be used, making gSTED compatible with live-cell studies both in vitro and in intravital imaging studies [91, 92].

STED has also been used to further examine the distribution of the tSNARE at the plasma membrane, revealing that the apparent size of SNARE "clusters" estimated using diffraction-limited imaging (~250 nm) was indeed bounded by the limit of resolution of the microscopes used. STED imaging revealed that "clusters" were in fact "clusters of clusters," of ~60 nm diameter, with each spot modelled to contain approximately 70 syntaxin molecules closely associated with each other via SNARE motif interactions [79]. The size of these immunodetected SNARE spots was also at the limit of resolution of the STED microscope, raising the possibility that the SNAREs are organized into smaller structures in situ or in fact that they are not clustered at all, with the apparent spots being due to diffracted and convolved image data arising from as few as single molecules. More recently, however, it has become possible to acquire image data describing the spatial localizations of very large cohorts of single molecules at or in the plasma membranes of cells: these approaches—termed single-molecule localization microscopies (SMLMs)—use a combination of modified photoactivatable or switchable fluorescent probes with computational processing to deliver rendered images describing the nanoscale locations of 100,000s of single molecules in each sample [83, 84].

1.3.2 Single-Molecule Localization Microscopy of Exocytotic Proteins

SMLM utilizes standard, diffraction-limited microscopes, typically under TIRF illumination, to permit the resolution of individual fluorescently labelled molecules on the plasma membrane. While STED achieves super-resolution through the reduction in size of the emitted point spread function, SMLM techniques permit the resolution of individual proteins by separating their emissions temporally. Under standard excitation, a fluorescently stained sample, such as an immunostained cell, will emit photons from each fluorophore at the same time. As the labelled proteins are typically at high density, their emissions will overlap and we will observe the cellular structures in which the proteins reside but be unable to resolve the constituent individual proteins. In contrast, SMLM techniques activate emission from a sparse subset of the fluorophores in a labelled sample at any one time. These emissions are sufficiently spatially resolved to permit their individual localization with a precision of tens of nanometers. By cycling through multiple activations, ultimately, the entire population of molecules can be localized with high precision.

The simplest way to perform SMLM is to use ground-state depletion with individual molecule return (GSDIM) [80] or its variant stochastic optical reconstruction microscopy (STORM or dSTORM [93]). In the case of GSDIM, fluorescent organic dyes (e.g., cy5, Alexa, or Atto), conjugated to antibodies, are driven into a long-lived dark state using a high-energy laser. This process can normally be enhanced by the use of reducing buffer conditions and oxygen scavengers. This approach has been used to examine the spatial organization of syntaxin on the plasma membrane concluding that, at the molecular level, syntaxin is indeed clustered. It is important to note, however, that due to the repeated cyclical emission of single fluorophores as they escape from the dark state and undergo fluorescence emission, apparent clustering of even single fluorophores can be observed.

An alternative approach in SMLM is to use photoactivation localization microscopy (PALM) [83, 84]. This techniques use photoactivatable fluorophores (commonly EGFP of mCherry variants) which initially exist in an inactive state. Upon illumination with high-energy UV light, the fluorophore matures and can now undergo fluorescence excitation and emission. Eventually this fluorophore will be permanently destroyed through photo-bleaching. This linear path of activation, excitation, and photo-bleaching permits counting of single molecules and more complex statistical analysis (photoblinking can be problematic unless minimized through experimental design). Recently we have applied PALM in neuroendocrine cells to examine the spatial organization of the tSNARE relative to secretory vesicles [16, 94]. Intriguingly secretory vesicles are positioned, at the plasma membrane, where tSNARE density is low.

Both PALM and GSDIM require the use of fixed cells as a result of the long data acquisition time. This freezes the position of molecules to permit the observation of many tens of thousands of molecules in the absence of any diffusion or large-scale movement. It is possible, however, to perform PALM in live cells and track the molecular movement of large cohorts of individual molecules [95]. Single-particle tracking PALM (sptPALM) of tSNARE has been performed and clearly shows that while they diffuse in the plasma membrane, they are not free and instead are contained within microdomains of, as yet, unknown mechanism or function. It will be important in the future to combine the SMLM techniques with other functional imaging modalities to probe in more detail the ultrastructural organization of the secretory machinery and its regulation.

1.3.3 Quantifying Protein–Protein Interactions In Situ

Even SMLM techniques, with ~10 nm localization certainty, cannot determine whether or not proteins interact with one another. The nature of the acquisition means that two protein molecules localized within a small distance of one another may have had their signals detected some time apart. Furthermore, although 2-color SMLM has been described (e.g., for PALM) [96], this is difficult and problematic to perform because photoactivatable GFP has a rather poor discrimination between "on" and "off" and tends to autoactivate leading to reduced signal to background (our observations).

Different biophysical approaches can be used to quantify protein–protein interactions in cells, however. Forster resonance energy transfer (FRET) is a physical effect where resonance energy is transferred in a non-radiative manner (i.e., no photon is emitted) between a higher-energy (so lower wavelength) so-called "donor" fluorophore and a longer-wavelength "acceptor" molecule. For this to occur, two criteria must be satisfied—the donor and acceptor must have substantially overlapping emission and excitation spectra, respectively, and they must be close together. As the distance dependency of energy transfer has a sixth power relationship, FRET occurs only over very short distances; the "Forster distance," where half-maximal energy transfer occurs, is in the range of a few nanometers for typical biological fluorophores.

Quantifying FRET in each pixel of an image offers a very attractive imaging tool for cell biologists, as if FRET can be measured; this is indicative of the two proteins under scrutiny being within interaction range. When combined with mutagenesis studies and other appropriate controls, therefore, the detection of FRET can be used to indicate protein interactions in a cell with high spatial resolution.

Unfortunately, quantifying FRET using fluorescence intensity imaging is very difficult, for several reasons. First, the necessity of overlapping spectra between the fluorophores under study means that separating spectrally the donor and acceptor signals is impossible. Several approaches have been described involving complex arithmetical manipulations of the intensity data to suppress bleed-through and cross talk between the fluorophores, which may help. Second, fluorescence intensity in images is linked to fluorophore concentration, which is impossible to control in cell expression experiments. Furthermore, as the arithmetical calculations mentioned above rely usually on ratios between corrected donor and acceptor signals, spatial heterogeneities in expression levels of either fluorophore can have strong effects on the end result without necessarily having an underlying biological driver. In addition, intensity can be affected by other confounding factors: light-path, scatter, detector nonlinearity, and photo-bleaching, for example. Several techniques involving deliberate photo-destruction of the acceptor molecule, in order to estimate the amount of FRET preceding the bleach, have also been described; these are in turn problematic because of the high excitation energies required as well as doubts over whether the bleaching effect itself can alter the spectral properties of the donor fluorophores. Finally, and importantly, even if FRET can be detected, it is impossible to determine whether spatial changes in the FRET signal are due to (a) changes in the inter-fluorophore distances (i.e., complex conformation) or (b) changes in the proportion of interacting versus noninteracting molecules in a given area of the cell and subsequent image. As each pixel of the image may contain dozens or hundreds of molecules, this becomes the major limiting factor in the interpretation of such data. For all these reasons, single-molecule spectroscopies have become the method of choice for quantifying protein–protein interactions in cells.

1.3.4 Fluorescence Lifetime Imaging Microscopy

Time-correlated single photon counting (TCSPC) approaches combined with pixel-by-pixel acquisition lead to directly quantitative data describing both fluorophore proximity and also the relative proportion in each pixel of an image [97]. Ultrashort laser pulses, either femtosecond or picosecond duration, are used to provide the fluorescence excitation. These pulses are synchronized using timing electronics to provide an accurate measurement of the arrival time of a single photon emitted from a fluorophore within each laser pulse—the inefficiency of fluorescence excitation combined with low-level intensity illumination ensures only single photons are emitted and acquired per laser pulse—hence "single photon counting." As the laser pulses are ultrafast and have high repetition rate (megahertz), it is feasible to acquire many 10,000s of single photons correlated with millions of laser pulses, in turn assigned to each pixel of an image. Subsequent mathematical fitting of the photon arrival time curves (which follow exponential decays) can deliver the time constant describing the mean excited state lifetime of fluorophores.

This approach has several important advantages. As FRET provides an efficient route to relaxation for excited state donor molecules, the excited state fluorescent lifetime (i.e., the time spent in the excited state after absorption of an exciting photon) is decreased if a suitable acceptor molecule comes close. At the half-maximal distance for FRET (the Forster distance), the donor fluorescence lifetime is reduced by 50 %. The measured fluorescence lifetime is independent of donor concentration (as it is a physical parameter characteristic of the molecule being studied) and also of light-path and scatter. Furthermore, no measurement of the acceptor is required as the information necessary to detect FRET is contained in the donor lifetime data. Finally, importantly, if a donor fluorophore with an inherently mono-exponential (i.e., single lifetime) fluorescence decay is used, this makes subsequent calculations simple. In the presence of a proximal acceptor, the mean lifetime per pixel will shorten. These data may be deconvolved, commonly revealing two independent fluorescence lifetimes—one not dissimilar from the "non-FRET" donor and one significantly shorter. The relative magnitudes of each of these exponents can in turn be used to estimate the proportion of interacting versus noninteracting molecules per pixel, and thus in different regions of an image.

We have applied TCSPC-FLIM extensively to describe the interactions of Munc18-1 and syntaxin1a as well as syntaxin1a and SNAP-25 (the tSNARE) [16, 52, 94, 98–101]. Combined with appropriate controls, in particular mutagenesis to disrupt potential interactions or binding sites, these approaches have proved invaluable and highly quantitative.

TCSPC-FLIM however does have some drawbacks. It is expensive and complex. Pulsed laser sources are also commonly expensive and difficult to tune to different wavelengths (although this has now changed with the advent of super-continuum laser sources). A failure to detect FRET does not necessarily mean that no interaction is present—just that the inter-fluorophore distance is too great or the sensitivity of the assay is too low. Importantly, a large number of photons need to be detected in order to make accurate fits to the decay data—in the order of 100,000 photons per pixel for a bi-exponential fit in a FRET experiment—and this takes several minutes to acquire. The latter issue in particular means that TCSPC-FLIM, although accurate, is not compatible with live-cell imaging or imaging dynamic events.

1.3.5 Fluorescence Correlation Spectroscopy

An extremely useful, high-content, and quantitative adjunct for FLIM comes in the form of fluorescence correlation spectroscopy (FCS). This is a long-standing technique originally described in the 1970s [102, 103]. Pioneering work since then, in particular from the laboratory of Petra Schwille, has led to the development of this approach for use in living cells in order to quantify protein–protein interactions with very high temporal resolution [75–77].

Briefly, in FCS, the excitation laser is focused to a spot using a high numerical aperture lens—typically this leads to an excitation volume within the spot in the order of femtoliters. In a living sample, fluorescent molecules (e.g., GFP-fusion proteins) will diffuse across this excitation spot, which is held very stationary at a point inside the cell, and as they pass will emit a burst of emission photons. If these photons can be measured at a rate substantially faster than the time it takes for a molecule to cross the excitation volume, then data are acquired describing the residence time of single molecules in an area of known dimensions and with a known volume. As long as sufficiently few molecules are present in the measurement volume at any given time (fewer than approximately ten molecules, requiring low expression levels in the nanomolar range), then it is simple to calculate the number of molecules, thus concentration as well as the rate of diffusion. This approach commonly uses modern single photon counting detectors to achieve microsecond measurement rates and can be performed in multiple channels simultaneously—this offers the potential to cross-correlate the emission data from differently labelled proteins simultaneously.

As the rate of diffusion is dependent mass, when a protein complex forms, both monomers decrease their diffusion rates when they bind; this provides a strong indication that an interaction has occurred.

We have applied FCS to the study of syntaxin1a and Munc18-1 inside single synaptic boutons, where spatial resolution is limited (see Sect. 1.3.1). This approach, combined with, for example, electrophysiology, provides multiple quantitative measurements of molecular behaviors with temporal resolution sufficiently high to be compatible with dynamic molecular cellular events.

2 Future Perspectives

Over recent decades the field of exocytosis has both benefitted from advances in microscopies and spectroscopies and been a driving influence in their development. Together, the technologies described in this chapter provide an optical toolbox for current researchers to probe, with unparalleled precision, the molecular organization and interactions of the secretory machinery. Many of these technologies are currently highly specialized, performed in a small number of leading laboratories in the world. However, as the technology becomes more widely available, it will undoubtedly drive new experimental innovations and findings.

References

1. Weber T et al (1998) SNAREpins: minimal machinery for membrane fusion. Cell 92:759–772
2. Sollner T et al (1993) SNAP receptors implicated in vesicle targeting and fusion. Nature 362:318–324
3. Clary DO, Rothman JE (1990) Purification of three related peripheral membrane proteins needed for vesicular transport. J Biol Chem 265:10109–10117
4. Melia TJ et al (2002) Regulation of membrane fusion by the membrane-proximal coil of the tSNARE during zippering of SNAREpins. J Cell Biol 158:929–940
5. Sollner T, Bennett MK, Whiteheart SW, Scheller RH, Rothman JE (1993) A protein assembly-disassembly pathway in vitro that may correspond to sequential steps of synaptic vesicle docking, activation, and fusion. Cell 75:409–418
6. Fasshauer D, Otto H, Eliason WK, Jahn R, Brunger AT (1997) Structural changes are associated with soluble N-ethylmaleimide-sensitive fusion protein attachment protein receptor complex formation. J Biol Chem 272:28036–28041
7. Mikoshiba K, Fukuda M, Ibata K, Kabayama H, Mizutani A (1999) Role of synaptotagmin, a Ca2+ and inositol polyphosphate binding protein, in neurotransmitter release and neurite outgrowth. Chem Phys Lipids 98:59–67
8. Shoji-Kasai Y et al (1992) Neurotransmitter release from synaptotagmin-deficient clonal variants of PC12 cells. Science 256:1821–1823
9. Sugita S, Sudhof TC (2000) Specificity of Ca2+-dependent protein interactions mediated by the C2A domains of synaptotagmins. Biochemistry 39:2940–2949
10. Tucker WC, Chapman ER (2002) Role of synaptotagmin in Ca2+-triggered exocytosis. Biochem J 366:1–13
11. Zhang X, Rizo J, Sudhof TC (1998) Mechanism of phospholipid binding by the C2A-domain of synaptotagmin I. Biochemistry 37:12395–12403
12. Garcia RA, Forde CE, Godwin HA (2000) Calcium triggers an intramolecular association of the C2 domains in synaptotagmin. Proc Natl Acad Sci U S A 97:5883–5888
13. Brose N, Petrenko AG, Sudhof TC, Jahn R (1992) Synaptotagmin: a calcium sensor on

the synaptic vesicle surface. Science 256:1021–1025

14. Rickman C et al (2006) Conserved perfusion protein assembly in regulated exocytosis. Mol Biol Cell 17:283–294
15. Rickman C, Hu K, Carroll J, Davletov B (2005) Self-assembly of SNARE fusion proteins into star-shaped oligomers. Biochem J 388:75–79
16. Rickman C et al (2010) tSNARE protein conformations patterned by the lipid microenvironment. J Biol Chem 285: 13535–13541
17. Fasshauer D, Margittai M (2004) A transient interaction of SNAP-25 and syntaxin nucleates SNARE assembly. J Biol Chem 279: 7613–7621
18. Fasshauer D, Antonin W, Subramaniam V, Jahn R (2002) SNARE assembly and disassembly exhibit a pronounced hysteresis. Nat Struct Biol 9:144–151
19. Sutton RB, Fasshauer D, Jahn R, Brunger AT (1998) Crystal structure of a SNARE complex involved in synaptic exocytosis at 2.4 A resolution. Nature 395:347–353
20. Ubach J et al (2001) The C2B domain of synaptotagmin I is a Ca2+-binding module. Biochemistry 40:5854–5860
21. Shao X et al (1997) Synaptotagmin–syntaxin interaction: the C2 domain as a Ca2+-dependent electrostatic switch. Neuron 18:133–142
22. Fernandez-Chacon R et al (2001) Synaptotagmin I functions as a calcium regulator of release probability. Nature 410:41–49
23. Otto H, Hanson PI, Jahn R (1997) Assembly and disassembly of a ternary complex of synaptobrevin, syntaxin, and SNAP-25 in the membrane of synaptic vesicles. Proc Natl Acad Sci U S A 94:6197–6201
24. Canaves JM, Montal M (1998) Assembly of a ternary complex by the predicted minimal coiled-coil-forming domains of syntaxin, SNAP-25, and synaptobrevin. A circular dichroism study. J Biol Chem 273:34214–34221
25. Hu K, Rickman C, Carroll J, Davletov B (2004) A common mechanism for the regulation of vesicular SNAREs on phospholipid membranes. Biochem J 377:781–785
26. Nickel W et al (1999) Content mixing and membrane integrity during membrane fusion driven by pairing of isolated v-SNAREs and tSNARE. Proc Natl Acad Sci U S A 96:12571–12576
27. Fukuda M, Aruga J, Niinobe M, Aimoto S, Mikoshiba K (1994) Inositol-1,3,4,5-tetrakisphosphate binding to C2B domain of IP4BP/synaptotagmin II. J Biol Chem 269:29206–29211
28. Mackler JM, Drummond JA, Loewen CA, Robinson IM, Reist NE (2002) The C(2)B Ca(2+)-binding motif of synaptotagmin is required for synaptic transmission in vivo. Nature 418:340–344
29. Yang B, Steegmaier M, Gonzalez LC Jr, Scheller RH (2000) nSecl binds a closed conformation of syntaxinlA. J Cell Biol 148:247–252
30. Rizo J, Sudhof TC (2002) Snares and Munc18 in synaptic vesicle fusion. Nat Rev Neurosci 3:641–653
31. Pevsner J, Hsu SC, Scheller RH (1994) n-Secl: a neural-specific syntaxin-binding protein. Proc Natl Acad Sci U S A 91:1445–1449
32. Dulubova I et al (1999) A conformational switch in syntaxin during exocytosis: role of munc18. EMBO J 18:4372–4382
33. Brose N, Hofmann K, Hata Y, Sudhof TC (1995) Mammalian homologues of *Caenorhabditis elegans* unc-13 gene define novel family of C2-domain proteins. J Biol Chem 270:25273–25280
34. Betz A et al (1998) Munc13-1 is a presynaptic phorbol ester receptor that enhances neurotransmitter release. Neuron 21:123–136
35. Betz A, Okamoto M, Benseler F, Brose N (1997) Direct interaction of the rat unc-13 homologue Munc13-1 with the N terminus of syntaxin. J Biol Chem 272:2520–2526
36. Stevens DR et al (2005) Identification of the minimal protein domain required for priming activity of Munc13-1. Curr Biol 15:2243–2248
37. Ashery U et al (2000) Munc13-1 acts as a priming factor for large dense-core vesicles in bovine chromaffin cells. EMBO J 19:3586–3596
38. de Wit H et al (2009) Synaptotagmin-1 docks secretory vesicles to syntaxin-1/SNAP-25 acceptor complexes. Cell 138:935–946
39. Comunanza V et al (2010) CaV1.3 as pacemaker channels in adrenal chromaffin cells: specific role on exo- and endocytosis? Channels (Austin, TX) 4:440–446
40. Düfer M et al (2011) BK channels affect glucose homeostasis and cell viability of murine pancreatic beta cells. Diabetologia 54:423–432
41. Yang S-N, Berggren P-O (2006) The role of voltage-gated calcium channels in pancreatic

beta-cell physiology and pathophysiology. Endocrine Rev 27:621–676

42. Charvin N et al (1997) Direct interaction of the calcium sensor protein synaptotagmin I with a cytoplasmic domain of the alpha1A subunit of the P/Q-type calcium channel. EMBO J 16:4591–4596
43. Park Y, Kim K-T (2009) Dominant role of lipid rafts L-type calcium channel in activity-dependent potentiation of large dense-core vesicle exocytosis. J Neurochem 110:520–529
44. Catterall WA (1999) Interactions of presynaptic Ca2+ channels and snare proteins in neurotransmitter release. Ann N Y Acad Sci 868:144–159
45. Rettig J et al (1996) Isoform-specific interaction of the alpha1A subunits of brain Ca2+ channels with the presynaptic proteins syntaxin and SNAP-25. Proc Natl Acad Sci U S A 93:7363–7368
46. Sheng ZH, Yokoyama CT, Catterall WA (1997) Interaction of the synprint site of N-type Ca2+ channels with the C2B domain of synaptotagmin I. Proc Natl Acad Sci U S A 94:5405–5410
47. Zhong H, Yokoyama CT, Scheuer T, Catterall WA (1999) Reciprocal regulation of P/Q--type Ca2+ channels by SNAP-25, syntaxin and synaptotagmin. Nat Neurosci 2:939–941
48. Mochida S, Sheng ZH, Baker C, Kobayashi H, Catterall WA (1996) Inhibition of neurotransmission by peptides containing the synaptic protein interaction site of N-type Ca2+ channels. Neuron 17:781–788
49. Rettig J et al (1997) Alteration of Ca2+ dependence of neurotransmitter release by disruption of Ca2+ channel/syntaxin interaction. J Neurosci 17:6647–6656
50. Betz WJ, Bewick GS (1992) Optical analysis of synaptic vesicle recycling at the frog neuromuscular junction. Science (New York, NY) 255:200–203
51. Betz WJ, Mao F, Bewick GS (1992) Activity-dependent fluorescent staining and destaining of living vertebrate motor nerve terminals. J neurosci 12:363–375
52. Betz WJ et al (2004) Multi-dimensional time-correlated single photon counting (TCSPC) fluorescence lifetime imaging microscopy (FLIM) to detect FRET in cells. J Microsc 215:1–12
53. Cousin MA, Held B, Nicholls DG (1995) Exocytosis and selective neurite calcium responses in rat cerebellar granule cells during field stimulation. Eur J Neurosci 7:2379–2388
54. Pyle JL, Kavalali ET, Piedras-Rentería ES, Tsien RW (2000) Rapid reuse of readily releasable pool vesicles at hippocampal synapses. Neuron 28:221–231
55. Goda Y, Stevens CF (1994) Two components of transmitter release at a central synapse. Proc Natl Acad Sci U S A 91:12942–12946
56. Gordon SL, Leube RE, Cousin MA (2011) Synaptophysin is required for synaptobrevin retrieval during synaptic vesicle endocytosis. J Neurosci 31:14032–14036
57. Degtyar VE, Allersma MW, Axelrod D, Holz RW (2007) Increased motion and travel, rather than stable docking, characterize the last moments before secretory granule fusion. Proc Natl Acad Sci U S A 104:15929–15934
58. Duncan RR et al (2003) Functional and spatial segregation of secretory vesicle pools according to vesicle age. Nature 422:176–180
59. Steyer JA, Horstmann H, Almers W (1997) Transport, docking and exocytosis of single secretory granules in live chromaffin cells. Nature 388:474–478
60. Jaiswal JK, Fix M, Takano T, Nedergaard M, Simon SM (2007) Resolving vesicle fusion from lysis to monitor calcium-triggered lysosomal exocytosis in astrocytes. Proc Natl Acad Sci U S A 104:14151–14156
61. Lang T et al (1997) Ca2+-triggered peptide secretion in single cells imaged with green fluorescent protein and evanescent-wave microscopy. Neuron 18:857–863
62. Miesenböck G, De Angelis DA, Rothman JE (1998) Visualizing secretion and synaptic transmission with pH-sensitive green fluorescent proteins. Nature 394:192–195
63. Balaji J, Ryan TA (2007) Single-vesicle imaging reveals that synaptic vesicle exocytosis and endocytosis are coupled by a single stochastic mode. Proc Natl Acad Sci U S A 104:20576–20581
64. Kishimoto T et al (2005) Sequential compound exocytosis of large dense-core vesicles in PC12 cells studied with TEPIQ (two-photon extracellular polar-tracer imaging-based quantification) analysis. J Physiol 568:905–915
65. Oheim M, Loerke D, Stühmer W, Chow RH (1998) The last few milliseconds in the life of a secretory granule. Docking, dynamics and fusion visualized by total internal reflection fluorescence microscopy (TIRFM). Eur Biophys J 27:83–98
66. Nofal S, Becherer U, Hof D, Matti U, Rettig J (2007) Primed vesicles can be distinguished

from docked vesicles by analyzing their mobility. J Neurosci 27:1386–1395

67. Wang DS, Miller R, Shaw R, Shaw G (1996) The pleckstrin homology domain of human beta I sigma II spectrin is targeted to the plasma membrane in vivo. Biochem Biophys Res Commun 225:420–426
68. Börnig H, Geyer G (1974) Staining of cholesterol with the fluorescent antibiotic "filipin". Acta Histochem 50:110–115
69. Parasassi T, De Stasio G, d' Ubaldo A, Gratton E (1990) Phase fluctuation in phospholipid membranes revealed by Laurdan fluorescence. Biophys J 57:1179–1186
70. Parasassi T, De Stasio G, Ravagnan G, Rusch RM, Gratton E (1991) Quantitation of lipid phases in phospholipid vesicles by the generalized polarization of Laurdan fluorescence. Biophys J 60:179–189
71. Yu W, So PT, French T, Gratton E (1996) Fluorescence generalized polarization of cell membranes: a two-photon scanning microscopy approach. Biophys J 70:626–636
72. Gaus K et al (2003) Visualizing lipid structure and raft domains in living cells with two-photon microscopy. Proc Natl Acad Sci U S A 100:15554–15559
73. Munro S (2003) Lipid rafts: elusive or illusive? Cell 115:377–388
74. Owen DM et al (2006) Fluorescence lifetime imaging provides enhanced contrast when imaging the phase-sensitive dye di-4-ANEPPDHQ in model membranes and live cells. Biophys J 90:L80–L82
75. Bacia K, Schuette CG, Kahya N, Jahn R, Schwille P (2004) SNAREs prefer liquid-disordered over "raft" (liquid-ordered) domains when reconstituted into giant unilamellar vesicles. J Biol Chem 279:37951–37955
76. Kim SA et al (2010) Quantifying translational mobility in neurons: comparison between current optical techniques. J Neurosci 30:16409–16416
77. Kim SA, Heinze KG, Schwille P (2007) Fluorescence correlation spectroscopy in living cells. Nat Methods 4:963–973
78. Lang T, Margittai M, Holzler H, Jahn R (2002) SNAREs in native plasma membranes are active and readily form core complexes with endogenous and exogenous SNAREs. J Cell Biol 158:751–760
79. Sieber JJ et al (2007) Anatomy and dynamics of a supramolecular membrane protein cluster. Science (New York, NY) 317:1072–1076
80. Fölling J et al (2008) Fluorescence nanoscopy by ground-state depletion and single-molecule return. Nat Methods 5:943–945
81. Willig KI et al (2006) Nanoscale resolution in GFP-based microscopy. Nat Methods 3:721–723
82. Willig KI, Rizzoli SO, Westphal V, Jahn R, Hell SW (2006) STED microscopy reveals that synaptotagmin remains clustered after synaptic vesicle exocytosis. Nature 440:935–939
83. Betzig E et al (2006) Imaging intracellular fluorescent proteins at nanometer resolution. Science (New York, NY) 313:1642–1645
84. Hess ST, Girirajan TPK, Mason MD (2006) Ultra-high resolution imaging by fluorescence photoactivation localization microscopy. Biophys J 91:4258–4272
85. Hell SW, Wichmann J (1994) Breaking the diffraction resolution limit by stimulated emission: stimulated-emission-depletion fluorescence microscopy. Opt Lett 19:780–782
86. Kamin D et al (2010) High- and low-mobility stages in the synaptic vesicle cycle. Biophys J 99:675–684
87. Opazo F et al (2010) Limited intermixing of synaptic vesicle components upon vesicle recycling. Traffic (Copenhagen, Denmark) 11:800–812
88. Westphal V et al (2008) Video-rate far-field optical nanoscopy dissects synaptic vesicle movement. Science (New York, NY) 320:246–249
89. Moffitt JR, Osseforth C, Michaelis J (2011) Time-gating improves the spatial resolution of STED microscopy. Opt Express 19:4242–4254
90. Vicidomini G et al (2011) Sharper low-power STED nanoscopy by time gating. Nat Methods 8:571–573
91. Testa I et al (2012) Nanoscopy of living brain slices with low light levels. Neuron 75:992–1000
92. Berning S, Willig KI, Steffens H, Dibaj P, Hell SW (2012) Nanoscopy in a living mouse brain. Science (New York, NY) 335:551
93. Rust MJ, Bates M, Zhuang X (2006) Sub-diffraction-limit imaging by stochastic optical reconstruction microscopy (STORM). Nat Methods 3:793–795
94. Smyth AM, Rickman C, Duncan RR (2010) Vesicle fusion probability is determined by the specific interactions of munc18. J Biol Chem 285:38141–38148
95. Manley S et al (2008) High-density mapping of single-molecule trajectories with

photoactivated localization microscopy. Nat Methods 5:155–157

96. Shroff H et al (2007) Dual-color superresolution imaging of genetically expressed probes within individual adhesion complexes. Proc Natl Acad Sci U S A 104: 20308–20313
97. Duncan RR, Bergmann A, Cousin MA, Apps DK, Shipston MJ (2004) Multi-dimensional time-correlated single photon counting (TCSPC) fluorescence lifetime imaging microscopy (FLIM) to detect FRET in cells. J Microsc 215:1–12
98. Valkonen M et al (2007) Spatially segregated SNARE protein interactions in living fungal cells. J Biol Chem 282:22775–22785
99. Duncan RR (2006) Fluorescence lifetime imaging microscopy (FLIM) to quantify protein-protein interactions inside cells. Biochem Soc Trans 34:679–682
100. Palmer ZJ et al (2008) S-nitrosylation of syntaxin 1 at Cys(145) is a regulatory switch controlling Munc18-1 binding. Biochem J 413:479–491
101. Medine CN, Rickman C, Chamberlain LH, Duncan RR (2007) Munc18-1 prevents the formation of ectopic SNARE complexes in living cells. J Cell Sci 120:4407–4415
102. Elson EL, Webb WW (1975) Concentration correlation spectroscopy: a new biophysical probe based on occupation number fluctuations. Annu Rev Biophys Bioeng 4:311–334
103. Magde D, Elson EL, Webb WW (1974) Fluorescence correlation spectroscopy. II. An experimental realization. Biopolymers 13:29–61

Chapter 6

Electrophysiologic Measurements of Membrane Capacitance in Hormone-Secreting Cells

Boštjan Rituper and Robert Zorec

Abstract

Many methods have been developed for the study of exocytosis and endocytosis. One of the most elegant consists of cellular membrane capacitance measurements which reflect membrane area changes due to fusion or fission of secretory vesicles. This parameter can be monitored directly by the electrophysiologic patch-clamp technique developed by Erwin Neher and Bert Sakmann in 1976. It is still widely used, primarily for the study of currents across ion channels but also for measurements of membrane capacitance as introduced by Erwin Neher and Alain Marty in 1982. In this chapter we provide an overview of the patch-clamp membrane capacitance measurement techniques used in our laboratory to study anterior pituitary hormone-secreting cells.

Key words Regulated exocytosis, Single vesicle, Membrane capacitance, Unitary exocytic events

Abbreviations

C_m	Membrane capacitance
C_{pa}	Patch capacitance
C_s	Stray capacitance
C_{sm}	Specific membrane capacitance
C_v	Vesicle capacitance
ECT	Extracellular fluid
E_{rev}	Reversal potential
G_a	Access conductance
G_m	Membrane conductance
G_p	Pore conductance
G_{pa}	Patch conductance
ICT	Intracellular fluid
Im	Imaginary part of complex admittance
R_a	Access resistance
Re	Real part of complex admittance
R_m	Membrane resistance
R_s	Leakage resistance

Peter Thorn (ed.), *Exocytosis Methods*, Neuromethods, vol. 83,
DOI 10.1007/978-1-62703-676-4_6,

T Attenuation factor
Y Admittance
Z Impedance
ω Angular frequency

1 Introduction

Exocytosis is a highly conserved and ubiquitous process of eukaryotic cells and consists of many steps including vesicle trafficking, docking, priming, and the merger between the vesicle membrane and the plasma membrane leading to the formation of a fusion pore, which can either fully expand or reversibly close [1]. There are two types of exocytosis: constitutive and regulated. Both varieties arguably utilize similar molecular exocytic machinery; the only difference is that regulated exocytosis requires activation by a stimulus, such as an increase in cytosolic $[Ca^{2+}]_{free}$ [2]. Regulated exocytosis thus requires an additional step in the exocytic cascade—a stimulus-sensing switch capable of initiating membrane fusion and widening a narrow, unproductive fusion pore [3]. Synaptotagmins are believed to mediate the switching and respond to increased cytosolic $[Ca^{2+}]_{free}$ (synaptotagmin-1/-2) or cAMP (synaptotagmin-12) [4]. Most of the cells in the human body are capable of constitutive exocytosis, but only few specialized cells, such as neurons, endocrine and neuroendocrine cells, can release their cargo in response to physiologic stimuli. Regulated exocytosis is not uniform in all cells. In fact, there are several types of regulated exocytosis.

Kiss-and-run exocytosis [5] is characterized by the formation of a transient fusion pore, during which vesicle content may be released, followed by closure of the fusion pore. This allows the vesicle to be reused in the next round of exocytosis. The fusion pore can also fluctuate between open and closed states in a process referred to as fusion pore flickering [6] or can retain the transient nature of rhythmic opening and closing for several minutes (pulsing pore) [7–9]. These types are collectively known as transient exocytosis. Secretory vesicles can also fuse with the plasma membrane in such a way that the fusion pore formed expands fully (full-fusion exocytosis) [10]. How the vesicle content is released seems to depend on the cell type. A typical example of full-fusion exocytosis can be recorded in pituitary melanotrophs [11], but the predominant mode of vesicular content release from pituitary lactotrophs is transient exocytosis [12, 13].

In the last few decades, many methods have been developed to monitor exocytosis. Amperometry is an electrophysiologic method whereby the carbon fiber electrode adjacent to the cells is used to electrochemically monitor the release of vesicular content. The electrooxidizing reaction taking place between the carbon electrode surface and the vesicle cargo molecules that are released generates current spikes that indirectly report the interaction between the vesicle and the plasma membrane. Only substances that are readily oxidized or reduced, such as catecholamines, can be detected with this method, therefore it is usually used when studying exocytosis in chromaffin cells [14]. The use of fluorescent markers has also proved to be of value. FM lipophilic styryl dyes [7, 15, 16] and pHluorins [8, 17], pH-sensitive mutants of green fluorescent protein, are popular choices.

These methods allow for indirect observation of unitary exocytic events. However, it is also possible to study the interaction of a single vesicle with the plasma membrane directly using the patch-clamp method. This electrophysiologic method was first introduced by Sakmann and Neher in 1976 and later improved [18]. It is still one of most widely used techniques for the study of ion channels in living cells. Erwin Neher and Bert Sakmann decisively demonstrated the existence of ion channels and explained how they function; they were awarded the Nobel Prize in 1991. Neher, together with Alain Marty, made another important advancement of this technique; they were the first to monitor unitary fusions of vesicles with the plasma membrane. Unitary fusions were recorded as minute discrete steps in membrane capacitance (C_m), a parameter linearly related to the area of the membrane, which is affected by exocytosis and endocytosis [19]. These small discrete steps in C_m were the first direct real-time measurement of unitary exocytic events.

It is possible to combine the patch-clamp technique with amperometry (patch amperometry [20]), fluorescence microscopy [21], and with caged-photolysis approaches whereby a strong ultraviolet light is used to increase $[Ca^{2+}]_{free}$ swiftly in a spatially homogeneous manner [22].

The aim of this chapter is to provide an overview of the patch-clamp membrane capacitance measurement techniques used in our laboratory to study anterior pituitary hormone-secreting cells.

2 A Historical Perspective

It is important to place the development of the patch-clamp technique, previously reviewed many times (e.g., refs. [23, 24]) into the modern era of electrophysiology, which began with Alan Hodgkin, Andrew Huxely, and Bernard Katz. Their collaboration culminated in direct measurements of the transmembrane potential of the giant squid axon in the resting and excited state—the first direct demonstration of the membrane reversal potential [25]. They were also able to explain the potential overshoot and attributed it to the selective permeability of potassium ions, and concluded that the membrane potential, which in the resting state depended on the permeability of potassium ions, is largely dominated by the permeability of sodium ions during excitation, and the passage from one condition to the other is a consequence of a specific change in membrane permeability induced by the electrical stimulus. The voltage-clamp apparatus, first devised by Cole and Mormont in 1949, is a device that enables the measurement of ion currents across the membranes of excitable cells while holding the membrane voltage at a set level. Using this technique, Hodgkin and Huxley [26] clarified the mechanisms underlying the generation of the action potential in nerve fibers and provided evidence that the nervous impulse depends on the previously existing electrical energy gradient accumulated between the interior and exterior of the nerve fiber and is thus an active process. This discovery brings with it a spectacular prediction—the existence of ion channels and gradient-restoring ion pumps. Hodgkin and Huxley were awarded the Nobel Prize in 1963. Attention now turned to the neuromuscular junction. It was known at that time that the end of the motor neuron axon and the motor end plate were bridged by a gap (similar to the synaptic cleft in neuronal synapses). How then, if there is a gap in the cable, can a signal be transferred from the nerve to the muscle? This puzzle was solved by Bernard Katz and colleagues. The resulting quantal theory brought significant progress in the understanding of signal transduction and set the foundation for electrophysiologic exploration of exocytosis. Katz was studying the release of neurotransmitter acetylcholine (ACh) from the axonal endings and was able to determine that after the ACh is released, it crosses the synaptic gap and binds to the receptors on the motor end plate, providing a stimulus that depolarizes muscle and leads to muscle contraction. He also realized that the ACh is released in quanta (small packets) and concluded that it must be packed in small synaptic vesicles that collide with the axonal membrane on stimulation, leading to temporary membrane fusion and an all-or-nothing discharge of vesicular content into the synaptic cleft. His findings are described in detail in the 10th edition of The Sherrington Lectures entitled *The Release of Neural Transmitter Substances* [27]. Katz was awarded the Nobel Prize in 1970. The next important advancement in electrophysiology was the development of patch-clamp techniques to study ion channels [28] and later to study membrane area dynamics, which reflects exocytosis and endocytosis [19].

3 Minimal Equivalent Circuit of a Small Spherical Cell

Capacitance is the ability of a body to store electric charge. Such a body is called a capacitor. A capacitor of C farads with V volts across its terminal is able to store Q coulombs of charge on one plate and $-Q$ on the other as described by

$$C = \frac{Q}{V} \tag{1}$$

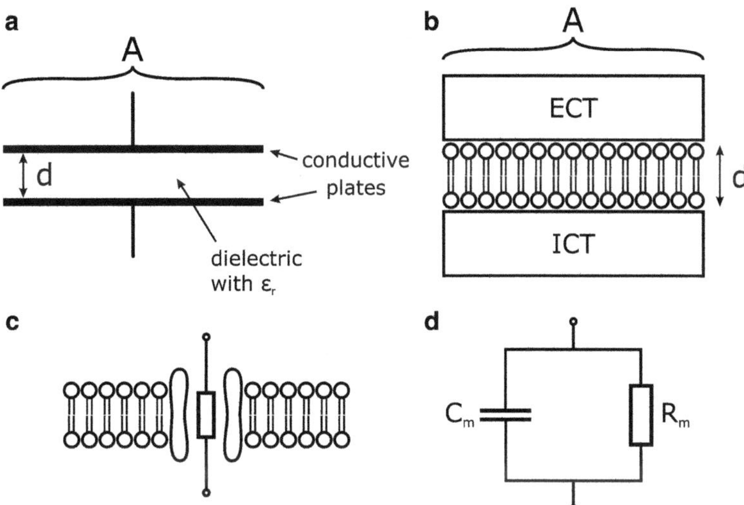

Fig. 1 (**a**) A parallel plate capacitor consisting of two conductive plates with surface area *A*, separated by a distance *d* with a dielectric with relative permittivity ε_r. (**b**) The plasma membrane of a cell electrically resembles a parallel plate capacitor. Extracellular (ECT) and intracellular (ICT) fluids, acting as conductive plates, are bridged by the cell plasma membrane with thickness *d* and surface area *A*. (**c**) An ion-channel spanning lipid bilayer acts as a resistor. (**d**) The minimal equivalent circuit, consisting of a capacitor (C_m) and resistor (R_m) in parallel, is an accurate and adequate representation of the electrical properties of a small spherical cell

The parallel plate capacitor is a common capacitor. It consists of two conductive plates separated by a dielectric material (Fig. 1a). The capacitance of this electrical component is directly proportional to the surface area of the plates and inversely proportional to the distance between the plates. It also depends on the permittivity of the dielectric between the plates and can be calculated from the following equation:

$$C = \varepsilon_0 \varepsilon_r \frac{A}{d} \tag{2}$$

where C denotes capacitance, ε_0 is the permittivity of free space, ε_r is the relative permittivity of the dielectric material between the plates, A is the surface area of the conductive plates, and d is the distance between the plates.

The current flowing through such a capacitor depends on its capacitance and the voltage change over time and can be calculated using (3):

$$I = C\frac{\mathrm{d}V}{\mathrm{d}t} \tag{3}$$

In electrical terms, the plasma membrane of a cell can be described by a parallel plate capacitor (Fig. 1a, b). Intracellular and extracellular fluids act as conductive plates and a lipid bilayer acts as a dielectric. If the voltage is applied across the plasma membrane, the amount of charge stored is proportional to the surface area of the cell membrane and inversely proportional to the membrane thickness. Because the membrane thickness usually remains fairly constant in all eukaryotic cells and is similar to that in purely lipidic liposomes [29], changes in capacitance reflect changes in the surface area of the cell membrane. Thus, in the case of biological membranes, (2) can be simplified to

$$C = C_{sm} A \quad (4)$$

where C_{sm} is the specific membrane capacitance, a constant with units of fF/μm^2. C_{sm} for mammalian pituitary cells is ~8 fF/μm^2 [30].

Cell membranes contain a wide variety of proteins in addition to lipids. These can be embedded in the outer or inner leaflet of the lipid bilayer, or can span the whole bilayer (transmembrane proteins) (Fig. 1c). Proteins seem to have a minimal influence on C_{sm}, but more importantly, several transmembrane protein types regulate the flow of ions across the membrane and thus influence membrane conductance (G_m). These proteins, acting as electrical resistors, are ion channels, (co)transporters and pumps; ion channels have by far the biggest impact on G_m. Another important property of ion channels is that they can exist in open or closed configurations. The overall conductance of a plasma membrane thus depends on the number and type of open channels in a given period of time.

Changes in cell membrane G_m follow Ohm's law. If the constant voltage (V, denoting the driving voltage, which also takes into account the reversal potentials of particular ionic species that are conducted via distinct ion channel types, but are combined here) is applied across the membrane, then changes in membrane resistance R_m ($R_m = 1/G_m$) are reflected in changes in the resulting current I_m across the membrane as follows:

$$I_m = \frac{V}{R_m} = VG_m \quad (5)$$

The minimal equivalent circuit consists of the smallest amount of connected electronic components effectively representing the electrical characteristics of the cell membrane of a spherical cell. In any given cell, large numbers of ion channels are embedded in the plasma membrane. The minimal equivalent circuit would get hopelessly complicated if every component of the cell had to be represented individually. However, considering the generalized Thévenin theorem, the equivalent circuit can be simplified. It consists of a capacitor and a resistor linked in parallel (RC circuit), and represents the electric properties of the plasma membrane of a single cell fairly accurately and adequately (Fig. 1d). However, this is only true as long as the cell is spherical and not too large (the conductance of the intracellular medium should be much higher than that of a cellular membrane). For cells with long thin protrusions (such as neurons), additional electronic components need to be considered in the electrical model in order to generate a more accurate representation.

According to Katz's quantal theory [27], neurotransmitters, which are stored in secretory vesicles, are released as quanta on stimulation in an all-or-nothing fashion. For this to be true, secretory vesicles must fuse with the plasma membrane. On fusion, the surface area of the plasma membrane is increased by the membrane area of the fusing vesicle. It can be seen from (2) that the capacitance of the plasma membrane is increased proportionally, provided there is no change in membrane thickness and/or permittivity of the dielectric properties of the lipid bilayer. In other words, changes in the cell membrane surface area can be measured as changes in C_m, reflecting area fluctuations due to fusion and fission of secretory vesicles (i.e., exocytosis and endocytosis). Although Del [31] proposed that the merger between the vesicle membrane and the plasma membrane is transient, the vesicle membrane is now usually considered to be fully integrated into the plasma membrane. The former mode is also termed transient fusion exocytosis; the latter is called full-fusion exocytosis (Fig. 2).

If the voltage (V) is applied across the minimal equivalent circuit of a spherical cell (Fig. 3a, b), the current (I_t) through such a circuit is given by the sum of two components:

$$\overrightarrow{I_t} = \overrightarrow{I_r} + \overrightarrow{I_c} \tag{6}$$

I_r is the ionic current that is carried by the flow of ions through channels in the membrane (resistive current), I_c is the capacitive current flow that results from a change in the amount of charge separated by the membrane; hence, from (3) and (5),

$$I = \frac{V}{R_m} + C_m \frac{dV}{dt} \tag{7}$$

where V is the membrane potential. If $dV/dt = 0$, which is the case when constant direct current (DC) voltage is applied, current will only flow through the resistive part of the circuit. This configuration is ideal for measuring changes in membrane conductance (i.e., current through ion channels) and is generally referred to as the voltage-clamping approach.

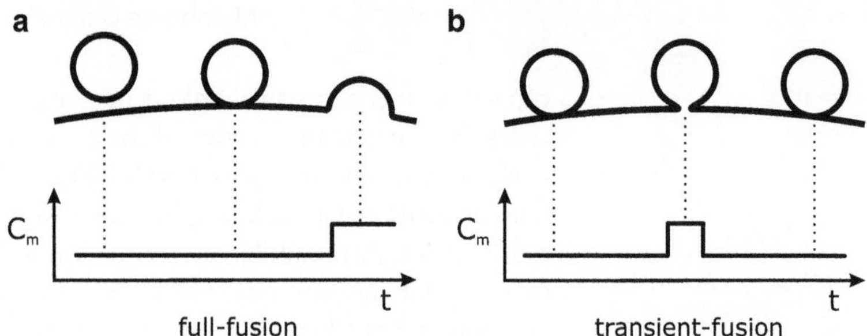

Fig. 2 Modes of exocytosis. (**a**) Full-fusion exocytosis is characterized by vesicular trafficking, docking, priming, and the merger of the vesicle with the plasma membrane, whereby the vesicle is fully integrated with the plasma membrane. When measuring changes in membrane capacitance (C_m) as a function of time (**a**, plot), full fusion of the vesicle with the membrane is seen as an upward step in capacitance, not subsequently followed by a down step. (**b**) The initial stages of transient fusion exocytosis are similar to the whole-cell mode. On merging with the membrane, a fusion pore is created, which can be dilated and subsequently narrowed/closed. Transient fusion exocytosis can be seen as an upward step in capacitance (**b**, plot) that is followed by a down step of similar amplitude

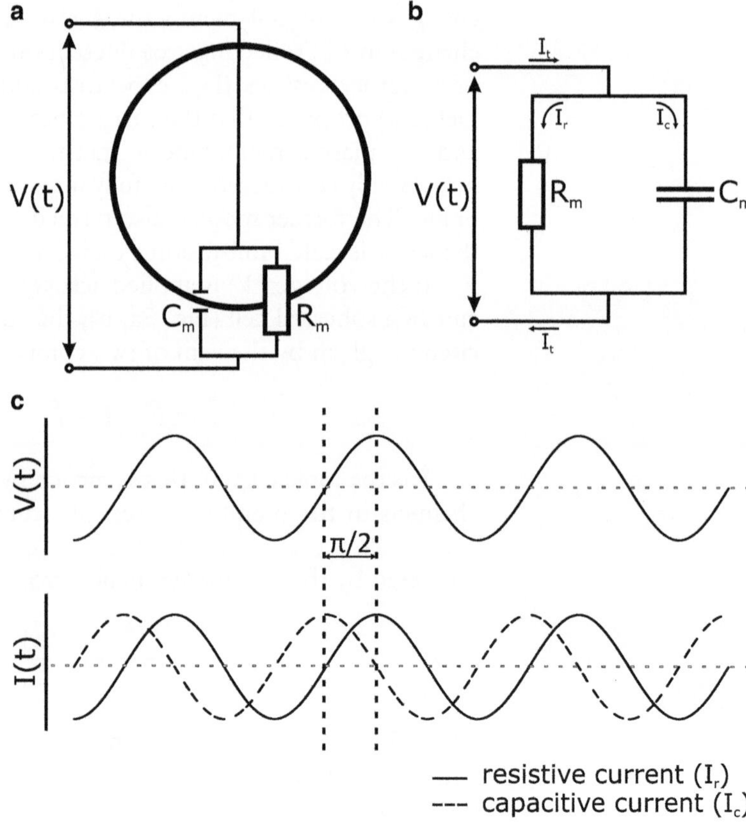

Fig. 3 (**a**) Minimal equivalent circuit of a small spherical cell consists of a capacitor (C_m) and resistor (R_m) in parallel. (**b**) If the voltage ($V(t)$) is applied across such a circuit, the total current ($I(t)$) is given by the sum of two components: I_r (current through the resistor) and I_c (current through the capacitor). If direct current voltage ($dV/dt = 0$) is applied, only I_r will be flowing (7). If sine voltage is applied (**c**), the resulting current will be the sum of two sinusoidal currents; I_r will be in phase with the driving voltage and I_c will precede the driving voltage by 90° or $\pi/2$ (8)

If, on the other hand, a voltage in the form of $V(t) = V_0 \sin(\omega t)$ (sine voltage) is applied to the same circuit, the resulting current will be of the form

$$I(t) = \frac{V_0 \sin(\omega t)}{R_m} + C_m V_0 \omega \cos(\omega t) \tag{8}$$

where ω represents the angular frequency (radians/second) of applied sine voltage and equals $2\pi f$ (f= frequency in Hz). The resulting current is thus the sum of two sinusoidal currents; the resistive (ionic) current (I_r) is in phase with the driving voltage and the capacitive current (I_c) leads the phase of the driving voltage by 90° or $\pi/2$ (Fig. 3c). Using this information and special hardware (a dual-phase lock-in amplifier integrated with the patch-clamp amplifier), changes in the membrane capacitance and conductance can be measured (see Fig. 7).

Another method of minimal equivalent circuit analysis is to apply voltage steps and measure current transients, from which C_m and G_m can be calculated (see Fig. 5b).

Application of a voltage step across an RC circuit provokes an abrupt increase in current ($dV/dT \ggg 0$, see (3)) which spikes at the amplitude of I_0. During this time, the capacitor charges and subsequently discharges through the resistor. The decrease in the current has the form

$$I = I_0 e^{-t/RC} \tag{9}$$

where e is the base of the natural logarithm and the product of RC is for the time constant (τ) of the decrease in the current. The time constant represents the time it takes for a capacitor (C) to discharge via resistance in series (R) for ($1 - 1/e$) or roughly 63.2 % of its initial charge. When the capacitor is discharged (remember that the voltage is constant at this point), only resistive current remains, which equals $I_r = V/R_m$. The C_m and G_m of the minimal equivalent circuit are then calculated from I_0, I_r, V, and τ (see Section 3.2).

3.1 The Experiment

Cell membranes are constantly remodeled by means of exocytosis and endocytosis. The whole plasma membrane can be recycled within a few hours [32]. On fusion and fission, the surface area of the plasma membrane is increased or decreased by the surface area of vesicles undergoing fusion or fission. Accordingly, the capacitance of the plasma membrane is either increased or decreased. Thus, by measuring changes in capacitance, changes in plasma membrane surface area can be measured, which reflect exocytosis and endocytosis.

How is this done in practice? The solution for this problem was engineered by Neher and Sakmann in 1981. The key was the

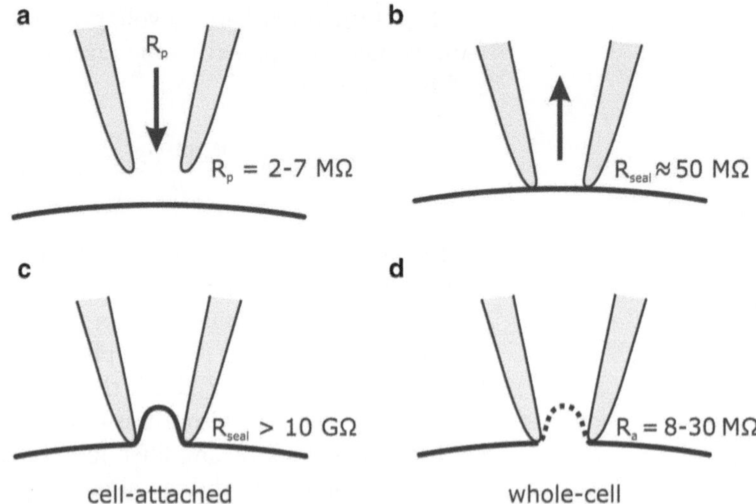

Fig. 4 Patch-clamp configurations. (**a**) Pipettes with the resistance (R_p) of 1–5 MΩ are used for experiments. (**b**) A contact between the pipette tip and the plasma membranes is registered as an increase in resistance (now termed seal resistance, R_{seal}) to approximately 50 MΩ. (**c**) Gentle suction is then applied, forming firm contact between the pipette tip orifice and the plasma membrane and increasing the seal resistance to >10 GΩ (i.e., giga seal), which indicates that a cell-attached configuration has been achieved. (**d**) Another brief pulse of negative pressure tears the membrane patch spanning the pipette opening and establishes a diffusional continuum between the pipette and the cell lumens, i.e., whole-cell configuration

use of relatively large-bore glass pipettes that form a very tight seal with the cell membrane. An experiment can then be performed in several ways, including cell-attached and whole-cell configurations [18] (Fig. 4).

Cells attached to a coverslip are mounted on the microscope (usually inverted microscopes are used) and bath solution is applied. The pipette, secured within a holder, is lowered into the bath solution and the pipette resistance ($R_{pipette}$) is measured using the square wave voltage pulse and the resulting pipette current step (Fig. 4a). Pipettes with resistance in the range of 1–5 MΩ are used. Next, the pipette is carefully placed on top of the cell surface and gentle contact is established. This can be seen as an increase in pipette resistance (to ~50 MΩ) (Fig. 5b). Gentle suction is then applied, forming firm contact between the pipette tip orifice and the plasma membrane, which increases pipette resistance steeply; then a giga seal is formed ($R > 10$ GΩ), usually abruptly. This configuration is also known as the cell-attached configuration and is used for high-resolution measurements of unitary fusion and fission events of a single vesicle (Fig. 4c). If another brief pulse of negative pressure is applied, part of the membrane spanning the pipette opening (i.e., the patch membrane) is torn and a diffusional continuum between the pipette lumen and the cytosol is

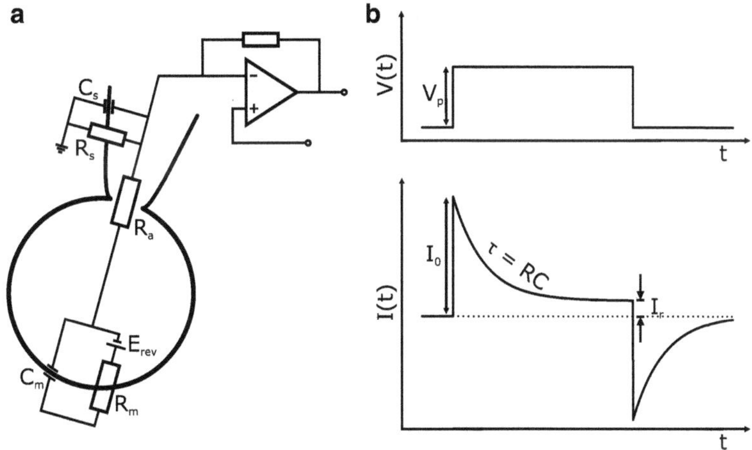

Fig. 5 Time-domain analysis of a minimal equivalent circuit. (**a**) Capacitance measurements complicate the minimal equivalent circuit by further introducing access resistance (R_a), leakage resistance (R_s), and stray capacitance (C_s). In a good patch, R_s is several orders of magnitude greater than R_a and can be neglected. C_s is cancelled out (i.e., subtracted) using a fast capacitance cancellation potentiometer built into the patch-clamp amplifier. (**b**) Application of a voltage step (**b**, *upper plot*) across such a circuit results in a capacitive current transient that can be used to measure I_0, τ, and I_r (**b**, *lower plot*), from which the remaining parameters (R_a, R_m, C_m) of a cell are calculated using (15)–(17)

established with the access resistance (R_a) in range of 8–30 MΩ. This configuration is known as the whole-cell configuration (Fig. 4d). It is ideal for measuring relatively large changes in C_m over time and has the advantage of being able to introduce potential activators and inhibitors of exocytosis/endocytosis directly into the cytosol via diffusion from the pipette [18, 33].

3.2 Time-Domain Analysis of the Minimal Equivalent Circuit

The time-domain technique is the simplest method for determining the passive circuit parameters (C_m, G_m, G_a; see below). This technique analyses the current relaxation in response to an instantaneous increase in transmembrane voltage, which is usually produced by a train of square wave pulses. The resulting dielectric polarization of the lipid bilayer leads to current transients in response to those pulses and parameters [34]. Time-domain analysis is usually performed in the whole-cell mode of recording.

The introduction of the glass pipette complicates the minimal equivalent circuit of a single spherical cell (Fig. 5a). In addition to C_m and G_m, access resistance (R_a), leakage resistance (R_s), and stray capacitance (C_s) have to be dealt with, the latter consisting of contributions from the head stage of the amplifier, the pipette holder, and the pipette tip. Moreover, the reversal potential (E_{rev}), which represents mechanisms that establish concentration gradients for different ions across the cell membrane, and the effect of different permeabilities for permeating ionic species across the plasma membrane (Fig. 5) has to be considered. The equivalent circuit is only valid if the parameters are independent of the voltage (and frequency). Because many cellular membranes contain voltage-dependent ion channels, care needs to be taken so that the driving voltage pulses do not interfere with ion channel activation. The nonlinearity of G_m can then be neglected [34]. C_s is cancelled out using fast capacitance cancellation circuits built into the amplifier in the cell-attached mode of recording. This procedure simply subtracts capacitive contributions from sources other than the cell membrane.

Application of a voltage steps results in capacitive current transients (Fig. 5b), which are fitted with an exponential function of the form

$$I(t) = (I_0 - I_r)^{-t/\tau} + I_r \tag{10}$$

where t is the time after the voltage step. The passive circuit parameters are then related to I_0, I_r, and τ by the following equations:

$$R_a = \frac{V_p R_s}{I_0 R_s - V_p} \tag{11}$$

$$G_m = \frac{1}{R_m} = \frac{1}{\dfrac{V_p R_s}{I_r R_s - V_p} - R_a} \tag{12}$$

$$C_m = \tau \left(\frac{1}{R_a} + G_m \right) \tag{13}$$

The solutions to these equations are not unique, because three measured parameters are not sufficient to calculate four elements of the equivalent circuit. Using a simple mathematical manipulation, (11) can be rewritten as

$$\frac{1}{R_a} = \frac{I_0}{V_p} - \frac{1}{R_s} \tag{14}$$

If $R_s \ggg R_a$ (in order of gigaohms), then $1/R_s$ is close to 0 and the following can be assumed:

$$R_a^* = \frac{V_p}{I_0} \tag{15}$$

Using the same reasoning, (12) and (13) reduce to

$$G_m^* = \frac{I_r}{V_p - R_a^* I_r} \tag{16}$$

$$C_m^* = \tau\left(\frac{1}{R_a^*} + G_m^*\right) \tag{17}$$

If a continuous train of voltage pulses is applied, C_m and G_m can be monitored as a function of time. The time interval between two consecutive pulses must be at least $5 \times \tau$, otherwise the current has not settled to a steady state value and measurements of I_r are not reliable. The time resolution limit for a typical neuroendocrine cell with $C_m = 5$ pF and $R_a = 10$ MΩ is then $5 \times C_m \times R_a = 0.25$ ms, provided that $R_a \lll R_m$ (see (17)).

The advantage of this method is that changes in G_m and R_a can be well separated and even large changes in G_m and reversal potential do not interfere with the determination of C_m. On the other hand, the total signal power of the method is fairly low resulting in high capacitance noise, therefore this method is not the method of choice for studying single fusion events, but it is sufficient for monitoring relatively large changes in C_m [35]. The resolution can be improved by using sine wave voltage stimulation for charging and discharging C_m and the resulting current can be analyzed by using a dual-phase lock-in amplifier (see next section).

To perform time-domain analysis, a patch-clamp amplifier, a square wave function generator, a set of analog or digital low-pass filters, a multichannel A/D converter, a computer, and specialist software dedicated to fitting exponential functions over the current transients, from which τ is calculated, are needed. Nevertheless, to estimate C_m, a patch-clamp amplifier with capacitance cancellation controls [30] is sufficient. Here, C_m is estimated by cancellation of capacitive current transients using a built-in analog cancellation circuit. This approach allows measurements only at relatively long time intervals.

3.3 Frequency-Domain Analysis of a Minimal Equivalent Circuit

By considering (7) and (8), instead of measuring current transients, frequency-domain analysis can be used, which is based on the fact that the currents flowing through the resistive part of the cell membrane (I_r) are in phase with the driving sine wave voltage and the capacitive currents (I_c) are phase shifted with respect to the driving voltage by 90° ($\pi/2$) (Fig. 3). The essence of the analysis is the separation of the two current components by means of a two-phase lock-in amplifier. This piece of equipment is used in standard electronic circuit analysis and can be connected to the standard patch-clamp amplifier. However, there are not many all-in-one amplifiers available commercially (one was developed in our laboratories [30]).

Frequency-domain analysis can be used for microscopic (compensated) or macroscopic (compensated and uncompensated) capacitance measurements. In the former, bulk (resting) C_m is cancelled and minor changes in the isolated cell-attached membrane patch are recorded. On the other hand, macroscopic measurements of C_m are used for monitoring relatively large excursions in C_m over time in a whole-cell mode recording configuration.

To fully understand the frequency-dependent analysis of a minimal equivalent circuit, one must be familiar with the physical basis of the analysis [36]. This analysis is uniquely possible because of a fundamental property of sine waves when applied to linear circuits. In a linear circuit, its output, when driven by the sum of two input signals, equals the sum of its individual outputs when driven by each signal in turn. Our minimal equivalent circuit fits this description perfectly as both capacitors and resistors are linear devices. The output of a linear circuit driven with a sine wave at frequency f is itself a sine wave with the same frequency; the only parameters that can change are the amplitude and phase of the output signal with regard to the input signal. No other driving signal has these properties.

3.3.1 Generalization of Ohm's Law

When dealing with currents driven by a sine wave voltage in an alternating current (AC) circuit, Ohm's law is no longer sufficient as it does not account for changes in the phase angle introduced by capacitors. However, for linear circuits, voltage, current, and resistance can be generalized in order to rescue Ohm's law. Although the amplitude and phase shifts of voltage or current can be specified by writing them out in form

$$V(t) = A\sin(\omega t + \theta) \tag{18}$$

where A is the peak amplitude, ω is the angular frequency of the sine wave, and θ is the phase angle introduced by the circuit, this type of analysis is seldom necessary as the calculation can be made much simpler by the use of complex numbers to represent changes in voltage and currents. Euler's formula is used for the conversion between both representations.

3.3.2 Impedance

Impedance (Z) is essentially generalized resistance and represents the complex ratio of the voltage to the current in a linear AC circuit. It consists of resistance and reactance:

$$Z = R(resistance) + jX(reactance)$$

and contains information on the magnitude and the phase of the sine wave. Impedance is generally represented as a complex number with resistance as its real part (Re) and reactance as its imaginary part (Im) and can be depicted in polar form (Fig. 6). The introduction of impedance in the analysis of AC circuits is necessary as devices such as inductors and capacitors (more relevant in our case)

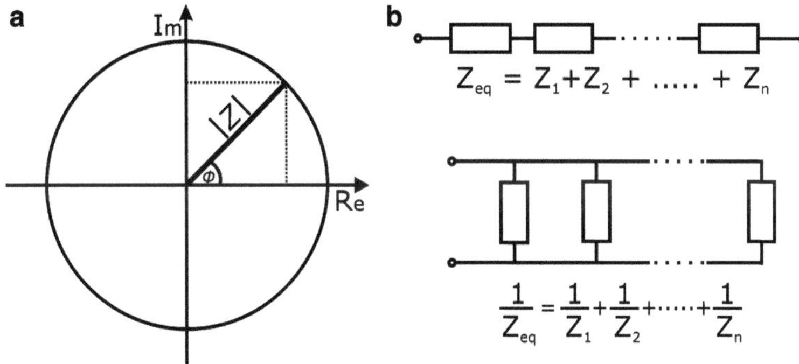

Fig. 6 (**a**) Polar representation of impedance Z, where $Z=|Z|(\cos\Phi+j\sin\Phi)$; also $Z=|Z|e^{j\Phi}$ and contains information on both the amplitude and phase change of the current when driven by a sine voltage over a linear circuit. (**b**) Rules for calculations with impedances. Impedance (Z_{eq}) of components in series equals the sum of the impedances of the individual components (**b**, *top*, (27)); the impedance (Z_{eq}) of components in parallel equals the inverse of the sum of the inverse impedances of the individual components of a circuit (**b**, *bottom*, (28))

impede the current flow differently from normal resistors; they have reactance, whereas resistors have resistance. It follows from this that the capacitors have purely imaginary impedance as they introduce 90° phase angle shifts and resistors have purely real impedance. This is discussed later.

Impedance is defined as

$$Z = \frac{V}{I} \tag{19}$$

where V is the driving voltage and I is the resulting current.

3.3.3 Impedance of a Capacitor

Equation (3) gives the current over a capacitor in relation to the changes in driving voltage. If we apply the sine voltage in the form of

$$V = V_0 \sin(\omega t) \tag{20}$$

then the current across the capacitor is of the form (see (8))

$$I(t) = V_0 C\omega \cos(\omega t) \tag{21}$$

If we insert (20) and (21) into (19), the impedance (Z_c) of a capacitor is in the form

$$Z_c = \frac{V_c(t)}{I_c(t)} = \frac{V_0 \sin(\omega t)}{V_0 C\omega \cos(\omega t)} = \frac{\sin(\omega t)}{\omega C \sin(\omega t + (\pi/2))} \tag{22}$$

where $\frac{1}{\omega C}$ represents the amplitude ratio and $\frac{\pi}{2}$ the phase shift. Since a capacitor introduces a 90° phase angle change, its impedance is purely imaginary (i.e., reactance). The impedance of a capacitor has a polar form of

$$Z_c = \frac{1}{\omega C} e^{-j(\pi/2)} \tag{23}$$

where e is the base of the natural logarithm and j is the imaginary unit. Using Euler's formula, this equation can be rewritten as

$$Z_C = \frac{1}{j\omega C} \tag{24}$$

3.3.4 Impedance of a Resistor

As considered for the capacitor, the current across a resistor when driven by a sine voltage is in the form

$$I(t) = \frac{V_0 \sin(\omega t)}{R} = I_0 \sin(\omega t) \tag{25}$$

The impedance of a resistor (Z_r) is then

$$Z_r = \frac{V_r(t)}{I_r(t)} = \frac{V_0 \sin(\omega t)}{I_0 \sin(\omega t)} = R \tag{26}$$

It is obvious from this equation that resistors do not introduce phase changes to the current and thus have purely real impedance (i.e., resistance).

3.3.5 Rules When Calculating with Impedances

The total impedance of a circuit consisting only of components connected in series can be calculated from

$$Z_{total} = Z_1 + Z_2 + \cdots + Z_n \tag{27}$$

where n is the number of components in the circuit.

Similarly, the total impedance of a circuit consisting only of components connected in parallel can be calculated from

$$\frac{1}{Z_{total}} = \frac{1}{Z_1} + \frac{1}{Z_2} + \cdots + \frac{1}{Z_n} \tag{28}$$

These rules should be familiar as they are very similar to the rules for calculating resistances for DC circuits (see Fig. 6b).

3.4 Admittance of a Minimal Equivalent Circuit and Calculation of Passive Cell Parameters in Uncompensated Whole-Cell Mode

Admittance (Υ) is defined as the inverse of impedance ($\Upsilon(\omega) = 1/Z(\omega)$) and consists of conductance (A) and susceptance (B), where A represents the real (Re) part of the admittance and B the imaginary (Im) part of the admittance:

$$\Upsilon(\omega) = A + iB \tag{29}$$

The minimal equivalent circuit of a small spherical cell in whole-cell configuration consists of a resistor (R_m) and capacitor (C_m), connected in parallel, together with a resistor (R_a), connected in series (Fig. 6a). The total admittance of such a circuit can be derived from (24) to (28) and is of the form [34]

$$Y(\omega) = \frac{1 + j\omega C_m R_m}{R_a + R_m + j\omega C_m R_m R_a} = A + iB \tag{30}$$

where R_a is the access resistance, R_m is the membrane resistance, and C_m is the membrane capacitance.

From ref. [33], we find that

$$A = \frac{R_a + R_m + \omega^2 C_m^2 R_m^2 R_a}{(R_a + R_m)^2 + \omega^2 C_m^2 R_m^2 R_a^2} \tag{31}$$

$$B = \omega C_m R_m \left[\frac{1 - AR_a}{R_a + R_m} \right] \tag{32}$$

A two-phase lock-in amplifier provides two outputs Y_1 and Y_2, which are proportional to A and B, respectively, provided that the phase angle settings are chosen correctly. From A and B and the DC conductance, which is given by

$$b = \frac{1}{R_a + R_m} = \frac{I_{DC}}{U_{DC} - E_{rev}} \tag{33}$$

three passive parameters R_a, R_m, and C_m can be calculated using the following equations:

$$R_a = \frac{A - b}{A^2 + B^2 - Ab} \tag{34}$$

$$R_m = \frac{1}{b} \frac{(A - b)^2 + B^2}{A^2 + B^2 - Ab} \tag{35}$$

$$C_m = \frac{1}{\omega B} \frac{(A^2 + B^2 - Ab)^2}{(A - b)^2 + B^2} \tag{36}$$

In comparison with time-domain analysis, frequency-domain analysis allows for a much better signal-to-noise ratio. It is possible to achieve a resolution of up to 40 aF, which translates to a vesicle diameter of ~40 nm (see Section 5.2), if the microscopic recording approach is used. On the other hand, absolute measurements of C_m are only accurate if the stray capacitance C_s is nullified correctly. C_s and its associated transient are seldom ideally fast; thus, complete cancellation of C_s may only be optimal and incomplete.

Moreover, this type of analysis is sensitive to nonlinear changes in G_m (due to voltage-activated channels) as well as possible changes in reversal potential during long recordings. Because the passive cell membrane parameter calculation includes the reversal potential, it is also of utmost importance to measure the reversal potential as accurately as possible. For pituitary lactotrophs and melanotrophs, E_{rev} was measured to be −50 mV.

3.5 High-Resolution Measurements in Compensated Whole-Cell Mode

To improve the signal-to-noise ratio even further, the bulk of C_m is cancelled and the patch-clamp amplifier is operated at high gain (typically 50 or 20 pA/V, with different feedback resistances in the head stage). Under these conditions the simple expressions for R_a, R_m, and C_m from (34) to (36) no longer apply. However, small changes in cell passive parameters can still be measured [34]. Apart from phase shift and an attenuation factor $|T^2|$, outputs Y_1 and Y_2 of the lock-in amplifier are proportional to

$$Y_1 \approx \Delta G_m - \Delta G_a \omega^2 C_m^2 / G_a^2 \tag{37}$$

$$Y_2 \approx \omega \Delta C_m \tag{38}$$

Special care needs to be given to the phase angle setting as changes introduced by access conductance (G_a) can be high. In addition, the attenuation factor $|T|$ can be significantly lower than 1; thus, the lock-in signal calibration should be recalculated by the following equation:

$$1/|T|^2 = k = (1 + G_m / G_a)^2 + (2\pi f C_m / G_a)^2 \tag{39}$$

where k is the scaling calibration factor and f is the frequency in Hz. In practice, both the phase angle and the scale factor can be obtained by means of the (hardware) calibrated capacitance neutralization feature of patch-clamp amplifiers and the use of calibration pulses during the measurement (see below).

3.6 High-Resolution Measurements in Compensated Cell-Attached Mode

The minimal equivalent circuit for cell-attached configuration is further complicated by the addition of two electronic elements representing patch membrane capacitance (C_{pa}) and patch membrane conductance (G_{pa}) (Fig. 7a). The admittance of such a circuit is given by the following equation:

$$Y(\omega) = \frac{G_a (G_{pa} + i\omega C_{pa})(G_m + i\omega C_m)}{G_a (G_{pa} + i\omega C_{pa}) + (G_{pa} + i\omega C_{pa} + G_a)(G_m + i\omega C_m)} \tag{40}$$

The solutions of (40) are not unique, because the three measured parameters (i.e., Y_1, Y_2, and DC conductance) are not sufficient to calculate five elements of the equivalent circuit. Nonetheless,

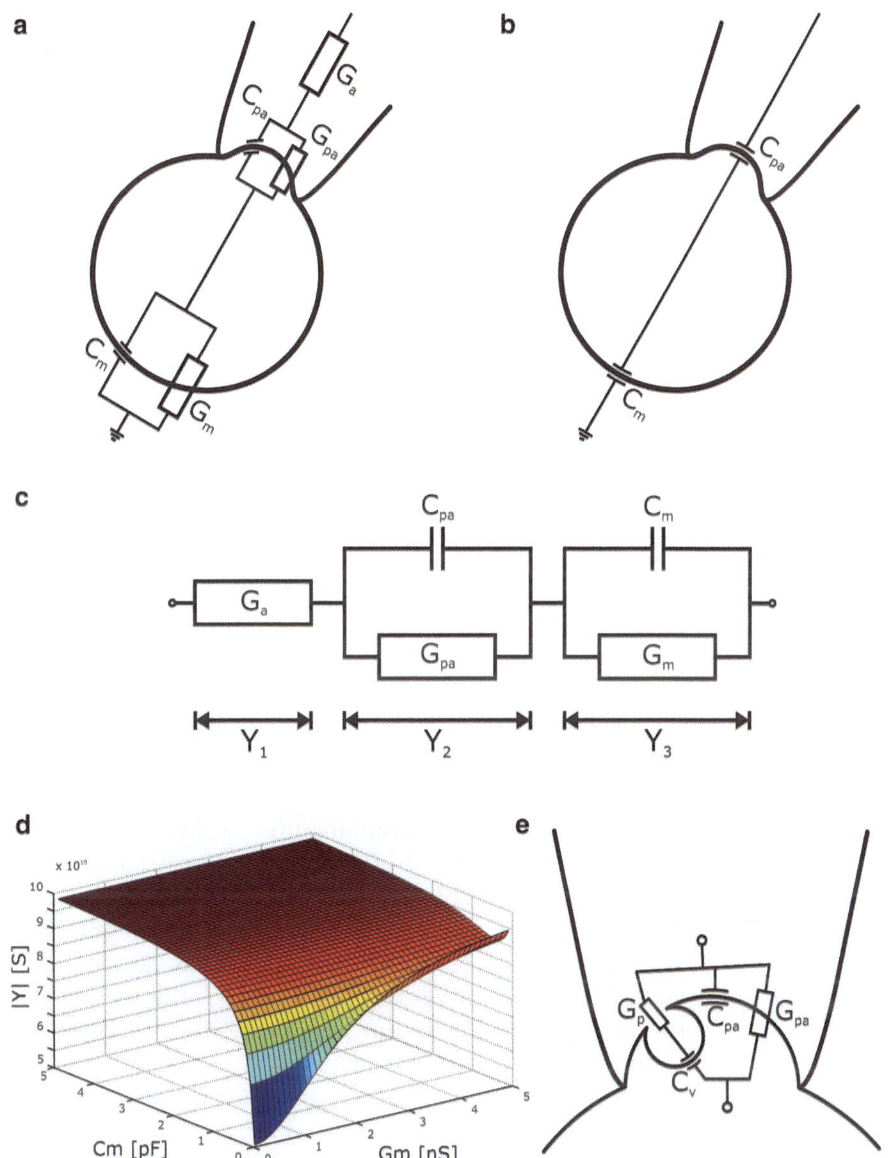

Fig. 7 Frequency-domain analysis of a minimal equivalent circuit. (**a**) The minimal equivalent circuit for the cell-attached configuration is further complicated by the addition of two electronic elements representing patch membrane capacitance (C_{pa}) and patch membrane conductance (G_{pa}). (**b**) Simplified representation of the cell-attached configuration, where only C_{pa} and C_m are depicted. Because C_{pa} and C_m are connected in series, changes in C_m have a negligible influence on C_{total}, whereas changes in C_{pa} have a much greater effect (41). Measured changes in C_{total} are thus an accurate representation of changes in C_{pa}. (**c**) Schematic representation of the minimal equivalent circuit for cell-attached mode; the total admittance can be calculated from (40). (**d**) Simulation of the effects of changes in C_m and G_m on the overall complex admittance (Y) of the minimal equivalent circuit in the cell-attached configuration. Realistic values for C_m and G_m are 2–5 pF and 0.5–3 nS, respectively. In that range, even large changes in C_m and G_m produce negligible changes in the overall complex admittance. (**e**) Minimal equivalent circuit of a fusing vesicle with membrane capacitance C_v, which forms the fusion pore conductance (G_p) in a stable patch with capacitance C_{pa} and conductance G_{pa} (43)–(46)

(37) and (38) are still valid if G_m and C_m are replaced with G_{pa} and C_{pa}, respectively. In that case, the Y_2 output of the lock-in amplifier provides a signal that is proportional to changes in C_{pa}. This becomes clearer if we look at the simplified schematic representation of the cell-attached configuration, where only C_{pa} and C_m are depicted (Fig. 7b). Exocytosis and endocytosis lead to changes in C_{pa} as well as C_m. Because C_{pa} and C_m are connected in series, the resulting capacitance (C_{total}) is given by (see also ref. [35])

$$\frac{1}{C_{total}} = \frac{1}{C_{pa}} + \frac{1}{C_m} \tag{41}$$

The diameter of the tip of the pipette used for capacitance measurements is approximately 1 μm. The capacitance of the patch membrane (i.e., C_{pa}) is typically <0.1 pF, whereas the typical C_m of pituitary lactotrophs ranges from 3 to 10 pF (up to 100 times larger). Due to the relationship shown in (41), changes in C_m have a negligible influence on C_{total}, whereas changes in C_{pa} have a much greater effect. The effects of ΔC_{pa} are so much larger than those of ΔC_m that the latter can simply be ignored. Consequently, we are left with the situation described in the paragraph above. Equation (40) can be used to simulate the effects of ΔC_m on the overall complex admittance of the minimal equivalent circuit in cell-attached mode (Fig. 7c, d).

The cell-attached configuration is ideally suited to high-resolution measurement of small capacitance changes in the patch membrane. The resolution limit is such that it allows the recognition of fusion of vesicles with diameters as small as 30–40 nm. The average secretory vesicle in lactotrophs has capacitance of ~1 fF and diameter of ~200 nm [12]. Moreover, this method also allows the measurement of fusion pore properties, such as fusion pore conductance, fusion pore diameter, and fusion pore dwell time. Figure 7e shows the minimal equivalent circuit of a fusing vesicle with membrane capacitance C_v, which forms the fusion pore conductance (G_p). Assuming a stable patch recording with a constant patch membrane capacitance (C_{pa}), patch conductance (G_{pa}), R_a, C_v, and correct phase settings, then the lock-in outputs Y_1 and Y_2 provide Re and Im components of complex admittance of the electric circuit in Fig. 7e, respectively (see refs. [35, 37, 38]):

$$\mathrm{Re} = \frac{(\omega C_v)^2 / G_p}{1 + (\omega C_v / G_p)^2} \tag{42}$$

$$Im = \frac{\omega C_v}{1 + (\omega C_v / G_p)^2} \tag{43}$$

From these, C_v and G_p can be calculated (see ref. [39]):

$$C_v = \frac{1}{\omega}\frac{\mathrm{Re}^2 + Im^2}{Im} \tag{44}$$

$$G_p = \frac{\mathrm{Re}^2 + Im^2}{\mathrm{Re}} \tag{45}$$

Thus, it is possible to measure changes in membrane conductance that are directly triggered by the creation of a fusion pore during a transient exocytic event.

4 Protocols and Examples

4.1 Whole-Cell Uncompensated Measurements

Uncompensated whole-cell measurements are usually used for time-dependent measurements in C_m and reflect changes in the membrane surface area that are due to exocytosis and endocytosis. Such recording setup allows various compounds to be introduced into the cytoplasm to study the effects on overall exocytosis and endocytosis. The signal-to-noise ratio is usually insufficient for single fusion or fission events to be distinguishable from the C_m trace.

Figure 8 shows the traces for changes in C_m, G_a, and G_m as a function of time. Exocytosis was stimulated with the dialysis of 1 μM $[Ca^{2+}]_{free}$. ΔC_m was ~40 %. Note that G_a was stable and well above 100 nS throughout the experiment and that there was a gradual increase in G_m throughout the experiment.

The correct phase setting of the lock-in amplifier is essential in patch-clamp experiments. The phase-sensitive lock-in amplifier provides two outputs Υ_1 and Υ_2 that are proportional to the sinusoidal component at two orthogonal phases Φ_1 and Φ_2. Φ_1 can be adjusted, whereas Φ_2 is always shifted by 90° (with respect to Φ_1) [34]. In theory, when the bulk capacitance is not compensated and the phase angle of the lock-in amplifier is set to 0, Υ_1 and Υ_2 provide Re and Im components of complex admittance directly (see (30)–(32)). However, in practice, small shifts in phase are introduced by the acquisition setup. This needs to be accounted for in the analysis, usually by minor changes in phase angle before recording from a cell. The problem is solved by application of purely capacitive perturbations (e.g., by repeated attaching of small-value capacitor in the circuitry) and the phase angle is then set so that the perturbations are only visible in the Υ_2 output. The choice of output filters makes a considerable difference, so once the phase angle introduced by the setup has been determined, the filter settings should not be changed. Phase angle settings generally do not need to be readjusted unless changes to the frequency of the sine wave or filter settings are made.

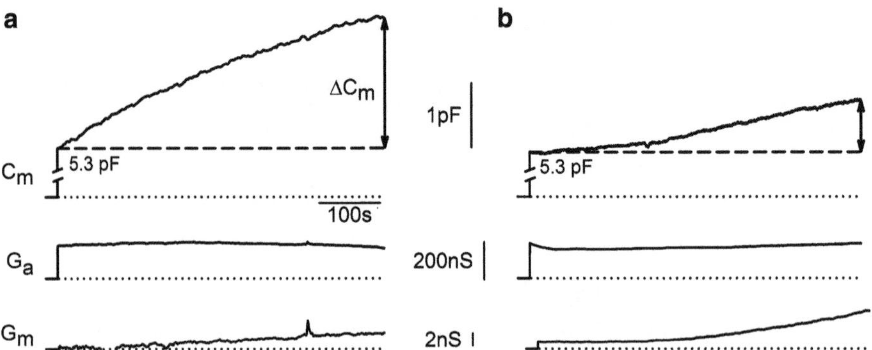

Fig. 8 Whole-cell compensated measurements of C_m. Dialysis of 1 μM $[Ca^{2+}]_{free}$ was used to stimulate exocytosis. G_a was stable and well above 100 nS throughout the experiment. Also note the gradual increase in G_m. The recording was obtained at a holding potential (V_{hold}) of −70 mV. Assuming reversal potential (V_{rev}) of −50 mV, a driving force of −20 mV was achieved. V_{hold} was superimposed on a 1.11-mV rms, 1,591-Hz sine voltage. The resulting outputs, Y_1, Y_2, current (I_m), and DC voltage (V_{DC}), were filtered using analog low-pass 4-pole Bessel filters (−3 dB) at 30 Hz for Y_1 and Y_2 and 10 Hz for I_m and V_{DC}. Signals were digitized at an acquisition rate of 50 Hz

The next step is the formation of the giga seal in the cell-attached configuration and compensation of the stray capacitance (C_s). C_s consists of contributions from mainly the head stage of the amplifier, pipette holder, and pipette tip, and it needs to be compensated if the absolute C_m is to be measured precisely. This is done with fast capacitance cancellation circuits that are built into the lock-in amplifier (C_{fast} potentiometer). The current output is observed and the C_{fast} is adjusted until the sine wave current is cancelled out as best as possible. The absolute values of C_s can reach well above 2 pF (almost 50 % of the total small pituitary cell C_m) but can be reduced using certain techniques, such as coating the pipette tips with a silicon resin (e.g., Sylgard) and reducing the volume of the bath solution into which the pipette tip is immersed. Once C_s has been compensated for, a brief pulse of negative pressure (i.e., suction) is applied to disrupt the patch membrane. At this point, the amplitude of the sinusoidal current increases and indicates that the whole-cell configuration has been attained. C_m, G_a, and C_m can now be measured.

The recordings from Fig. 8 were obtained at a holding potential (V_{hold}) of −70 mV. Assuming a reversal potential (V_{rev}) of −50 mV (for lactotrophs and melanotrophs), a driving force of −20 mV was achieved. It is important that V_{hold} is not selected so that $V_{hold} - V_{rev} = 0$ as this would create infinities when calculating DC conductance (see (31)). V_{hold} was superimposed on a 1.11-mV rms, 1,591-Hz sine voltage. The resulting outputs Υ_1, Υ_2, current (I_m), and DC voltage (V_{DC}) were filtered using analog low-pass 4-pole Bessel filters (−3 dB) at 30 Hz for Υ_1 and Υ_2 and 10 Hz for I_m and V_{DC}. Signals were digitized at an acquisition rate of 50 Hz.

Calculated passive equivalent cell parameters C_m, G_m, and G_a were digitally filtered if necessary.

The quality of a recording can be assessed during experiments by monitoring the outputs of the lock-in amplifier. This can be seen if (32) is rewritten and solved for G_a:

$$G_a = A + \left[\frac{B(G_a + G_m)}{\omega C_m} \right] \tag{46}$$

If $G_m \lll G_a$, which is the case in our experiments, then (47) reduces to

$$G_a = A + \left[\frac{BG_a}{\omega C_m} \right] \tag{47}$$

The magnitudes of A and B are thus determined by G_a, and a low noise recording, where $\omega C_m < G_a$, is indicated by $B < (G_a - A)$. An experiment can be rejected if G_a is too low ($G_a < \omega C_m$) [30]. Generally, recordings are rejected if G_a falls below 50 nS; however, it is good practice to maintain G_a above 100 nS. This can be facilitated by very gentle reductions of the pressure inside the recording pipette during the experiment; however, care needs to be taken not to disrupt the patch or affect the washout of the cytoplasm.

4.2 Cell-Attached Compensated Measurements

In a cell-attached compensated configuration, a signal-to-noise ratio can be achieved such that fusion and fission of individual vesicles of a diameter as small as 40 nm can be resolved. If one wants to achieve such resolution, the minimization of C_s becomes even more important, because it exerts a great deal of influence over the level of noise in the recording (see Section 5.2). A cell-attached configuration is achieved exactly as described in the previous section. However, outputs Υ_1 and Υ_2, although the latter is proportional to the patch capacitance, are now out of phase and attenuated by the factor $|T|$ (see (37)–(39)). It is possible to recalculate the recorded signals offline (see below); however, this problem can be solved easily by application of purely imaginary signal perturbations by means of hardware-determined calibration pulses (Fig. 9a, b). Υ_1 and Υ_2 outputs are directly monitored and purely capacitive calibration steps are applied to the signal (before the signal phase separation). The phase angle is then set such that the resulting discrete steps are only visible in the Υ_2 output (Im trace, Fig. 9b). The amplitude of these capacitance steps can be set (10 fF in our case), and when applied later in the experiment, they serve as calibration for the capacitive (Im) trace (Fig. 9c, d, asterisks). Changes in the Re trace can be measured reliably when the Im trace is calibrated. It is also possible to use suction pulses to calibrate the signal [40]. Phase angle adjustments need to be done for each recording.

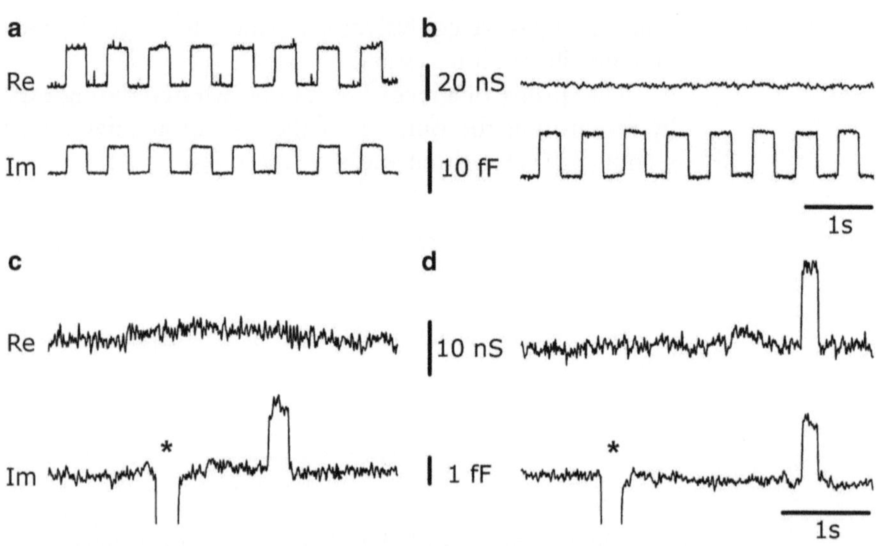

Fig. 9 High-resolution compensated measurements of capacitance in cell-attached mode. (**a**) Purely imaginary perturbations with an amplitude of 10 fF are applied before phase separation and the phase is adjusted until the resulting discrete steps (**b**) are only visible in Y_2 output (Im). This is done online for every cell tested. (**c**) Representation of a unitary exocytic event without projection on the Re. The correct phase angle setting is tested by the means of purely capacitive hardware-determined calibration pulses (denoted by an *asterisk*). The phase is set correctly if there is no projection of the calibration pulse on the Re component of the signal. (**d**) Roughly 35 % of unitary exocytic events are projected on the Re despite correct phase setting. These projections represent changes in patch conductance that are due to a narrow fusion pore (exhibiting relatively high resistance and therefore shifting the phase). The recordings above were obtained at a holding potential (V_{hold}) of 0 mV. V_{hold} was superimposed on a 111-mV rms, 1,591-Hz sine voltage. The resulting outputs Y_1, Y_2, current (I_m), and DC voltage (V_{DC}) were filtered using analog low-pass 4-pole Bessel filters (−3 dB) at 30 Hz for Y_1 and Y_2 and 10 Hz for I_m and V_{DC}. Signals were digitized at an acquisition rate of 200 Hz. Calibration pulses were applied approximately every 20 s. *Asterisks* denote calibration pulses

The recordings from Fig. 9 were obtained in rat lactotrophs at a holding potential (V_{hold}) of 0 mV. V_{hold} was superimposed on a 111-mV rms, 1,591-Hz sine voltage. The resulting outputs Y_1, Y_2, current (I_m), and DC voltage (V_{DC}) were filtered using analog low-pass 4-pole Bessel filters (−3 dB) at 30 Hz for Y_1 and Y_2 and 10 Hz for I_m and V_{DC}. Signals were digitized at an acquisition rate of 200 Hz. Calibration pulses were applied approximately every 20 s and were used to determine putative phase angle changes and for signal calibration. A typical capacitance resolution of ~300 aF was easily achieved with these settings, which exceeded the required resolution when considering the average lactotroph vesicle capacitance of ~1 fF [12]. As can be observed in Fig. 9c, d (Im trace), the transient fusion events are seen as upward steps and are defined by the amplitude and duration. From these parameters, several others can be calculated, including vesicle size, frequency of fusion events, and probability of an open fusion pore. Approximately 35 % of events are projected on the Re trace despite correct phase settings. These projections represent changes in patch conductance that are due to a narrow fusion pore (exhibiting relatively high resistance and therefore shifting the phase). The fusion pore conductance can be calculated (see (46)) and, from this, the fusion pore radius (r) is given by

$$r = \sqrt{\frac{G_p \rho \lambda}{\pi}} \tag{48}$$

where G_p is the pore conductance, ρ is the resistivity of saline (100 Ω cm), and λ is the estimated length of a gap junction channel (15 nm) [37] (see also ref. [38]).

The limitations of the acquisition setup do not allow all fusion pores to be detected on the Re trace. The upper and lower limits of G_p resolution depend on the angular frequency of the applied sine wave voltage and the diameter of the fusing vesicle (i.e., vesicle C_v; see ref. [40]):

$$G_p^{upper} \approx 3\omega \Delta C \tag{49}$$

$$G_p^{lower} \approx \frac{1}{3}\omega \frac{C_v^2}{\Delta C} \tag{50}$$

where ΔC is the capacitance noise (in rms) (see Section 5.2), C_v is the fusing vesicle capacitance, and a signal-to-noise ratio of 3 is taken into account. The theoretical upper and lower limits for the detection of the fusion pore, with $C_v = 1$ fF, $\omega = 10{,}000$ (1.6 kHz sine wave), and $\Delta C = 100$ aF, are 3 pS and 33.3 pS [41], which translate to fusion pore diameters of 0.24 and 0.8 nm, respectively. Detailed protocols and examples can be found in ref. [42].

5 Special Considerations

5.1 Offline Correction of Phase Error

The projections of capacitive (Im) steps on the Re trace are due to inaccurate phase angle settings (i.e., phase error). This error can be corrected online (as described above), or signals can be recalculated offline using the following set of equations (also see ref. [39]):

$$\Upsilon_c = \Upsilon_c^0 \cos(\Delta\varphi) + \Upsilon_g^0 \sin(\Delta\varphi) \tag{51}$$

$$\Upsilon_g = \Upsilon_g^0 \cos(\Delta\varphi) - \Upsilon_c^0 \sin(\Delta\varphi) \tag{52}$$

where Υ_c and Υ_g are the phase-shifted traces, Υ_g^0 and Υ_c^0 are the originally recorded traces, and $\Delta\varphi$ is the phase shift (in degrees).

The phase error can be calculated from the ratio of the step amplitudes in the $\mathrm{Im}(\Upsilon_c)$ and $\mathrm{Re}(\Upsilon_g)$ traces [40], provided that the deflections were caused by the purely capacitive perturbations:

$$\Delta\varphi = \tan^{-1}\left(\frac{\Delta\Upsilon_g^0}{\Delta\Upsilon_c^0}\right) \tag{53}$$

5.2 Noise

To reliably measure unitary exocytic events of small secretory vesicles, noise has to be minimized. In whole-cell measurements, the thermal (Johnson's) noise of the access resistance (R_a) is the dominant noise source and instrumental noise is minimal [43]. In the cell-attached configuration, several noise sources must be considered, including noise contributions from the patch-clamp amplifier, seal conductance, the patch itself, as well as the recording pipette and pipette solution [40].

Nevertheless, the recordings in cell-attached mode show considerably lower noise levels compared with whole-cell measurements. The cell-attached configuration thus allows measurements with much higher resolution. Instrumental noise, determined by capacitance measurement in the open head-stage configuration, is frequency dependent. While the current noise spectral density increases with ω [44], the amplitude of the capacitance signal increases in proportion to the sine wave frequency, and this increase outweighs the increase in noise. Thus, the signal-to-noise ratio improves with increasing frequency. Experimental data [40] show that the instrumental capacitance noise decreases exponentially up to a sine frequency of ~10 to ~20 kHz.

Figure 10 shows instrumental capacitance noise measurements in the open head-stage configuration at 1.6 and 6.4 kHz with a 1-GΩ feedback resistor in the head stage. Instrumental noise contributions to the overall noise (rms) are 38 aF at 1.6 kHz and 14 aF at 6.4 kHz in our system. Note that the data were acquired with a 10-kHz output filter switched in. This filter, known to increase the noise at higher frequencies, is used to prevent saturation of the patch-clamp amplifier and thus increases the capacitance measuring range.

Capacitance noise in the cell-attached configuration can further be reduced by increasing the amplitude of the sine wave [40]. We routinely apply a voltage of 111 mV (rms), although amplitudes as high as 200 mV (rms) can be used in stable patches. The patch-clamp amplifier is, however, easily saturated at these amplitudes, especially at higher frequencies; thus, the reduction of noise is usually a trade-off between increases in sine wave frequency and amplitude. In our case, the use of a 111-mV rms sine wave at 6.4 kHz usually results in noise levels of 20–30 aF (rms), which allows for routine observation of vesicles with a capacitance of 40–90 aF and diameter of 40–60 nm, although vesicles with C_v as

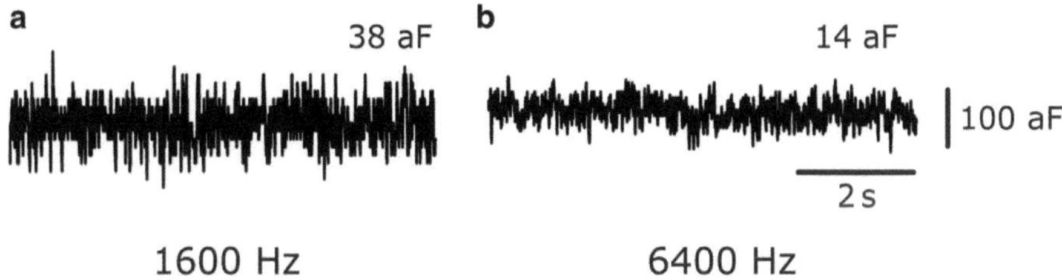

Fig. 10 Instrumental capacitance noise recorded in the open head-stage configuration at 1.6 (**a**) and 6.4 kHz (**b**) with a 1-GΩ feedback in the head stage. Noise (in rms) was calculated at 3-s intervals and was 38 aF at 1.6 kHz and 14 aF at 6.4 kHz driving sine voltage

small as 35 aF were observed (patch-clamp instrument with a 1-GΩ feedback resistor in the head stage).

To obtain such resolution, the minimization of C_s is vital. One of the main contributors to this is the pipette tip. A silicon resin (Sylgard) coating is usually used to reduce the pipette tip stray capacitance (see below) before C_s is cancelled (by C_{fast} potentiometer).

5.3 Pipettes

Pipettes are made from borosilicate or quartz glass capillaries by local melting and pulling in a micropipette puller. For easier filling, capillaries contain filament within the lumen. For experimentation, we use thin-walled borosilicate capillaries with an outer diameter of 1.5 mm and inner diameter of 0.86 mm. The puller can be programmed for various pull velocities and melting temperatures, which results in pipettes with various shapes and resistances. Pipettes with straight tips and resistances of 1–5 MΩ are generally produced. Before patching, the tips are fire-polished to straighten jagged edges resulting from braking in the final pulling cycle. This procedure allows for much better contact between the plasma membrane and the pipette surface and reduces the risk of membrane rupture. The next step is the coating of the pipette tip with Sylgard, a silicone polymer with excellent dielectric properties, which can reduce stray capacitance of the tip by 1–2 pF if applied correctly. The pipettes are then ready for use.

6 Conclusion

The methods described in this chapter have allowed us and others to gain new insights into the mechanisms of regulated exocytosis in anterior pituitary and other secretory and nonsecretory cells. In the early studies, we used whole-cell recording techniques to monitor changes in membrane capacitance. The results have revealed that an increase in $[Ca^{2+}]_{free}$ is an essential stimulus for an increase in C_m of bovine lactotrophs [45–47] and rat melanotrophs [11]. By using these techniques in combination with molecular biology, we were able to show that small GTPase Rab3B is an important regulator of

Ca^{2+}-dependent regulated exocytosis in bovine lactotrophs [48]; a few years later, the results have shown that in rat melanotrophs, the Rab3A is the key small GTPase isoform [49], indicating the molecular heterogeneity of the exocytic apparatus in different secretory cell types. By combining whole-cell recording and caged Ca^{2+} photolysis studies, the results have shown that the rapid and slow phases of regulated exocytosis have distinct molecular and physiologic characteristics in the same cell [50]. To further study the nature of the distinct mechanisms of exocytosis in a single cell, the unitary exocytic events that were observed in whole-cell recordings of bovine lactotrophs were measured [47]. In more detailed studies conducted on rat melanotrophs [32, 51, 52], predominantly full-fusion events were recorded. In contrast, when we started to use rat lactotrophs as a new secretory cell model, abundant transient exocytic events were recorded [7], which allowed more detailed study of the nature of these events [8, 9, 12, 41, 53]. The results have revealed that transient exocytic events represent a stable intermediate stage that can precede full-fusion events. Similar conclusions were obtained earlier on giant vesicles [37]. Moreover, the diameter of the fusion pore at resting conditions was shown to depend on the vesicle size: the smaller the vesicle diameter, the narrower the equilibrium fusion pore diameter [53]. Subsequent experiments have shown that this relationship can be changed by the manipulation of fusion proteins [41], electrostatic properties of membrane surfaces [54, 55], and likely the lipid composition [13]. In the current model, when the vesicle membrane and plasma membrane merge, a stable, very narrow fusion pore is formed, the stability of which is determined by a high local concentration of anisotropic membrane components in the membrane of the neck of the exocytic fusion pore [56]. Influences via SNARE proteins, likely due to electrostatic interactions, mediate changes in this local anisotropicity and hence drive fusion pore widening and prolongation of dwell time, eventually leading to full-fusion exocytosis [3].

Patch-clamp capacitance measurements have also been used to study regulated exocytosis in plants [57–59]. It was demonstrated for the first time that Ca^{2+}-dependent regulated exocytosis operates in plant as well as in animal cells. Furthermore, the recorded unitary exocytic events were similar to those present in animal cells [60, 61]. These methods have also been used on brain astrocytes, which, for about a century, were thought not to exhibit regulated exocytosis. Caged Ca^{2+} photolysis combined with membrane capacitance measurements revealed that these cells, like neurons, also exhibit this process [62] at about two orders of magnitude slower rate than neurons [63].

Patch-clamp capacitance measurement methods have provided a new platform to study the mechanisms of regulated exocytosis in a number of cell types and will likely be a key technique in future studies of the detailed mechanisms of the function and structure of fusion pores, the nanoarchitecture of which remains to be discovered.

Acknowledgements

This work was supported by the Slovenian Research Agency under grant P3 301 and projects J3 4051, J3 4146, and J3 632. We are indebted to Alenka Guček who contributed the noise measurement data.

References

1. Rituper B, Davletov B, Zorec R (2010) Lipid–protein interactions in exocytotic release of hormones and neurotransmitters. Clin Lipidol 5:747–761
2. Burgess TL, Kelly RB (1987) Constitutive and regulated secretion of proteins. Annu Rev Cell Biol 3:243–293
3. Vardjan N, Jorgacevski J, Zorec R (2013) Fusion pores, SNAREs, and exocytosis. Neuroscientist 19:160–174
4. Vardjan N, Stenovec M, Jorgacevski J, Kreft M, Grilc S, Zorec R (2009) The fusion pore and vesicle cargo discharge modulation. Ann N Y Acad Sci 1152:135–144
5. Ceccarelli B, Hurlbut WP, Mauro A (1973) Turnover of transmitter and synaptic vesicles at the frog neuromuscular junction. J Cell Biol 57:499–524
6. Fernandez JM, Neher E, Gomperts BD (1984) Capacitance measurements reveal stepwise fusion events in degranulating mast cells. Nature 312:453–455
7. Stenovec M, Kreft M, Poberaj I, Betz WJ, Zorec R (2004) Slow spontaneous secretion from single large dense-core vesicles monitored in neuroendocrine cells. FASEB J 18:1270–1272
8. Vardjan N, Stenovec M, Jorgacevski J, Kreft M, Zorec R (2007) Elementary properties of spontaneous fusion of peptidergic vesicles: fusion pore gating. J Physiol 585:655–661
9. Vardjan N, Stenovec M, Jorgacevski J, Kreft M, Zorec R (2007) Subnanometer fusion pores in spontaneous exocytosis of peptidergic vesicles. J Neurosci 27:4737–4746
10. Heuser JE, Reese TS (1973) Evidence for recycling of synaptic vesicle membrane during transmitter release at the frog neuromuscular junction. J Cell Biol 57:315–344
11. Rupnik M, Zorec R (1992) Cytosolic chloride ions stimulate Ca^{2+}-induced exocytosis in melanotrophs. FEBS Lett 303:221–223
12. Jorgacevski J, Stenovec M, Kreft M, Bajic A, Rituper B, Vardjan N, Stojilkovic S, Zorec R (2008) Hypotonicity and peptide discharge from a single vesicle. Am J Physiol Cell Physiol 295:C624–C631
13. Rituper B, Flasker A, Gucek A, Chowdhury HH, Zorec R (2012) Cholesterol and regulated exocytosis: a requirement for unitary exocytotic events. Cell Calcium 52:250–258
14. Wightman RM, Jankowski JA, Kennedy RT, Kawagoe KT, Schroeder TJ, Leszczyszyn DJ, Near JA, Diliberto EJ Jr, Viveros OH (1991) Temporally resolved catecholamine spikes correspond to single vesicle release from individual chromaffin cells. Proc Natl Acad Sci U S A 88:10754–10758
15. Stenovec M, Solmajer T, Perdih A, Vardjan N, Kreft M, Zorec R (2007) Distinct labelling of fusion events in rat lactotrophs by FM 1-43 and FM 4-64 is associated with conformational differences. Acta Physiol 191:35–42
16. Betz WJ, Mao F, Bewick GS (1992) Activity-dependent fluorescent staining and destaining of living vertebrate motor nerve terminals. J Neurosci 12:363–375
17. Miesenbock G, De Angelis DA, Rothman JE (1998) Visualizing secretion and synaptic transmission with pH-sensitive green fluorescent proteins. Nature 394:192–195
18. Hamill OP, Marty A, Neher E, Sakmann B, Sigworth FJ (1981) Improved patch-clamp techniques for high-resolution current recording from cells and cell-free membrane patches. Pflugers Arch 391:85–100
19. Neher E, Marty A (1982) Discrete changes of cell membrane capacitance observed under conditions of enhanced secretion in bovine adrenal chromaffin cells. Proc Natl Acad Sci U S A 79:6712–6716
20. Dernick G, Gong LW, Tabares L, Alvarez de Toledo G, Lindau M (2005) Patch amperometry: high-resolution measurements of single-vesicle fusion and release. Nat Methods 2:699–708
21. Smith CB, Betz WJ (1996) Simultaneous independent measurement of endocytosis and exocytosis. Nature 380:531–534

22. Neher E, Zucker RS (1993) Multiple calcium-dependent processes related to secretion in bovine chromaffin cells. Neuron 10:21–30
23. Piccolino M (1998) Animal electricity and the birth of electrophysiology: the legacy of Luigi Galvani. Brain Res Bull 46:381–407
24. Verkhratsky A, Krishtal OA, Petersen OH (2006) From Galvani to patch clamp: the development of electrophysiology. Pflugers Arch 453:233–247
25. Hodgkin AL, Huxley AF, Katz B (1952) Measurement of current-voltage relations in the membrane of the giant axon of Loligo. J Physiol 116:424–448
26. Hodgkin AL, Huxley AF (1952) Currents carried by sodium and potassium ions through the membrane of the giant axon of Loligo. J Physiol 116:449–472
27. Katz B (1969) The release of neural transmitter substances (Vol. 10). Liverpool University Press.
28. Neher E, Sakmann B (1976) Single-channel currents recorded from membrane of denervated frog muscle fibres. Nature 260:799–802
29. Solsona C, Innocenti B, Fernandez JM (1998) Regulation of exocytotic fusion by cell inflation. Biophys J 74:1061–1073
30. Zorec R, Henigman F, Mason WT, Kordaš M (1991) Electrophysiological study of hormone secretion by single adenohypophyseal cells. Methods Neurosci 4:194–210
31. Del Castillo J, Katz B (1957) La base "quantale" de la transmission neuromusculaire. Microphysiologie comparee des elements excitables. Coll. Internat. CNRS Paris, 67:245–258
32. Kreft M, Zorec R (1997) Cell-attached measurements of attofarad capacitance steps in rat melanotrophs. Pflugers Archiv 434:212–214
33. Pusch M, Neher E (1988) Rates of diffusional exchange between small cells and a measuring patch pipette. Pflugers Arch 411:204–211
34. Lindau M, Neher E (1988) Patch-clamp techniques for time-resolved capacitance measurements in single cells. Pflugers Arch 411: 137–146
35. Lollike K, Lindau M (1999) Membrane capacitance techniques to monitor granule exocytosis in neutrophils. J Immunol Methods 232: 111–120
36. Horowitz P, Hill W (1989) The art of electronics. Cambridge University Press, Cambridge
37. Breckenridge LJ, Almers W (1987) Currents through the fusion pore that forms during exocytosis of a secretory vesicle. Nature 328: 814–817
38. Spruce AE, Breckenridge LJ, Lee AK, Almers W (1990) Properties of the fusion pore that forms during exocytosis of a mast cell secretory vesicle. Neuron 4:643–654
39. Lindau M (1991) Time-resolved capacitance measurements: monitoring exocytosis in single cells. Q Rev Biophys 24:75–101
40. Debus K, Lindau M (2000) Resolution of patch capacitance recordings and of fusion pore conductances in small vesicles. Biophys J 78:2983–2997
41. Jorgacevski J, Potokar M, Grilc S, Kreft M, Liu W, Barclay JW, Buckers J, Medda R, Hell SW, Parpura V, Burgoyne RD, Zorec R (2011) Munc18-1 tuning of vesicle merger and fusion pore properties. J Neurosci 31:9055–9066
42. Rituper B, Guček A, Jorgačevski J, Flašker A, Kreft M, Zorec R (2013) High-resolution membrane capacitance measurements for the study of exocytosis and endocytosis. Nature Protocols 8(6):1169–1183
43. Gillis KD (1995) Techniques for membrane capacitance measurements. In: Bert Sackmann EN (ed) Single-channel recording, 2nd edn. Plenum Press, New York
44. Sigworth FJ (1995) Electronic design of the patch clamp. In: Sakmann B, Neher E (eds) Single-channel recording, 2nd edn. Plenum Press, New York
45. Sikdar SK, Zorec R, Brown D, Mason WT (1989) Dual effects of G-protein activation on Ca-dependent exocytosis in bovine lactotrophs. FEBS Lett 253:88–92
46. Sikdar SK, Zorec R, Mason WT (1990) cAMP directly facilitates Ca-induced exocytosis in bovine lactotrophs. FEBS Lett 273:150–154
47. Zorec R, Sikdar SK, Mason WT (1991) Increased cytosolic calcium stimulates exocytosis in bovine lactotrophs. Direct evidence from changes in membrane capacitance. J Gen Physiol 97:473–497
48. Lledo PM, Vernier P, Vincent JD, Mason WT, Zorec R (1993) Inhibition of Rab3B expression attenuates Ca^{2+}-dependent exocytosis in rat anterior pituitary cells. Nature 364:540–544
49. Rupnik M, Kreft M, Nothias F, Grilc S, Bobanovic LK, Johannes L, Kiauta T, Vernier P, Darchen F, Zorec R (2007) Distinct role of Rab3A and Rab3B in secretory activity of rat melanotrophs. Am J Physiol Cell Physiol 292:C98–C105
50. Rupnik M, Kreft M, Sikdar SK, Grilc S, Romih R, Zupancic G, Martin TF, Zorec R (2000) Rapid regulated dense-core vesicle exocytosis requires the CAPS protein. Proc Natl Acad Sci U S A 97:5627–5632

51. Zupancic G, Kocmur L, Veranic P, Grilc S, Kordas M, Zorec R (1994) The separation of exocytosis from endocytosis in rat melanotroph membrane capacitance records. J Physiol 480(Pt 3):539–552
52. Sikdar SK, Kreft M, Zorec R (1998) Modulation of the unitary exocytic event amplitude by cAMP in rat melanotrophs. J Physiol 511(Pt 3):851–859
53. Jorgacevski J, Fosnaric M, Vardjan N, Stenovec M, Potokar M, Kreft M, Kralj-Iglic V, Iglic A, Zorec R (2010) Fusion pore stability of peptidergic vesicles. Mol Membr Biol 27:65–80
54. Calejo AI, Jorgacevski J, Silva VS, Stenovec M, Kreft M, Goncalves PP, Zorec R (2012) Aluminium-induced changes of fusion pore properties attenuate prolactin secretion in rat pituitary lactotrophs. Neuroscience 201:57–66
55. Kabaso D, Calejo AI, Jorgacevski J, Kreft M, Zorec R, Iglic A (2012) Fusion pore diameter regulation by cations modulating local membrane anisotropy. ScientificWorldJournal 2012:983138
56. Jorgacevski J, Kreft M, Vardjan N, Zorec R (2012) Fusion pore regulation in peptidergic vesicles. Cell Calcium 52:270–276
57. Tester M, Zorec R (1992) Cytoplasmic calcium stimulates exocytosis in a plant secretory cell. Biophys J 63:864–867
58. Zorec R, Tester M (1993) Rapid pressure driven exocytosis-endocytosis cycle in a single plant cell. Capacitance measurements in aleurone protoplasts. FEBS Lett 333:283–286
59. Thiel G, Rupnik M, Zorec R (1994) Raising the cytosolic Ca^{2+} concentration increases the membrane capacitance of maize coleoptile protoplasts—evidence for Ca^{2+}-stimulated exocytosis. Planta 195:305–308
60. Thiel G, Kreft M, Zorec R (1998) Unitary exocytotic and endocytotic events in Zea mays L. coleoptile protoplasts. Plant J 13:117–120
61. Thiel G, Kreft M, Zorec R (2009) Rhythmic kinetics of single fusion and fission in a plant cell protoplast. Ann N Y Acad Sci 1152:1–6
62. Kreft M, Stenovec M, Rupnik M, Grilc S, Krzan M, Potokar M, Pangrsic T, Haydon PG, Zorec R (2004) Properties of Ca^{2+}-dependent exocytosis in cultured astrocytes. Glia 46:437–445
63. Kreft M, Krizaj D, Grilc S, Zorec R (2003) Properties of exocytotic response in vertebrate photoreceptors. J Neurophysiol 90:218–225

Part II

Model Systems for the Study of Exocytosis

Chapter 7

Measuring Exocytosis in Endocrine Tissue Slices

Maša Skelin Klemen, Jurij Dolenšek, Andraž Stožer, and Marjan Slak Rupnik

Abstract

A number of approaches have been developed during the last decades to assess exocytosis in neuronal, endocrine, and other secretory cells. A detailed description of a selection of the methods that have been adapted to study regulated exocytosis in endocrine tissue slices is provided in this chapter. Such adaptation of the methods has been associated with a plethora of open issues that required innovative solutions, but innovation has been and shall be in the future plentifully rewarded by discoveries of novel, often emergent principles of cells' operation within an intact tissue. The use of tissue slices has enabled us to reduce negative effects of cell culturing. It helped us circumvent serious experimental challenges associated with the small size and enhanced sensitivity of embryonic endocrine cells from wild-type animals and those lacking important functional proteins and thus associated with perinatal lethality. More recent methods to study exocytosis specifically in the intact tissues include polar tracers and lipophilic dyes that can be used to directly visualize exocytosis with multiphoton excitation microscopy. A hallmark of the microscopic approach is that it offers positional information and the exocytic process can simultaneously be studied in a large number of cells. Optical approaches alone and in combination with electrophysiology hold promise for future excitement about the mechanism of regulated exocytosis.

Key words Regulated exocytosis, Endocrine cells, Tissue slices, Capacitance measurements, Depolarization protocols, Photo-release, Two-photon microscopy, Polar tracer

1 Introduction: The Mechanism of Regulated Exocytosis

Exocytosis is a process by which a cell secretes the content of its secretory vesicles into the extracellular space and is crucial for normal cellular function of all eukaryotic cells. Until now, we have been able to identify two major types of exocytosis. The first, termed constitutive exocytosis, is important for the renewal of plasma membrane and its proteins and, in some cell types, also for secretion of various extracellular molecules, such as extracellular matrix components and antibodies [1, 2]. The second, coined regulated exocytosis, requires a signal for triggering and, as its name suggests, therefore seems more tightly regulated. The regulated exocytosis is particularly well studied in neuronal synapses where

Peter Thorn (ed.), *Exocytosis Methods*, Neuromethods, vol. 83,
DOI 10.1007/978-1-62703-676-4_7,

neurotransmitters are released very rapidly, maintaining neurotransmission and further propagation of the signal [3]. Besides neurons, the regulated exocytosis is essential for normal cell function also in a wide range of nonneuronal cells. In these cells, regulated exocytosis is responsible for secreting large amounts of secretory products. For example, hormones like epinephrine, secreted by the adrenal chromaffin cells, and the peptide α-melanocyte-stimulating hormone (α-MSH), secreted by the melanotrophs of the intermediate lobe of pituitary, and insulin, secreted by the beta cells of the endocrine pancreas, are released via regulated exocytosis into the extracellular space and reach circulation. Additionally, digestive enzymes are exocytosed from acinar cells of the exocrine pancreas upon nervous and hormonal stimulation, reach the lumen of the gastrointestinal tract, and carry out breakdown of macromolecules in the food.

Regulated exocytosis has been described as a multistep process with several sites of regulation and which is strongly dependent upon calcium ions (Ca^{2+}) [4]. Exocytosis involves several biochemically distinct steps: vesicle biogenesis, content loading, vesicle translocation, tethering, docking, priming, and fusion of the vesicle with the plasma membrane, which results in release of vesicle content into the extracellular space [5, 6]. After biogenesis, secretory vesicles are recruited to the vicinity of the plasma membrane in an ATP-dependent manner and can be morphologically and biochemically docked to the plasma membrane. In this manner, they become less mobile compared to non-docked granules [7–9]. Priming of vesicles is the next step that renders docked vesicles competent for Ca^{2+}-triggered fusion. In adrenal chromaffin cells, only a small portion of vesicles undergo Ca^{2+}-triggered exocytosis in the absence of ATP [7, 10]. Similarly, in pancreatic beta cells, the second exocytic phase is observed only in the presence of ATP [11]. On the other hand, in melanotrophs primed vesicles appear stable, and acute ATP depletion does not prevent a significant exocytosis [12]. The primed vesicles are believed to represent a population of ready releasable vesicles (RRP) that are capable of immediate fusion with the plasma membrane after an increase in $[Ca^{2+}]_c$. Conversely, the majority of vesicles (slowly releasable vesicles, SRP) must undergo a reversible and ATP-dependent priming or even repriming reaction prior to fusion [7, 13].

Molecular details of the exocytic fusion process are well studied and are only briefly introduced here. During fusion of a secretory vesicle, formation of the so-called trans-SNARE complex is needed for bringing the opposite lipid bilayers of cell and vesicle membrane in proximity and to induce their fusion. The SNARE complex is formed by proteins synaptobrevin, SNAP-25, and syntaxin [14]. The latter accommodates synaptotagmin, a $[Ca^{2+}]_c$ sensor [15]. An elevation of $[Ca^{2+}]_c$ is required to trigger the regulated fusion of vesicles with the plasma membrane, and the Ca^{2+} fluxes into the cell can be through voltage-activated calcium channels (VDCC) or from intracellular stores or both. The fusion proceeds through the formation of the fusion pore, which allows the release of vesicle content into the extracellular space [16]. On some occasions, vesicles may not completely fuse with the plasma membrane. This type of exocytosis, also called "kiss and run exocytosis," has been shown to be present in neuroendocrine cells [17]. If the fusion pore fully expands and the vesicle membrane gets incorporated into the plasma membrane, full exocytosis occurs.

2 Materials

2.1 Solutions for Pituitary Slices

All chemicals are typically obtained from Sigma-Aldrich (St. Louis, Missouri, USA), unless otherwise specified:

- Phosphate-buffered saline (PBS) solution—dissolve one phosphate-buffered saline tablet (2.0055 g) in 200 ml of dd H_2O to obtain (in mM): 10 phosphate buffer, 2.7 KCl, and 137 NaCl.
- External solution 1 composed of the following (in mM): 125 NaCl, 2.5 KCl, 1.25 NaH_2PO_4, 2 Na pyruvate, 3 myoinositol, 0.5 ascorbic acid, 10 glucose, 26 $NaHCO_3$, 3 $MgCl_2$, 0.1 $CaCl_2$, and 6 lactate.
- External solution 2 composed of the following (in mM): 2.5 KCl, 1.25 NaH_2PO_4, 2 Na pyruvate, 3 myoinositol, 0.5 ascorbic acid, 250 sucrose, 10 glucose, 26 $NaHCO_3$, 3 $MgCl_2$, 0.1 $CaCl_2$, and 6 lactate.
- External solution 3 composed of the following (in mM): 125 NaCl, 2.5 KCl, 1.25 NaH_2PO_4, 2 Na pyruvate, 3 myoinositol, 0.5 ascorbic acid, 10 glucose, 26 $NaHCO_3$, 1 $MgCl_2$, 2 $CaCl_2$, and 6 lactate.

The osmolality of the extracellular solution shall be 300 ± 10 mOsm/kg, and it shall be bubbled continuously with 95 % O_2/5 % CO_2 to keep the pH stable at 7.3.

2.2 Solutions for Pancreatic and Adrenal Slices

- External solution 1 is composed of (in mM) 125 NaCl, 2.5 KCl, 26 $NaHCO_3$, 1.25 Na_2HPO_4, 2 sodium pyruvate, 0.25 ascorbic acid, 3 myoinositol, 6 lactate, 1 $MgCl_2$, 2 $CaCl_2$, and 6 glucose.

The osmolality of the extracellular solution shall be 300 ± 10 mOsm/kg, and it shall be bubbled continuously with 95 % O_2/5 % CO_2 to keep the pH stable at 7.4.

- HEPES-buffered saline (HBS) is composed of (in mM) 150 NaCl, 10 HEPES, 6 glucose, 5 KCl, 2 $CaCl_2$, and 1 $MgCl_2$ with pH adjusted to 7.4 by titration using 1 M NaOH.

2.3 Intracellular Solutions for Capacitance Measurements

- Intracellular solution used for depolarization-induced capacitance measurement is composed of (in mM) 135 CsCl, 10 HEPES, 20 TEA-Cl, 2 $MgCl_2$, 4 ATP Na_2, and 0.05 EGTA.
- Intracellular solution used for photolysis-induced capacitance measurement is composed of (in mM) 5 mM NP-EGTA, 2.8 mM $CaCl_2$, 0.1 mM fura-6 F together with 125 CsCl, 40 HEPES, 2 $MgCl_2$, 20 TEA-Cl, 2 Na_2ATP, and 0.05 EGTA at pH adjusted to 7.2 with CsOH and osmolality 300 ± 10 mOsm/kg. The intracellular solution with NP-EGTA and fura-6 F is used only for the tip of the pipette (1.5 μl). The rest of the pipette is filled with regular intracellular solution.

3 Methods for Studying Exocytosis

During the last three decades, several approaches have emerged that measure exocytic events at different levels of organization, the most reductionist being isolated cells [18, 19]. An important step toward systems level biology and conditions that more closely resemble the ones in vivo has been the introduction of acute tissue slices. At any of the organizational levels, exocytosis can be quantified employing one of the following experimental techniques: whole-cell patch-clamp recording, confocal imaging, and direct measurements of hormones using amperometry, cell sniffing, or biochemical assays (ELISA, RIA). The latter are beyond the scope of this chapter and should be referenced elsewhere [20]. The basic characteristics of each of the individual techniques presented here are summarized in Fig. 1.

3.1 Membrane Capacitance Approaches

In addition to measuring electrical currents and transmembrane potential changes, the patch-clamp technique enables us to measure exo- and endocytic activity by monitoring cell membrane capacitance [18, 21] (Fig. 1). In electrotechnical terms, cell membrane can be regarded as an electrical capacitor: insulator between the two plates of a capacitor being the phospholipid bilayer and the electrically conducting plates of the capacitor being cytosol on the one and extracellular fluid on the other side. To measure vesicle fusion events, we make use of the fact that membrane capacitance (i.e., charge on the cytosolic or extracellular side, which is of the same magnitude, divided by membrane potential difference across membrane) is a parameter that is proportional to the membrane surface area [21]. A measured change in membrane capacitance corresponds to a change in membrane area that is defined by the membrane capacitance constant. The value of the latter is constant for cells of the same type and practically identical across different types of cells, the range being 0.7–1.3 μF per cm^2 of membrane [22]. For high-resolution membrane capacitance recordings, a sinusoidal voltage

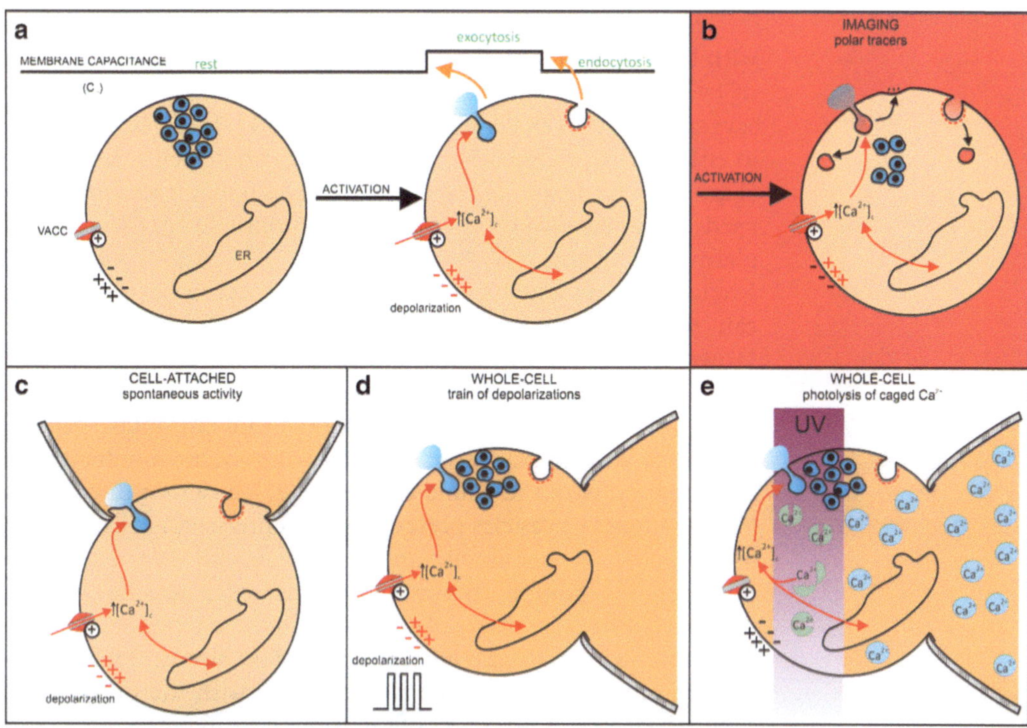

Fig. 1 Methods to study exocytosis. (**a**) Basic steps in the transduction pathway leading to regulated vesicle fusion in an endocrine cell. Upon activation of a cell, membrane depolarization with opening of voltage-activated calcium channels (VACC), Ca^{2+} release from the intracellular stores, or both contribute to an increase in $[Ca^{2+}]_c$. Ca^{2+} serves as a trigger for vesicle fusion with the plasma membrane. (**b**) Fusion of secretory vesicles can be detected optically by employing fluorescence imaging of extracellularly applied polar tracers. Fusion of a vesicle with the plasma membrane results in formation of a fusion pore through which the fluorescent dye enters the vesicle which then becomes visible in the cytosol. Subsequent complete fusion of such a vesicle depletes this signal. In an alternative scenario, upon fusion, the vesicle rapidly loses contact with the plasma membrane ("kiss and run" exocytosis), and the fluorescently labeled vesicle remains at least until it exits the observed focal plane. In the third possible scenario, endocytosis occurs after full fusion, and the fluorescence signal inside the cell increases. This method is discussed in detail in Section 2.2. (**c**) Capacitance measurement combined with the patch-clamp technique, used to measure both exo- and endocytosis. A vesicle fused with the plasma membrane increases the total surface area of the membrane. Since the surface area of plasma membrane is directly proportional to membrane capacitance, this increment can be detected as an increase in membrane capacitance. Alternatively decrease in membrane capacitance follows endocytic fission as depicted in **a**. In the cell-attached patch-clamp configuration, it is only realistic to measure spontaneous exocytic activity. (**d**) The whole-cell configuration of the patch-clamp technique enables one to study the exocytic activity of the majority plasma membrane. Depolarization of the plasma membrane using the patch pipette triggers stimulated exocytosis via activation of VACCs. (**f**) UV photolysis of caged Ca^{2+} increases $[Ca^{2+}]_c$ directly, dodging activation of the VACCs

at sufficient frequencies is applied to the holding potential, and the resulting current is analyzed. The amplifier outputs are proportional to small changes in cell membrane capacitance (C) and combined conductance (G) [18, 21]. The former output is used to investigate exocytosis.

Patch-clamp membrane capacitance measurement is a tool with high temporal resolution and high accuracy over a wide range of events, from single vesicle fusions to fusion of large pools of several thousands of vesicles [18, 21, 23, 24]. It can be performed in either the so-called cell-attached or the whole-cell configuration (Fig. 1). The whole-cell membrane capacitance measurement is used when one wants to monitor response from the total cell plasma membrane. With the classical whole-cell configuration (excluding perforated configuration), the cytosolic composition can readily be controlled through the patch pipette dialysis. However, this method comes with some inherent disadvantages that have to be mentioned. Namely, we measure only the net change in cell surface area and therefore cannot distinguish between exocytosis and endocytosis if both processes occur simultaneously. Additionally, with the whole-cell method, we cannot discriminate between types of vesicles present [3]. In chromaffin cells as well as in beta cells, synaptic-like microvesicles (SLMV) have been observed in addition to the large dense-core vesicles (LDCV) which represent the majority of vesicles in these types of cells [25].

In the cell-attached configuration, a tight electrical seal around a patch of plasma membrane is established with the tip of a glass capillary electrode. In this configuration, we can detect single exocytic events in the patched membrane under non-stimulated conditions likely recording only spontaneous activity [23, 26]. Due to the distinctly high temporal resolution of this method, we can separate endocytosis from exocytosis and also distinguish between the "kiss and run" and full-fusion events.

3.1.1 Depolarization Protocols

Depolarization protocols are used to mimic naturally occurring electrical activation of cells and have been used in several types of tissues [24]. During these protocols, the holding potentials are transiently changed from resting membrane potential values to depolarized values for varying periods of time, ranging from a few to several hundreds of milliseconds [27–29]. Each depolarization step is used to trigger regulated exocytosis; consecutive change in membrane capacitance changes can then be measured. To assess changes in exocytic kinetics related to different pools of vesicles, a train of depolarization pulses is used. As an alternative to a train of depolarizations, dual depolarizing pulses of different durations are used to deplete IRP and RRP in mouse chromaffin cells [24]. Varying intervals between the depolarization pulses can be used to assess the kinetics of RRP refilling [30].

3.1.2 Photo-Release of Caged Compounds

A very often used method for measuring exocytosis is the patch-clamp-based membrane capacitance measurement combined with photolysis of caged compounds, e.g., caged Ca^{2+} (Fig. 1). In this approach, an inactive caged compound is photo-cleaved into an active one by means of ultraviolet light illumination.

Caged compounds can be physiologically important molecules or their chelators rendered biologically inert by chemical modification. In the process of producing inactive caged compounds, cellular messengers, such as nucleotides [31], neurotransmitters [32], peptides [33], enzymes [34], or Ca^{2+} [35], are conjugated with a photolabile compound, usually a nitrobenzyl group. The conjugation makes these biologically important cellular messenger inactive or at least orders of magnitude less active than the photolyzed product [36]. In the case of Ca^{2+} ions, its high-affinity calcium chelators like BAPTA, EGTA, or EDTA are covalently modified [37]. The photolabile cages decrease their affinity to Ca^{2+} upon light illumination. In the case of NP-EGTA, K_d is increased from 0.080 μM to over 1,000 μM [35]. With this approach, the diffusion delays due to progressive opening of VACC and consecutive calcium-induced calcium release are avoided. Because of the very high temporal resolution of this method, it represents a major improvement in studies of cellular functions, particularly when it comes to studying cellular kinetics of processes like stimulus-secretion coupling. Therefore, it is very important that the rate of the photolysis is fast enough as to yield a biologically effective concentration in the time scale that corresponds to the investigated biological process [36].

The major advantage in using photo-release is a uniform increase in $[Ca^{2+}]_c$ upon UV illumination. By slow photolysis, we can avoid the problem of Ca^{2+} and Na^+ channel loss during tissue preparation or channel inactivation or rundown during prolonged stimulation. Photolysis also bypasses any modulatory effects that act on electrical excitability and Ca^{2+} influx [38] and enables a precise study of Ca^{2+} sensitivity of the exocytic process [11].

3.2 Two-Photon Extracellular Polar-Tracer Imaging

The major obstacle to studying exocytosis in electrically excitable cells, such as neurons and beta cells of the endocrine pancreas, has been the small size of the secretory vesicles. In neurons, where the vesicles are 30–50 nm in diameter, multiphoton microscopy cannot resolve single vesicles [39]. In beta cells where insulin granules have a mean diameter of 250 nm, multiphoton microscopy has been successfully employed to study exocytosis [19, 40–42].

Exocytic vesicles in non-excitable cells, such as the acinar cell of the exocrine pancreas, are typically larger than vesicles in excitable cells, measuring in diameter 150–2,000 nm [43–45]. Additionally, $[Ca^{2+}]_c$ required to trigger exocytosis is lower, the latency between the rise in $[Ca^{2+}]_c$ and the onset of vesicle fusion is longer, and the exocytosis is slower in non-excitable cells [44, 46, 47]. Due to these features, non-excitable cells have naturally been more easily accessible to optical techniques measuring exocytosis. The larger size of the vesicles proves especially important for imaging with a light microscope. Differential interference microscopy is able to identify individual fusion events and their spatial position, but cannot capture the behavior of the vesicles upon fusion [48]. To be able to follow the granules during fusion as well as after fusion, fluorescence techniques have been developed. So far, we have been able to apply to acinar cells in pancreas tissue slices the two-photon extracellular polar-tracer (TEP) fluorescence imaging, introduced by Kasai et al. [44]. With this approach, one is able to visualize and temporally as well as spatially quantify all exocytic events in the plane of focus within secretory tissues by postfusion labeling of secretory vesicles with an extracellular fluorescent tracer [45]. To visualize exocytosis with this technique, cells are immersed in an extracellular solution that contains a fluid-phase polar tracer, most commonly sulforhodamine-B, that enters the lumen of the vesicle upon its fusion with the plasma membrane. The basic idea of this technique is one of the simplest ones in the field, and it has been used with one-photon excitation imaging of cells possessing very large secretory vesicles [49]. In pancreatic acini, the lumen of the acinus cannot be readily visualized with one-photon confocal microscopy due to extensive light scattering by the secretory vesicles below apical plasma membrane [45]. Two-photon excitation, however, is intrinsically resistant to light scattering and enables clear representation of acinar lumens. Additionally, due to polar-tracer diffusion into the very thin (20–40 nm in diameter) and clean intercellular spaces, one can study exocytosis with great precision at the site where it occurs physiologically. Two-photon excitation excites only the tracer molecules in the focal plane. Therefore, high concentrations of polar tracers can be used without the danger of overheating or phototoxic effects. In addition, the dye molecules that are photo-bleached in the focal plane can effectively be replaced by diffusion of molecules from outside the focal plane, and due to absence of beam absorption except in the focal plane, the internal filter effect—an important drawback of one-photon confocal microscopy—is practically negligible. TEP imaging has been successfully used on different representative secretory tissues, enabling, among others, visualization of full-fusion exocytosis in beta cells of the endocrine pancreas, sequential exocytosis in acinar cells of the exocrine pancreas, and vacuolar sequential exocytosis in chromaffin cells of the adrenal medulla [19, 44, 50]. An extensive review of TEP imaging is provided elsewhere [45].

4 Adaptation of Methods to Tissue Slices

First, we describe preparation of acute tissue slice from three different rodent tissues, the pituitary, the pancreas, and the adrenal gland. In the second part, we discuss methods to measure membrane capacitance changes in conjunction with depolarization protocols and photo-release of caged Ca^{2+}. In the third part, fluorescence imaging using polar tracers to measure exocytosis is discussed.

4.1 Preparation of Tissue Slices

Since its introduction, the tissue slice technique has contributed immensely to the physiological studies of neuronal, neuroendocrine, and endocrine cells [51–54]. Tissue slices are nowadays used routinely for investigating the electrical properties of neurons and for analysis of synaptic transmission in central nervous system, since in slices neurons remain healthy, cell-to-cell contacts intact and therefore intercellular communication and tissue architecture are well preserved [55]. Almost a decade ago, the tissue slice technique was first applied to the pancreas, which enables one to record physiological parameters from a single cell or even from many cells at a time while preserving the anatomy of the pancreas [54].

For slice preparations rodents are killed by exposure to an atmosphere containing pure CO_2 and decapitation for pituitary and adrenal gland slices. Cervical dislocation is used in preparation of whole pancreas slices. It is critically important that the time from decapitation until the immersion of extracted tissue in cold external solution is kept as short as possible to avoid anoxia.

4.2 Pituitary Gland Slices

Pituitary glands should be immediately rinsed with ice-cold extracellular solution 1 for approximately 2 min. Subsequently, the glands shall be embedded in 2.5 % low-melting point agarose (Lonza Rockland Inc., Rockland, Maine, USA) in 1× PBS solution. The hardened agarose cubes are glued with cyanoacrylate glue (Super Attak, Henkel Slovenija d.o.o., Maribor, Slovenia) onto specimen disk of vibratome VT1000S (Leica Microsystems GmbH, Nussloch, Germany) and irrigated with ice-cold external solution 2. To preserve the pituitary gland morphology, 80 μm slices should be cut at 55–60 Hz with a velocity of 0.1 mm/s. Commercially available razor blades are used as cutting blades. Fresh slices are immediately transferred into a 500 ml chamber containing the oxygenated extracellular solution 1 where they can be kept on 32 °C for up to 8 h [56]. If slices are needed for a longer time period, they can be stored in cold extracellular solution. In this case, incubate slices on 32 °C for at least 20 min prior to recording. During recording, slices are continuously perifused at a rate of 1–2 ml/min with external solution 3.

4.3 Whole Pancreas Slices

Tissue slices are typically prepared from pancreata of 10–20-week-old NMRI or C57/B6 mice of either sex as described previously [54], although pancreata from younger animals (from P17 on) have been used successfully [57, 58]. In brief, after killing the animal, we access the abdomen by laparotomy and inject low-melting point 1.9 % agarose (Lonza Rockland Inc., Rockland, Maine, USA) dissolved in extracellular solution 1 at 40 °C into the proximal common bile duct clamped distally at the major duodenal papilla. After injection, the pancreas is cooled with an ice-cold extracellular solution 1. We then extract the agarose-permeated pancreas and gently wash it in ice-cold V. Small blocks of tissue

(0.1–0.2 cm^3 in size) are cut from the organ, cleared of connective tissue, and transferred to a 5 ml Petri dish filled with agarose at 40 °C and immediately cooled on ice. Individual cubes containing tissue blocks are then cut from hardened agarose and glued (Super Attak) onto the sample plate of the VT 1000 S vibratome (Leica Microsystems GmbH). The tissue is cut at 0.05 mm/s at 70 Hz into 140-μm-thick slices of a surface area of 20–100 mm^2. Throughout preparation and during slicing, the tissue is being held in an ice-cold extracellular solution 1 continuously bubbled with a gas mixture containing 95 % O_2 and 5 % CO_2 at barometric pressure to ensure oxygenation and a pH of 7.4. After cutting, slices are collected in 30 ml of HBS. Before measurements, the slices are kept in 30 ml of fresh HBS for up to 12 h at room temperature. HBS is exchanged every 2 h. The slices are held on the bottom of the chamber by a nylon-fiber net spread across a U-shaped platinum-wire weight. The total preparation time of whole pancreas tissue slices is typically up to 1 h. Throughout the whole preparation time, slices are kept in 6 mM glucose.

4.4 Adrenal Slices

The abdominal cavity is exposed and immediately cooled with several ml of ice-cold extracellular solution 1. Both kidneys together with suprarenal glands contained in the surrounding fat tissue are eviscerated. Care shall be taken not to mechanically damage the glands during isolation. Both organs are then transferred into an ice-cold extracellular solution, and several minutes shall be dedicated to remove all the surrounding fat tissue. Presence of any remains of the adipose tissue severely hinders the subsequent cutting. Isolated glands are embedded into 1.9 % agarose blocks (Lonza Rockland Inc.) and glued (Super Attak) onto specimen disk of the VT1000S vibratome (Leica Microsystems GmbH). Tissue is preserved best when cutting at a speed of 1.5 min per slice (0.05 mm/s) and frequency of 70 Hz. Slices of 110 μm thickness are used. This is thin enough as to enable light passage in the microscope setup to visualize cells during patch-clamping procedure and thick enough as to preserve ultrastructural organization of the tissue. Commercially available cutting blades are wiped with a mixture of ethanol and acetone (50:50 v/v) to clear the blade and improve cutting. The buffer tray of the vibratome shall be filled with ice-cold extracellular solution 1, continuously bubbled with carbogen to keep the slices during slicing cold, oxygenated, and at a pH of 7.4. Slices are temporarily stored on a nylon mesh in a 500 ml chamber in ice-cold extracellular solution 1, continuously bubbled with carbogen. The preparation of slices shall take no more than 1 h.

For the patch-clamp studies, prior to experimentation, individual slices are transferred into the perifused recording chamber. The agarose-embedded slices are held at the bottom of the recording chamber with a nylon mesh spread over a platinum U-shaped weight. If necessary, individual slices are flipped over to access the cells on the other side of the slice.

4.5 Membrane Capacitance Measurements Using the Patch-Clamp Technique

Cells are visualized on an upright Nikon Eclipse E600 FN microscope (Nikon, Tokyo, Japan) using differential interference contrast. Patch pipettes are pulled from borosilicate glass capillaries (1.5 mm outer diameter, 0.86 mm inner diameter, GC150F-15, Harvard Apparatus, USA) using a horizontal pipette puller (P-97, Sutter Instruments, USA). The pipette resistance shall be 2–3 MΩ in the cesium-based solution. Fast pipette capacitance (C_{fast}) is compensated for in the cell-attached configuration.

In the classical whole-cell configuration, compensated membrane capacitance measurements are obtained by using a SWAM IIc patch-clamp two-phase lock-in amplifier (Fig. 2; Celica, Ljubljana, Slovenia). After establishing the whole-cell configuration, slow membrane capacitance (C_{slow}) and series conductance (Gs)

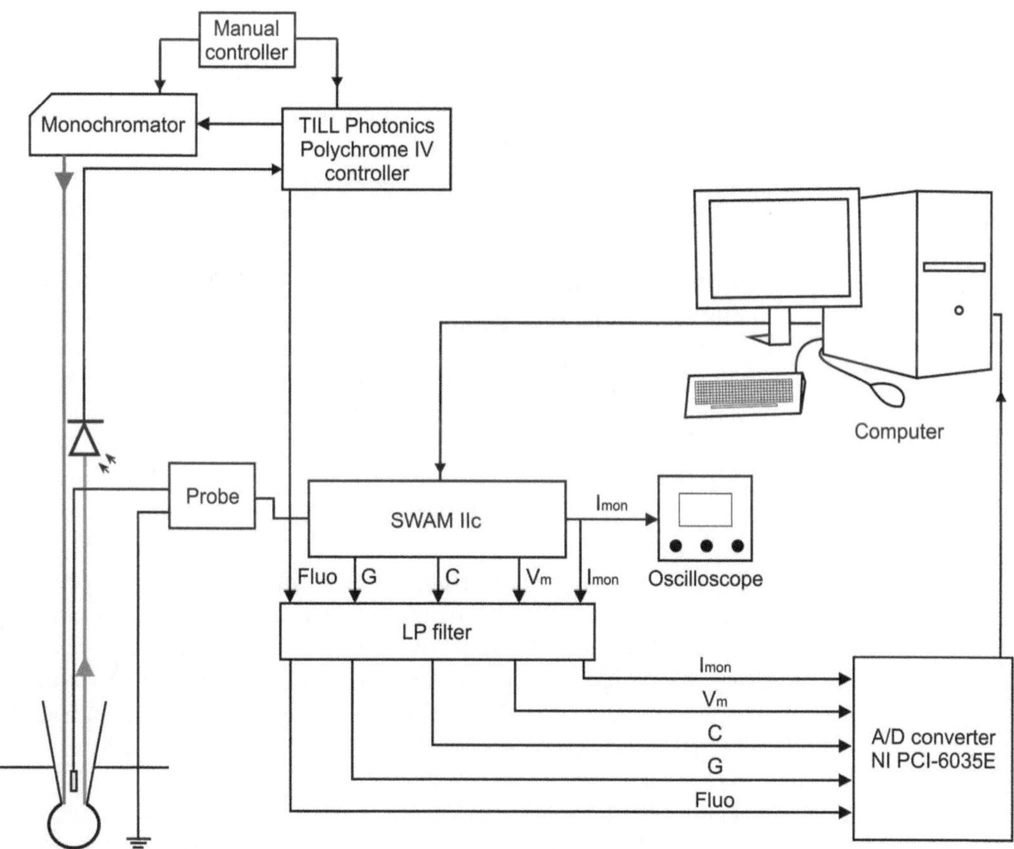

Fig. 2 A schematic representation of the patch-clamp setup for capacitance measurements used in both the depolarization protocol experiments and experiments involving photo-release of caged Ca^{2+}. See text for details

are compensated for by compensation controls. Only experiments with Gs > 50 nS are processed. To measure plasma membrane capacitance, a 1.6 kHz sinusoidal voltage (11.1 mV rms) is superimposed onto a holding potential of −80 mV. Recordings are low-pass filtered, transferred to a PC via an A/D converter (12 bit, Nidaq PCI-6035E, National Instruments), and recorded to the hard disk using WinWCP software (John Dempster, University of Strathclyde, UK). Only experiments not displaying crosstalk between membrane capacitance and membrane conductance monitored in parallel are used for analysis. All electrophysiological experiments are performed at 32 °C.

The procedure to measure membrane capacitance in a single cell within the tissue slice is similar to the one described for cultured isolated cells. Visualization is done by differential interference contrast rather than a phase contrast. A positive pressure has to be applied to the pipette to prevent collection of the debris during penetration of the pipette tip toward the deeper layers of cells. Individual cells may appear leaky in the whole-cell configuration due to the presence of gap junctions. In addition, postsynaptic currents are often detected already in a cell-attached configuration [59].

4.6 Depolarization of the Plasma Membrane by Current Injections

Sequential depolarizations that mimic naturally occurring activation by action potentials are used in endocrine cells to elucidate different pools of excitable vesicles [24, 60, 61]. To elicit exocytosis, we use a train of square pulse depolarizations. Our stimulus protocol consisted of 50 depolarizing pulses from −80 to 10 mV. Intervals between depolarizing pulses were coupled to a sine wave. This protocol enabled induction of exocytosis by depolarizing pulses during 60 ms long periods between depolarizations; capacitance changes induced by exocytosis were recorded. Specifically, in the WinWCP protocol editor, we used two digital output channels. One was used to control the holding potential of the amplifier and was therefore put to alternate between −80 and 10 mV in a square-like manner. The second channel was used to trigger sine wave generator of the amplifier during inter-depolarization periods by TTL pulses.

A representative recording of a capacitance change induced by the depolarizing train in chromaffin cell from acute mouse adrenal slice is depicted in Fig. 3. Each depolarization pulse triggered exocytosis that was measured during the inter-depolarization periods as a capacitance increase over the resting membrane capacitance. Altogether 50 depolarizations triggered an increase in excess of 800 fF.

4.7 Photolysis of Caged Ca^{2+}

The secretory activity can be triggered by photolysis of caged Ca^{2+} that results in an increase in $[Ca^{2+}]_c$. Fura-6F (0.1 mM in pipette solution; Molecular Probes, USA) is used to measure the $[Ca^{2+}]_c$ simultaneously with the whole-cell patch-clamp recordings. After dialyzing the cells with intracellular solution containing fura-6F and

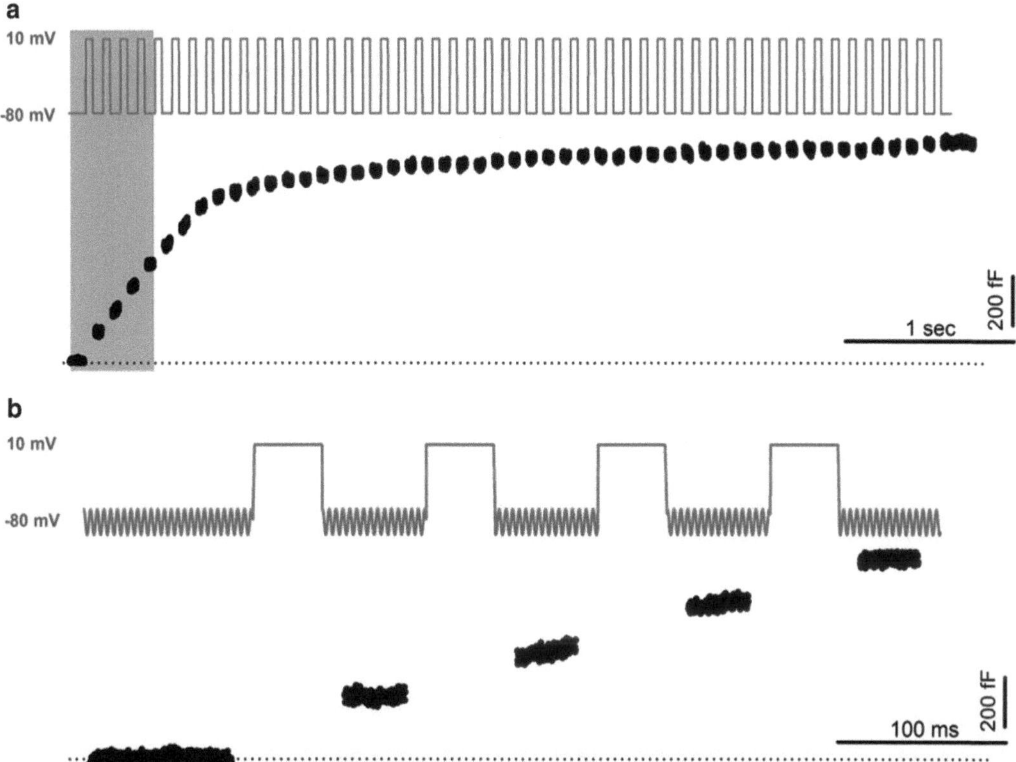

Fig. 3 Exocytosis in adrenal chromaffin cells elicited by a train of depolarization pulses. (**a**). A train of 50 depolarization pulses from −80 to 10 mV was used to stimulate the cell in the whole-cell patch-clamp configuration. The depolarization protocol is indicated in the *upper panel*; the result of the capacitance measurement is drawn in the *lower panel.* Every depolarization pulse induced an increase in membrane capacitance. (**b**) A close-up of the first four depolarizations shown in panel **a**. During inter-depolarization periods, a sine wave was superimposed on the holding potential (11.1 mV rms at 1.6 kHz) and the capacitance measured

Ca^{2+} bound to NP-EGTA (i.e., after whole-cell configuration is established, a few minutes are needed to allow for exchange of solution between the cell and the pipette), fura-6F is excited at 380 nm with a monochromator (Fig. 2, Polychrome IV; TILL Photonics, Germany). A long-pass dichroic mirror reflects the monochromatic light to the perfusion chamber and transmits the emitted fluorescence above 400 nm, which is further filtered through a 450-nm LP filter. The fluorescence intensity is measured with a photodiode (TILL Photonics). Upon excitation of fura-6F, Ca^{2+} is released from NP-EGTA, resulting in an increase in $[Ca^{2+}]_c$ and triggering Ca^{2+}-induced exocytosis. $[Ca^{2+}]_c$ was calculated by [56]:

$$\left[Ca^{2+}\right]_c = K_d \times \frac{F - F_{\min}}{F_{\max} - F}$$

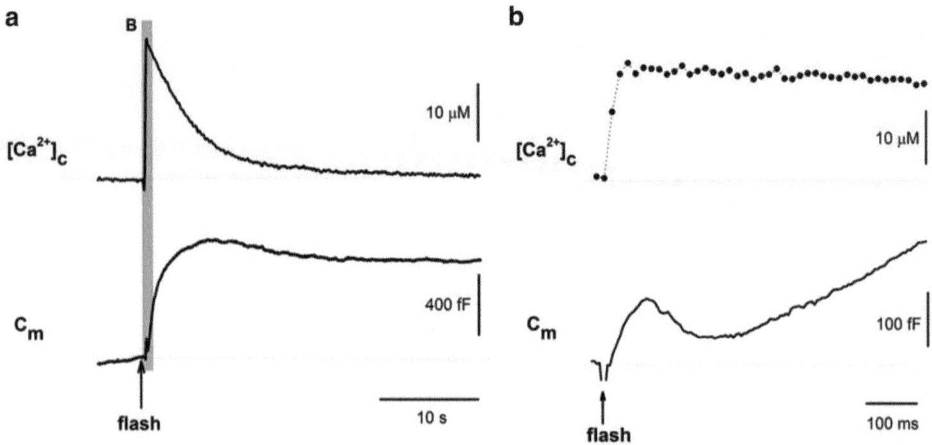

Fig. 4 Flash photolysis in pituitary melanotrophs. (**a**) Ultraviolet light flash application (*arrow*) induced changes in $[Ca^{2+}]_c$ (*upper panel*) which triggered change in membrane capacitance (C_m, *lower panel*). (**b**) Expanded time resolution of $[Ca^{2+}]_c$ and C_m trace from **a**

where K_d is the dissociation constant for fura-6F (5.3 μM [62]), F_{min} is the maximal fluorescence of fura-6F during recording, F_{max} is the autofluorescence in the cell-attached configuration, and F is the measured fluorescence.

We can increase $[Ca^{2+}]_c$ rapidly and to a large amplitude if using a high-intensity flash of ultraviolet light, as is the case in flash-photolysis experiments. In these experiments, a single value of Ca^{2+} concentration is sampled per cell, and all the kinetic phases may be triggered simultaneously, preventing their discrimination. Flash photolysis has been widely used in several types of secretory cells including pancreatic beta cells [63–65], chromaffin cells [66–68], and pituitary melanotrophs [69]. In the flash-photolysis experiments, $[Ca^{2+}]_c$ was increased to several tens of μM (Fig. 4). This concentration is high enough to deplete all releasable pools of granules enabling effectively measuring the size of releasable pools. However, since the pools with different kinetics and sensitivities are triggered simultaneously, Ca^{2+} dependencies cannot be properly separated.

Due to a typical use of an upright microscope in association with the tissue slices, flash-photolysis experiments where single cells can be exposed to a flash of ultraviolet light to produce massive uncaging cannot be performed. Instead we can use low-power ultraviolet illumination through the objective lens to release caged Ca^{2+} in a slow manner, resulting in a ramp-like increase in $[Ca^{2+}]_c$ which closely resembles changes in $[Ca^{2+}]_c$ as measured after physiological stimulation, e.g., glucose in pancreatic beta cells. Even this ramp-like slow photo-release of Ca^{2+} leads to a robust secretion (Fig. 5) in pancreatic beta cells, pituitary melanotrophs, as well as in adrenal chromaffin cells [70–72]. If we express the rate of the

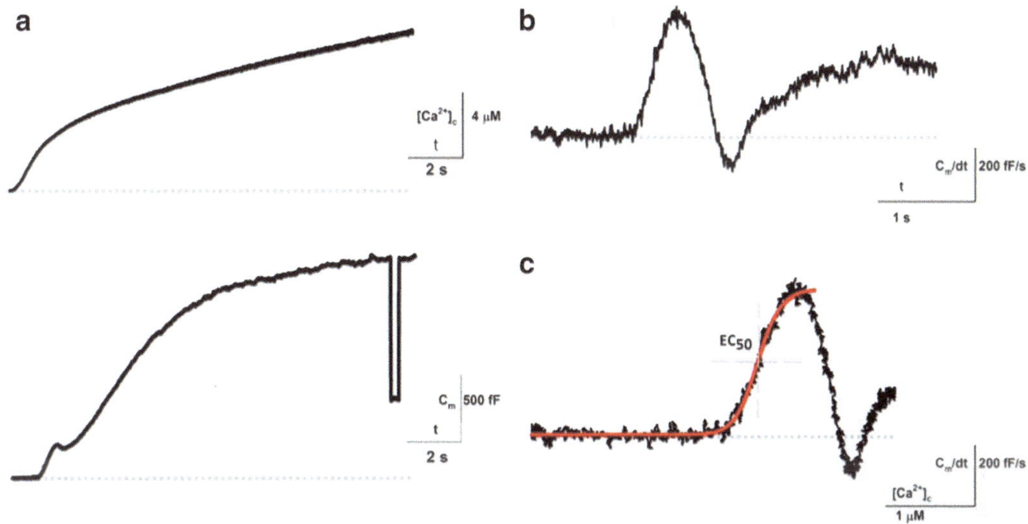

Fig. 5 Slow photo-release of caged Ca^{2+} in mouse beta cells from acute pancreas slice preparation. (**a**) Slow photo-release of caged Ca^{2+} produced a ramp-like increase in $[Ca^{2+}]_c$ (*upper panel*) which triggered an increase in membrane capacitance (C_m, *lower panel*). A square-like indent in the record indicated 1 pF calibration pulse. (**b**) Time derivative of C_m change presented in **a**. (**c**) The rate of the C_m change shows saturation kinetics when plotted versus $[Ca^{2+}]_c$ (*black line*), with high cooperativity and half-effective $[Ca^{2+}]_c$ (EC_{50}) of about 2.2 μM. A Hill function can be fitted through the data (*red line*)

release as a function of $[Ca^{2+}]_c$, we can observe saturation kinetics with high cooperativity [11].

Exocytic machinery in different cell types is sensitive to calcium in concentration ranges that differ for the types of cells studied [72]. Using slow photo-release, a slowly changing $[Ca^{2+}]_c$ can reveal the triggering of several exocytic kinetic phases, each with a different Ca^{2+} sensitivity. It should be noted that in the tissue slice preparation, even in whole-cell patch-clamp configuration, substantial endocytosis is much less likely to occur and contaminate the analysis of the exocytic processes compared to similar experiment in isolated cells.

4.8 Two-Photon Extracellular Polar-Tracer Imaging in Pancreas Tissue Slices

TEP imaging was performed on a Leica TCS SP5 AOBS Tandem II upright confocal system (Fig. 6) using a Leica HCX APO L water immersion objective (20×, NA 1.0). Exocytosis was visualized using sulforhodamine B (20 mg/ml, Molecular Probes, Eugene, Oregon, USA) as a membrane-impermeant fluorescent extracellular marker excited by femtosecond laser pulses at 800 nm. Fluorescent light between 550 and 700 nm was detected by a non-descanned detector (Leica Microsystems). Exocytosis of secretory vesicles occurred selectively at the apical plasma membrane. The Ω-shaped profile of the vesicle that fused first was stable long enough as to enable sequential exocytosis of the more deeply situated vesicles on the

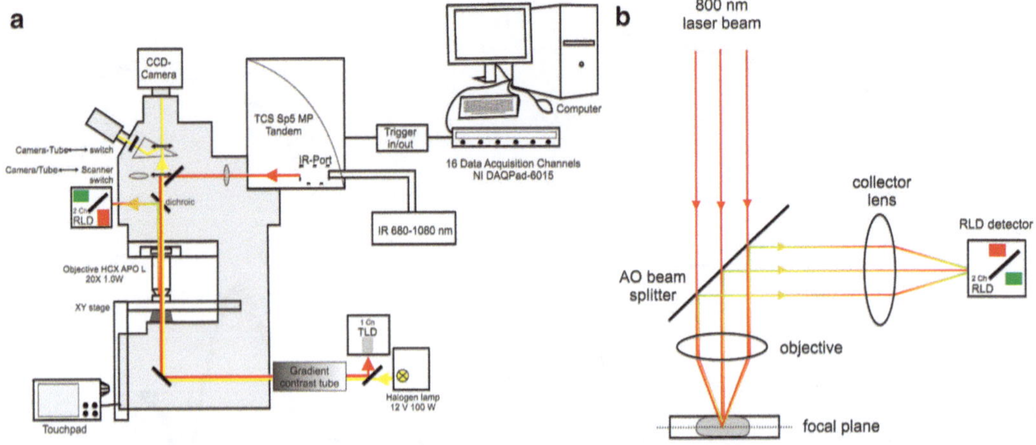

Fig. 6 A schematical summary of the main technical characteristics of the microscope used for TEP imaging. (**a**) For robust slice positioning under control of the eye and for finding a region for later recording with the camera, white light is used coming from a halogen lamp and lighting the slice in its chamber from below. The transmitted white light can be directed toward the objective tube or the CCD camera. For sulforhodamine B excitation and TEP imaging of exocytosis, infrared light with a wavelength of 800 nm coming from a Titan Sapphire pulsed laser is directed onto the slice from above. The transmitted infrared light is directed through a gradient contrast tube and collected by the so-called transmitted light detector (TLD). This pathway serves to depict a contrast image. The emitted fluorescent light is directed to the reflected light detector (RLD). Due to the ability to collect the focused as well as the scattered emitted light, this photomultiplier is also called the non-descanned detector. Besides the sulforhodamine B fluorescence, a 2-channel RLD can detect a second fluorescent signal, for example, a Fura-2 calcium-dependent fluorescent signal. To separate the signals, a filter cube is used. (**b**) A close-up of the pathway of the excitatory infrared light to the tissue slice and the reflected fluorescence to the 2-channel RLD. An acousto-optical beam splitter (AOBS) serves to select the wavelength of the incident light

already fused vesicle. This type of secondary exocytosis is named sequential exocytosis and is the predominant type of exocytosis in the acinar cells of the exocrine pancreas [44, 45]. It was first observed in 1965 in a static analysis performed with electron microscopy [73]. Its dynamics were first described in [44].

In the example shown, we applied TEP imaging to the exocrine part of our acute in situ preparation in order to be able to combine the advantages of TEP imaging with recordings of exocytosis in acinar cells under conditions that due to fast isolation and absence of enzymatic digestion during preparation more closely resemble the ones in vivo. Figure 7 summarizes a typical recording of sequential exocytosis upon stimulation with carbamylcholine.

5 Perspectives

The methodological approaches to study regulated exocytosis have been successfully implemented in several hallmark endocrine and exocrine tissues: pancreatic beta cells [11, 30, 57, 61, 74],

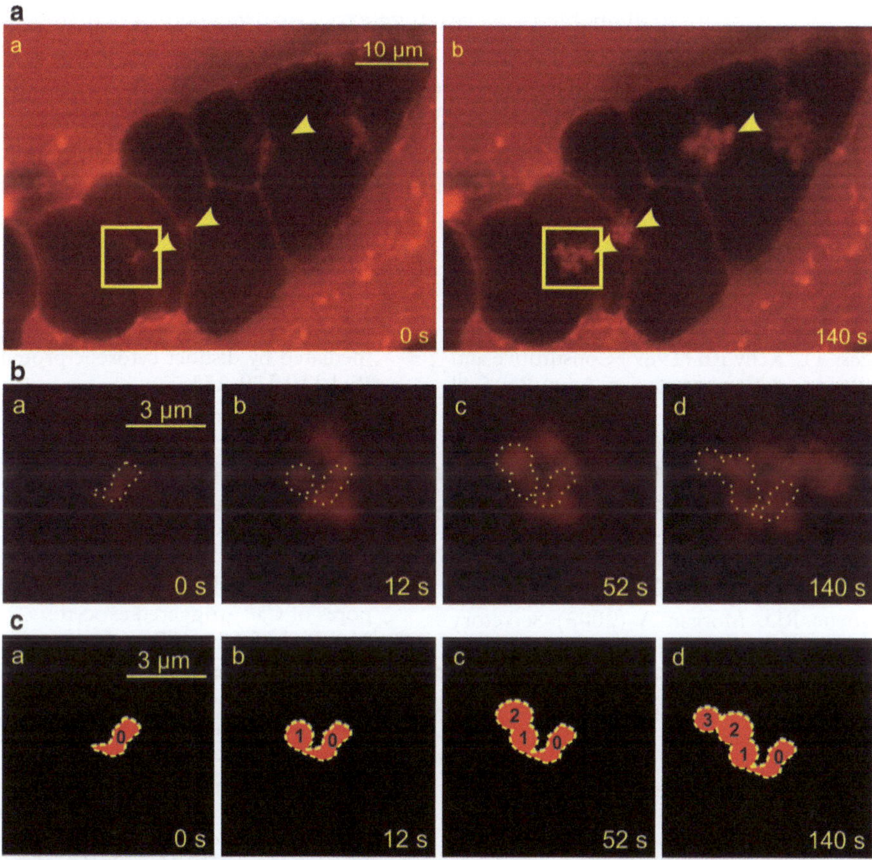

Fig. 7 A representative exocytic response elicited by 200 μM carbamylcholine and recorded via entry of the polar fluid-phase extracellular fluorescent tracer sulforhodamine B into fused secretory vesicles. (**a**) The appearance of three lumina before (*a*) and after stimulation with 200 μM carbamylcholine (*b*). The lumen enclosed by the full-lined square is shown in **b** and **c**. (**b**) Acinar lumen between apical poles of two acinar cells (*a*) to which three secretory vesicles fuse sequentially, as shown in figures *b*, *c*, and *d*. (**c**) A schematic representation of **b**, with sequential exocytosis of the three secretory vesicles, indicated by numbers 1, 2, and 3 with the lumen indicated by 0

pancreatic acinar cells [19, 75], adrenal chromaffin cells [24, 68], and pituitary gland melanotrophs [12, 59]. In a slice, tissue morphology, patterns of cell-to-cell connectivity, the vascular and ductal tree, the sensitivity of the membrane receptors to pharmacological stimulation, as well as parts of innervation remain well preserved, notably increasing the value of the preparation [76]. Many other nonclassical endocrine tissue preparations, containing scarcer endocrine cell populations (kidney, gastrointestinal tract), can be well studied using this approach. Additionally, the tissue-slice technology is ideally suited to the study of mutant animals and also human specimens.

In addition, we are facing an exciting period with the emergence of novel optical approaches with further increased temporal and spatial resolution even below the diffraction limit of light that shall enable us to visualize exocytic processes in real-time 3D space in addition to multiple spectral dimensions in a living system that closely resembles the in vivo conditions. There is no one single technique that would be ideally suited to reveal all aspects of regulated exocytosis. Conversely, all together they hold promise to deepen our understanding of this essential process with many clinical applications.

References

1. Burgess TL, Kelly RB (1987) Constitutive and regulated secretion of proteins. Annu Rev Cell Biol 3:243–293
2. Kelly RB (1985) Pathways of protein secretion in eukaryotes. Science 230:25–32
3. Kelly RB (1993) Storage and release of neurotransmitters. Cell 72(Suppl): 43–53
4. Martin TF (1997) Stages of regulated exocytosis. Trends Cell Biol 7:271–276
5. Burgoyne RD, Morgan A (2003) Secretory granule exocytosis. Physiol Rev 83:581–632
6. Burgoyne RD, Morgan A (1993) Regulated exocytosis. Biochem J 293(Pt 2):305–316
7. Holz RW, Bittner MA, Peppers SC, Senter RA, Eberhard DA (1989) MgATP-independent and MgATP-dependent exocytosis. Evidence that MgATP primes adrenal chromaffin cells to undergo exocytosis. J Biol Chem 264:5412–5419
8. Oheim M, Loerke D, Stuhmer W, Chow RH (1998) The last few milliseconds in the life of a secretory granule. Docking, dynamics and fusion visualized by total internal reflection fluorescence microscopy (TIRFM). Eur Biophys J 27:83–98
9. Martin TF, Kowalchyk JA (1997) Docked secretory vesicles undergo Ca^{2+}-activated exocytosis in a cell-free system. J Biol Chem 272:14447–14453
10. Parsons TD, Coorssen JR, Horstmann H, Almers W (1995) Docked granules, the exocytic burst, and the need for ATP hydrolysis in endocrine cells. Neuron 15:1085–1096
11. Skelin M, Rupnik M (2011) cAMP increases the sensitivity of exocytosis to Ca^{2+} primarily through protein kinase a in mouse pancreatic beta cells. Cell Calcium 49(2):89–99
12. Sedej S, Rose T, Rupnik M (2005) cAMP increases Ca^{2+}-dependent exocytosis through both pka and Epac2 in mouse melanotrophs from pituitary tissue slices. In, pp 799–813
13. Hay JC, Martin TF (1992) Resolution of regulated secretion into sequential MgATP-dependent and calcium-dependent stages mediated by distinct cytosolic proteins. J Cell Biol 119:139–151
14. Sudhof TC, Rothman JE (2009) Membrane fusion: grappling with SNARE and SM proteins. Science 323:474–477
15. Pang ZP, Sudhof TC (2010) Cell biology of Ca^{2+}-triggered exocytosis. Curr Opin Cell Biol 22:496–505
16. Jackson MB, Chapman ER (2008) The fusion pores of Ca^{2+} -triggered exocytosis. Nat Struct Mol Biol 15:684–689
17. Siegel GJ, Albers RW, Brady ST, Price DL (2006) Basic neurochemistry: molecular, cellular, and medical aspects. Elsevier Academic Press, Amsterdam
18. Lindau M, Neher E (1988) Patch-clamp techniques for time-resolved capacitance measurements in single cells. Pflugers Archiv-Eur J Physiol 411:137–146
19. Takahashi N, Kishimoto T, Nemoto T, Kadowaki T, Kasai H (2002) Fusion pore dynamics and insulin granule exocytosis in the pancreatic islet. Science 297:1349–1352
20. Willenborg M, Schumacher K, Rustenbeck I (2012) Determination of beta-cell function: insulin secretion of isolated islets. Methods Mol Biol 933:189–201
21. Neher E, Marty A (1982) Discrete changes of cell membrane capacitance observed under conditions of enhanced secretion in bovine adrenal chromaffin cells. Proc Natl Acad Sci U S A 79:6712–6716
22. Gentet LJ, Stuart GJ, Clements JD (2000) Direct measurement of specific membrane capacitance in neurons. Biophys J 79: 314–320
23. Vardjan N, Stenovec M, Jorgacevski J, Kreft M, Zorec R (2007) Subnanometer fusion pores in spontaneous exocytosis of peptidergic vesicles. J Neurosci 27:4737–4746
24. Voets T, Neher E, Moser T (1999) Mechanisms underlying phasic and sustained secretion in chromaffin cells from mouse adrenal slices. Neuron 23:607–615

25. Thomas-Reetz AC, De Camilli P (1994) A role for synaptic vesicles in non-neuronal cells: clues from pancreatic beta cells and from chromaffin cells. Faseb J 8:209–216
26. Jorgacevski J, Potokar M, Grilc S et al (2011) Munc18-1 tuning of vesicle merger and fusion pore properties. J Neurosci 31:9055–9066
27. Barg S, Ma X, Eliasson L et al (2001) Fast exocytosis with few Ca^{2+} channels in insulin-secreting mouse pancreatic β cells. Biophys J 81:3308–3323
28. Barg S, Eliasson L, Renstrom E, Rorsman P (2002) A subset of 50 secretory granules in close contact with L-type Ca^{2+} channels accounts for first-phase insulin secretion in mouse beta-cells. Diabetes 51:S74–S82
29. Gopel S, Zhang Q, Eliasson L et al (2004) Capacitance measurements of exocytosis in mouse pancreatic alpha-, beta- and delta-cells within intact islets of Langerhans. J Physiol-London 556:711–726
30. Rose T, Efendic S, Rupnik M (2007) Ca^{2+}-secretion coupling is impaired in diabetic Goto Kakizaki rats. J Gen Physiol 129:493–508
31. Kaplan JH, Forbush B 3rd, Hoffman JF (1978) Rapid photolytic release of adenosine 5'-triphosphate from a protected analogue: utilization by the na: Kpump of human red blood cell ghosts. Biochemistry 17: 1929–1935
32. Walker JW, McCray JA, Hess GP (1986) Photolabile protecting groups for an acetylcholine receptor ligand. Synthesis and photochemistry of a new class of o-nitrobenzyl derivatives and their effects on receptor function. Biochemistry 25:1799–1805
33. Walker JW, Gilbert SH, Drummond RM et al (1998) Signaling pathways underlying eosinophil cell motility revealed by using caged peptides. Proc Natl Acad Sci U S A 95: 1568–1573
34. Ghosh M, Song X, Mouneimne G, Sidani M, Lawrence DS, Condeelis JS (2004) Cofilin promotes actin polymerization and defines the direction of cell motility. Science 304: 743–746
35. Ellis-Davies GC, Kaplan JH (1994) Nitrophenyl-egta, a photolabile chelator that selectively binds Ca^{2+} with high affinity and releases it rapidly upon photolysis. Proc Natl Acad Sci U S A 91:187–191
36. Kaplan JH, Somlyo AP (1989) Flash photolysis of caged compounds: new tools for cellular physiology. Trends Neurosci 12:54–59
37. Ellis-Davies GCR (2007) Caged compounds: photorelease technology for control of cellular chemistry and physiology. Nat Meth 4: 619–628
38. Neher E (2006) A comparison between exocytic control mechanisms in adrenal chromaffin cells and a glutamatergic synapse. Pflügers Archiv Eur J Physiol 453:261–268
39. Thorn P (2012) Measuring calcium signals and exocytosis in tissues. Biochimica et Biophysica Acta (BBA) 1820:1179–1184
40. Takahashi N, Nemoto T, Kimura R et al (2002) Two-photon excitation imaging of pancreatic islets with various fluorescent probes. Diabetes 51:S25–S28
41. Takahashi N, Hatakeyama H, Okado H et al (2004) Sequential exocytosis of insulin granules is associated with redistribution of SNAP25. J Cell Biol 165:255–262
42. Kasai H, Hatakeyama H, Ohno M, Takahashi N (2010) Exocytosis in islet β-cells the islets of Langerhans. In: Islam MS (ed). Springer Netherlands, pp 305–338
43. Jena BP (2008) Chapter 2: Intracellular organelle dynamics at nm resolution. In: Bhanu PJ (ed) Methods in cell biology. Academic Press, Amsterdam, pp 19–37
44. Nemoto T, Kimura R, Ito K et al (2001) Sequential-replenishment mechanism of exocytosis in pancreatic acini. Nat Cell Biol 3: 253–258
45. Kasai H, Kishimoto T, Nemoto T, Hatakeyama H, Liu T-T, Takahashi N (2006) Two-photon excitation imaging of exocytosis and endocytosis and determination of their spatial organization. Adv Drug Deliv Rev 58:850–877
46. Low JT, Shukla A, Behrendorff N, Thorn P (2010) Exocytosis, dependent on Ca^{2+} release from Ca^{2+} stores, is regulated by Ca^{2+} microdomains. J Cell Sci 123:3201–3208
47. Thorn P, Fogarty KE, Parker I (2004) Zymogen granule exocytosis is characterized by long fusion pore openings and preservation of vesicle lipid identity. Proc Natl Acad Sci U S A 101:6774–6779
48. Warner JD, Peters CG, Saunders R et al (2008) Visualizing form and function in organotypic slices of the adult mouse parotid gland. Am J Physiol 295:G629–G640
49. Bi GQ, Alderton JM, Steinhardt RA (1995) Calcium-regulated exocytosis is required for cell membrane resealing. J Cell Biol 131: 1747–1758
50. Kishimoto T, Kimura R, Liu T-T, Nemoto T, Takahashi N, Kasai H (2006) Vacuolar sequential exocytosis of large dense-core vesicles in adrenal medulla. EMBO J 25:673–682
51. Schneggenburger R, López-Barneo J (1992) Patch-clamp analysis of voltage-gated currents in intermediate lobe cells from rat pituitary thin slices. Pflügers Archiv Eur J Physiol 420:302–312

52. Moser T, Neher E (1997) Rapid exocytosis in single chromaffin cells recorded from mouse adrenal slices. J Neurosci 17:2314–2323
53. Edwards FA, Konnerth A, Sakmann B, Takahashi T (1989) A thin slice preparation for patch clamp recordings from neurones of the mammalian central nervous system. Pflügers Archiv Eur J Physiology 414: 600–612
54. Speier S, Rupnik M (2003) A novel approach to in situ characterization of pancreatic β-cells. Pflügers Archiv Eur J Physiol 446:553–558
55. Sakmann B, Neher E, SpringerLink (Online service) (2009) Single-channel recording. In: Springer Science+Business Media, LLC,, Boston, MA, p 1 online resource
56. Sedej S, Tsujimoto T, Zorec R, Rupnik M (2004) Voltage-activated Ca^{2+} channels and their role in the endocrine function of the pituitary gland in newborn and adult mice. J Physiol 555:769–782
57. Meneghel-Rozzo T, Rozzo A, Poppi L, Rupnik M (2004) In vivo and in vitro development of mouse pancreatic β-cells in organotypic slices. Cell Tissue Res 316:295–303
58. Rozzo A, Meneghel-Rozzo T, Delakorda SL, Yang S-B, Rupnik M (2009) Exocytosis of insulin: in vivo maturation of mouse endocrine pancreas. Ann N Y Acad Sci 1152:53–62
59. Dudanova I, Sedej S, Ahmad M, et al. (2006) Important contribution of α±-neurexins to Ca^{2+}-triggered exocytosis of secretory granules. In, 10599–10613
60. Marengo FD (2005) Calcium gradients and exocytosis in bovine adrenal chromaffin cells. Cell Calcium 38:87–99
61. Chow RH, Klingauf J, Heinemann C, Zucker RS, Neher E (1996) Mechanisms determining the time course of secretion in neuroendocrine cells. Neuron 16:369–376
62. Gee KR, Archer EA, Lapham LA et al (2000) New ratiometric fluorescent calcium indicators with moderately attenuated binding affinities. Bioorg Med Chem Lett 10:1515–1518
63. Ge Q, Dong Y-m H, Z-t W, Z-x XT (2006) Characteristics of Ca^{2+}-exocytosis coupling in isolated mouse pancreatic β cells. Acta Pharmacol Sin 27:933–938
64. Takahashi N, Kadowaki T, Yazaki Y, Miyashita Y, Kasai H (1997) Multiple exocytotic pathways in pancreatic beta cells. J Cell Biol 138:55–64
65. Wan QF, Dong Y, Yang H, Lou X, Ding J, Xu T (2004) Protein kinase activation increases insulin secretion by sensitizing the secretory machinery to Ca^{2+}. J Gen Physiol 124:653–662
66. Heinemann C, Chow RH, Neher E, Zucker RS (1994) Kinetics of the secretory response in bovine chromaffin cells following flash photolysis of caged Ca^{2+}. Biophys J 67: 2546–2557
67. Neher E, Zucker RS (1993) Multiple calcium-dependent processes related to secretion in bovine chromaffin cells. Neuron 10:21–30
68. Voets T (2000) Dissection of three Ca^{2+}-dependent steps leading to secretion in chromaffin cells from mouse adrenal slices. Neuron 28:537–545
69. Rupnik M, Kreft M, Sikdar SK et al (2000) Rapid regulated dense-core vesicle exocytosis requires the CAPS protein. Proc Natl Acad Sci U S A 97:5627–5632
70. Liu Y, Schirra C, Stevens DR et al (2008) CAPS facilitates filling of the rapidly releasable pool of large dense-core vesicles. J Neurosci 28:5594–5601
71. Paulmann N, Grohmann M, Voigt J-P et al (2009) Intracellular serotonin modulates insulin secretion from pancreatic β-cells by protein serotonylation. PLoS Biol 7:e1000229
72. Dolensek J, Skelin M, Rupnik MS (2011) Calcium dependencies of regulated exocytosis in different endocrine cells. Physiol Res 60(Suppl 1):S29–S38
73. Ichikawa A (1965) Fine structural changes in response to hormonal stimulation of the perfused canine pancreas. J Cell Biol 24:369–385
74. Speier S, Gjinovci A, Charollais A, Meda P, Rupnik M (2007) Cx36-mediated coupling reduces beta-cell heterogeneity, confines the stimulating glucose concentration range, and affects insulin release kinetics. Diabetes 56:1078–1086
75. Low JT, Shukla A, Thorn P (2010) Pancreatic acinar cell: new insights into the control of secretion. Int J Biochem Cell Biol 42: 1586–1589
76. Rupnik M (2009) The physiology of rodent beta-cells in pancreas slices. Acta Physiol (Oxf) 195:123–138

Chapter 8

Intravital Microscopy and Its Application to Study Regulated Exocytosis in the Exocrine Glands of Live Rodents

Oleg Milberg, Natalie Porat-Shliom, Muhibullah Tora, Laura Parente, Andrius Masedunskas, and Roberto Weigert

Abstract

Regulated exocytosis is a fundamental event in specialized secretory organs that has been primarily studied in in vitro and ex vivo model systems. The recent application of intravital microscopy to image subcellular structures in vivo has enabled researchers to investigate the machinery controlling regulated exocytosis in live rodents. Here, we describe selected experimental models that have been used to investigate the dynamics of the secretory granules after their initial fusion their with the plasma membrane. Specifically, we used rodent salivary glands, an established model for exocrine secretion. Our goal is to provide the reader with guidelines on how to apply both qualitative and quantitative intravital microscopy to study regulated exocytosis and to highlight advantages and limitations of this approach.

Key words Regulated exocytosis, Exocrine glands, Actin cytoskeleton, Myosin, Intravital microscopy, Salivary glands

1 Introduction: Regulated Exocytosis in Exocrine Glands

Molecules destined for the cell surface are packed into membranous carriers that are transported to the cell periphery where they fuse with the plasma membrane and release their content into the extracellular space. This process is called exocytosis and can occur either in a constitutive or a regulated fashion [1–3]. Constitutive exocytosis occurs in every cell type and is fundamental for maintaining membrane homeostasis and cell polarity [4, 5]. On the other hand, regulated exocytosis occurs in specialized secretory cells and is elicited by extracellular stimuli that activate one or more receptors at the cell surface, which ultimately trigger an intracellular signaling cascade [6–8]. During regulated exocytosis transport intermediates are generated constitutively from intracellular

Peter Thorn (ed.), *Exocytosis Methods*, Neuromethods, vol. 83,
DOI 10.1007/978-1-62703-676-4_8,

organelles, such as the trans-Golgi network, and transported to the cell periphery where they accumulate under resting conditions [6–8]. Most of the stimuli elicit an increase in either intracellular Ca^{++} or cAMP that activates a series of processes which include the docking and the fusion of transport intermediates with the plasma membrane [6–8].

Regulated exocytosis occurs with different modalities and is regulated by various machineries depending on the secretory system. Indeed, secretory cells can be roughly divided in four categories: neuron, endocrine, hematopoietic, and exocrine. In the first three systems, molecules are transported in small vesicles (secretory vesicles that have diameters between 50 and 300 nm), which release their contents into the extracellular space, ultimately reaching the circulation [3, 9, 10]. In exocrine organs, molecules are stored in large vesicles (secretory granules that have diameters larger than 1 μm) and are released at specialized sites of the plasma membrane that form ductal structures which are in direct continuity with the external environment [1, 2]. After fusion, individual secretory granules remain connected to the extracellular space via the fusion pore that allows for their contents to be secreted. In some secretory systems, the membranes of the secretory granules are gradually integrated into the plasma membrane in a process that is coupled to the expansion of the fusion pore during the release of cargo molecules [11–13]. In other secretory systems, the fused secretory granules serve as docking and fusion sites for other secretory granules, which leads to the formation of strings of fused granules whose lumen is in continuity with the extracellular space (i.e., compound exocytosis) [11, 14, 15]. Regardless of the modalities of post-fusion events, the membranes of the secretory granules that are added to the plasma membrane are retrieved by the stimulation of endocytic events (i.e., compensatory endocytosis) [16–18].

The machinery controlling regulated exocytosis in exocrine glands has been extensively investigated by using three main models: pancreas, lacrimal glands, and salivary glands (SGs) [2, 19, 20]. Furthermore, several molecules have been identified to play fundamental roles in this process both in terms of signaling components and membrane regulators, such as Rab GTPase, lipid kinases, SNAREs, and the cytoskeleton [1]. The actin cytoskeleton, in particular, plays a fundamental role in controlling regulated exocytosis in exocrine glands, and several cytoskeletal components have also been shown to control both pre-fusion (docking, priming, and fusion) and post-fusion events (maintenance of the fusion pore, compound exocytosis, gradual collapse, and compensatory endocytosis) [2, 13, 19–22].

1.1 Methods to Study Regulated Exocytosis in Exocrine Glands

In exocrine glands, regulated exocytosis has been investigated by using various model systems and experimental approaches. In terms of model systems, this field has primarily relied on preparations derived from explanted organs rather than cell cultures. Indeed, cells isolated from exocrine organs and cultured on solid substrates rapidly dedifferentiate and within a few hours lose their polarity and secretory granules [23]. Instead, intact acinar preparations from pancreas, lacrimal glands, and SGs have been quite successful models and provided groundbreaking information on the kinetics of exocytosis and the molecular components regulating this process [11, 24]. In addition, lobule preparations and tissue slices have also been extensively used as model systems. The former, developed in David Castle's lab, have been instrumental in elucidating the complex nature of the secretory pathways in SGs, leading to the discovery of the minor regulated pathways [25, 26]. The latter, originally developed and used in the neurosciences, have been particularly useful for studying multiple aspects of regulated exocytosis both in the pancreas and in SGs [27, 28]. In these systems cell architecture is preserved compared to the native tissue since they retain several structural components, such as extracellular matrix, surrounding supporting cells, and ductal structures. Finally, regulated exocytosis has been studied in live rodents, although to a limited extent, due to challenges in the ability to manipulate these models [29].

In terms of experimental approaches, regulated exocytosis has been primarily investigated through the use of biochemical, immunological, or electrochemical assays. Although these methods permit an accurate determination of the amount of secreted molecules per cell, they do not reveal any information on the exocytic steps or the morphology and fate of individual exocytic vesicles. In this respect, assays based on measurements of membrane capacitance have provided detailed information on single fusion events at the plasma membrane, with the caveat of the inability to distinguish between endocytic and exocytic events. Electron microscopy has also been used to characterize the structure of the secretory granules at the plasma membrane before and after the fusion steps.

Finally, time-lapse light microscopy has elucidated several aspects of the dynamics of exocytosis in exocrine glands. Several studies were performed by bathing ex vivo preparations in small fluorescent dyes, such as sulforhodamine B or lucifer yellow, which accumulate in the ducts and upon the opening of the fusion pore access the lumen of the secretory granules. This approach has revealed that in most exocrine glands, secretory granules primarily undergo compound exocytosis [11, 14, 30–33]. Furthermore, it enabled measuring the kinetics of the pore expansion and the diffusion of proteins and lipids from the PM into the membrane of the granules [15, 34, 35]. More recently, laser-spinning disk has been successfully used in isolated pancreatic acini where the exocytosis of large secretory granules containing the cargo molecule syncollin fused with the pH-sensitive fluorescent dye pHluorin has been studied under both physiological and pathological conditions [36].

2 Materials

2.1 Microscope and Preparation of the Stage

To image the SGs in live rats and mice, we used an inverted microscope (Olympus IX81 equipped with a Fluoview 1000 scanner [37]). The stage was equipped with an insert designed to accommodate a 35-mm petri dishes (Olympus America, Center Valley, PA) that was closed with a 40-mm glass coverslip (# 1.5, from Bioptechs, Butler, PA). To image regulated exocytosis, we used either a Plan-Apo 60x/1.2 NA water-immersion objective or a Plan-Apo 60x/1.42 NA oil objective (Olympus). The objectives were connected to an objective heater (Bioptechs, Butler, PA), whereas the stage was preheated with heated pads (Foot Warmers, Heat Factory, Vista, CA). Confocal microscopy was performed by exciting the fluorophores as follows: 488 nm for GFP (0.5–1 % laser power); 561 nm for td-Tomato, Texas Red, and Alexa 594 (10 % laser power); and 633 nm for Alexa 647 (20–25 % laser power). We mostly used scanning speeds ranging from 200 ms up to 1 s per frame. When high spatial and temporal resolution scans are required, high zoom of the scan area is used. For example, using a 60× objective, a 16× scan area zoom, and 128 × 128 pixel image will yield 0.103 μm/pixel spatial scale. With these settings, when pixel dwell time is set to 4 μs/pixel, the resultant scan speed is close to 200 ms per frame.

2.2 Animal Preparations

For intravital microscopy we use Sprague–Dawley rats (150–250 g, Harlan Laboratories, Frederick, MD), FVB mice expressing EGFP (GFP mice), and C57BL/6 mice expressing the membrane-targeted td-Tomato peptide (mTomato mice). Both transgenic lines were purchased from the Jackson Laboratory and bred as homozygous in the local animal facility. The optimal weight for imaging is 20–25 g. The first step in imaging the dynamics of regulated exocytosis in vivo is the preparation of the animals. Mice and rats need to acclimate for 2–3 days in the area where they are going to be imaged, since any sudden change in the environment results in various degrees of degranulation, as reported by others [38]. Interestingly, regardless of the dietary regimen (fed ad libitum or fasting), the number of the secretory granules is not significantly affected. Although we did not observe any difference in secretory response related to the gender of the animals, we perform all the experiments using male mice or rats.

However, it is crucial to minimize any distress related to the anesthesia. To this aim, we devised a procedure that involves the preinduction of anesthesia by using isoflurane inhalation in a vaporizer chamber (Braintree Scientific, Braintree, MA, 800–900 mmHg oxygen, 2–3 % isoflurane) followed by the intraperitoneal injection of a freshly prepared mixture of ketamine and xylazine (100 mg/kg and 20 mg/kg, respectively, using a 25-gauge needle). After the animals are anesthetized, particular care has to be taken to maintain the body temperature in a range between 35 and 38 °C. The temperature is constantly measured by using a rectal thermometer (MicroTherma 2 T Thermometer with RET-3 and IT-21 thermocouple probes, Braintree Scientific, Braintree, MA), and it is maintained by using heated pads (Foot Warmers, Heat Factory, Vista, CA) or heated lamps (model HL1, Braintree Scientific, Braintree, MA).

2.3 Surgical Procedures to Expose the Submandibular Salivary Glands and Positioning on the Microscope Stage

The neck area of the anesthetized animal was cleaned with a gauze sponge soaked in 70 % ethanol. A small piece of skin at the midline was lifted up with dull tweezers and cut with clean scissors (0.5–1 cm). The skin was moved away from the underlying tissue and excised in order to leave a 0.5-cm-wide and 2-cm-long opening. The connective tissue was separated away from the submandibular gland that is then constantly washed with saline or covered with optical coupling gel to prevent dehydration (see [39, 40]). The animal was positioned on its side onto the preheated stage with the exposed salivary gland placed in the center of the coverslip. Before proceeding with the immobilization of the gland, the animal was secured and stabilized in this position by using masking tape. A small piece of lens cleaning tissue was then sandwiched between the gland and a small plastic square that was gently pressed to stabilize the gland. Particular care was taken to ensure that the blood flow was properly maintained. To this aim, the blood and the extent of the motion were quickly evaluated by using epi-fluorescence microscopy. Once the proper stabilization was achieved, the plastic square was secured to the stage using masking tape.

2.4 Retro-Diffusion of Fluorescent Dyes into the Wharton's Duct

The anesthetized animal was secured into a previously described stereotactic device [39] with the mandibles wide open and the cheeks extended to the sides. The tongue was folded towards the back of the mouth to expose the ductal orifices without obstructing the airways. The stereotactic device was held at about 45° and positioned under a stereomicroscope (SZX7, Olympus America, Center Valley, PA) to visualize the area below the tongue and locate the two orifices of the Wharton's ducts. A 30–40-cm PE-5 cannula (Strategic Applications, Libertyville, IL) was gently pushed into the orifice and introduced into the duct using bent sharp tweezers. For mice, the cannula was thinned by carefully stretching it close to the open flame of a Bunsen burner. To seal the cannula a small drop of Histoacryl tissue glue was applied to the orifice. The cannula was connected to a 1-ml syringe (via a 30- or a 33-GG needle), which contains the fluorescent dye (1–10 μg/ml in saline) and placed 25–30 cm above the animal to allow diffusion by gravity. After 20 min the acinar canaliculi should be clearly highlighted and visible by using epi-fluorescence microscopy. If not, a slight pressure can be applied to the plunger of the syringe to force the diffusion of the dye, but care must be taken to not disrupt the epithelial integrity.

3 Methods

3.1 Intravital Microscopy Permits the Visualization of Subcellular Structures In Vivo

Time-lapse light microscopy and the development of GFP technology have contributed to significant breakthroughs in understanding the dynamics and the mechanisms of several membrane-trafficking events, including regulated exocytosis. However, light microscopy has been primarily applied to in vitro and ex vivo models. In the early nineties, the development of two-photon microscopy [41] made deep tissue imaging possible and opened the door for the establishment of a new field called intravital microscopy (IVM). IVM has enabled imaging biological processes in several multicellular organisms, including rodents [42]. Initially, IVM has been applied primarily in the neurosciences to study synaptic plasticity and calcium signaling [43]. Afterwards, IVM was used to image the dynamics of cells involved in the immune response [44, 45] or to investigate cell migration and lympho-angiogenesis in tumor models [46–48]. Only recently, IVM has been applied to study the dynamics of subcellular structures in live animals. Indeed, the endocytosis of fluorescently labeled dextrans and folate has been imaged in the kidney of live rats and mice [49, 50].

Our group has developed a robust system to study membrane trafficking in vivo in live rodents using IVM. The ability to stabilize the SGs preparation, and minimizing motion artifacts due to heartbeat and respiration, combined with the use of selected transgenic mouse models has enabled us to investigate in detail the modality and the kinetics of regulated exocytosis in vivo [12, 13, 17, 51, 52]. We have also been able to study the endocytosis of selected molecules in rodent SGs and to follow their trafficking through the endo-lysosomal systems [40]. Furthermore, we have investigated the dynamics and modality of the internalization of naked DNA from the apical surface of the SG epithelium [18, 53]. Here, we provide a summary of the methods and tools that we have developed to carry out these studies. Since a detailed description of the procedures for animal handling has been recently published [39], we will focus primarily on the description of the models and the quantitative aspects of this approach.

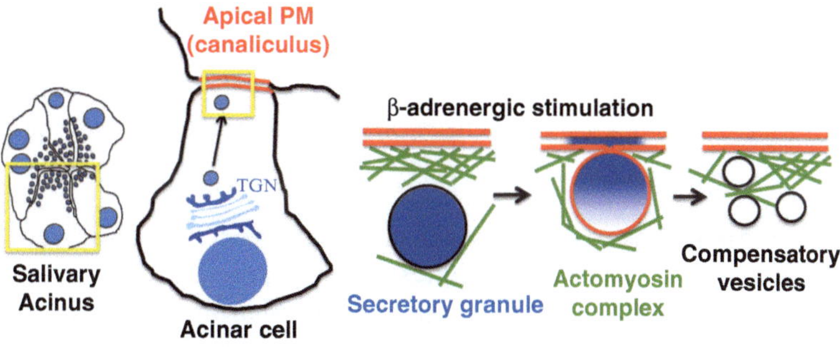

Fig. 1 Regulated exocytosis in salivary glands. In SGs, acini are the major secretory units and are composed of 9–10 polarized acinar cells [13]. The apical plasma membranes of adjacent acinar cells form the acinar canaliculi, which are very narrow ducts (0.3–0.4 μm in diameter) where both proteins and fluid are secreted. Secretory granules containing salivary proteins are formed from the trans-Golgi network and are transported to the apical plasma membrane. In resting conditions, secretory granules are prevented to fuse with the plasma membrane by a thick F-actin meshwork that acts as a functional barrier [2]. In both parotid and submandibular glands, the stimulation with beta-adrenergic receptor elicits docking and fusion of the secretory granules with the plasma membrane, the opening of the fusion pore, and the release of the cargo molecules into the canaliculi. The membranes of the secretory granules are first integrated in the apical plasma membrane via a process that requires the assembly of an actomyosin complex and successively retrieved by compensatory endocytosis

3.2 Tools and Procedures to Image the Secretory Granules at the Apical Plasma Membrane in the Salivary Glands of Live Rodents

In SGs, regulated exocytosis occurs primarily in the acini that are formed by polarized epithelial cells. Secretory granules are transported to the apical plasma membrane, where they fuse upon stimulation of the appropriate G protein-coupled receptor (Fig. 1). In order to investigate regulated exocytosis in vivo, we have developed a series of complementary approaches that make possible the visualization of the secretory granules, the apical plasma membrane, and other molecules that are implicated in this process. Specifically, these approaches are geared towards the elucidation of the steps that occur after the membrane bilayer of a secretory granule fuses with that of the plasma membrane.

In rodents, there are three major SGs: the parotid, which secretes primarily digestive enzymes, such as amylase; the submandibular, which secretes several molecules involved in defense against pathogens, such as peroxidase, lactoferrin, and immunoglobulins; and finally, the sublingual glands, which secrete primarily mucins [23]. In both the parotid and submandibular glands, regulated exocytosis is triggered by the stimulation of beta-adrenergic receptors, whereas in the sublingual gland, it is triggered by muscarinic stimulation [23]. As a model to study exocytosis, we selected the submandibular glands for two main reasons: first, they are located in the neck area and can be easily exposed and immobilized to minimize the motion artifacts, as previously described [39, 40], and second, the excretory duct of the submandibular glands (Wharton's duct) is easily accessible and can be used to inject fluorescent molecules, pharmacological agents, or to transfect genes [18], whereas the parotid excretory duct (Stensen's duct) is not.

3.3 The GFP Mouse as a Model to Quantitatively Study Regulated Exocytosis

To visualize the secretory granules, we used a mouse strain (FVB) that expresses GFP in the cytoplasm and in the nuclei (GFP mouse) [13, 54]. After translation, GFP is not imported into the ER and thus is excluded from the lumen of most of the organelles in the secretory pathway. For this reason, in the acini of the submandibular glands, large structures, such as the secretory granules, appear as well-defined circular profiles that can be easily visualized by either two-photon (not shown) or confocal microscopy (Fig. 2b, arrows). Since the average diameter of the secretory granules is 1.5 μm (Fig. 2c), they can be optimally resolved by using confocal microscopy and adjusting the size of the pinhole (0.8–0.9 μm), as previously reported [13, 39]. Only a small percentage of the structures that exclude GFP are mitochondria or lysosomes that can be identified by using specific vital dyes such as mitotracker (Fig. 2d) and lysotracker (not shown). This is further confirmed by the fact that stimulating exocytosis with subcutaneous (SC) injections of isoproterenol significantly reduces the number of structures, which exclude the GFP (Fig. 2e).

In order to visualize the exocytic events, it is essential to first locate the apical plasma membrane (or canaliculi). In the GFP mouse canaliculi can be identified since they have some unique characteristics, such as the following: (1) They are shared by two acinar cells, (2) they exhibit an enrichment in the GFP levels that correlate with the enrichment in F-actin, and (3) they are surrounded by secretory granules (Fig. 2b, arrowheads and insets). The best fields of view are usually 15–30 μm below the surface of the glands, as determined by visualizing the collagen fibers by second harmonic generation (Fig. 2a, between dotted lines). Once the optimal imaging area has been determined, time-lapse imaging is performed by setting the acquisition speed between 0.2 and 1 s/frame. A typical imaging session comprises 3–4 time-lapse sequences (10 min each) that are acquired in the same area. Under resting conditions secretory granules are stationary and exhibit minimal motion, whereas they slightly increase their mobility after 1 min from the SC injection of isoproterenol [13].

Due to the limit of resolution of light microscopy and to the small shifts in the *z*-axis that occur during imaging, the fusion between the secretory granules and the canaliculi cannot be directly imaged. However, post-fusion steps can be precisely monitored. Indeed, after fusion with the plasma membrane and the opening of the fusion pore, the secretory granules recruit an actomyosin scaffold that nonspecifically binds GFP and other cytoplasmic proteins [13].

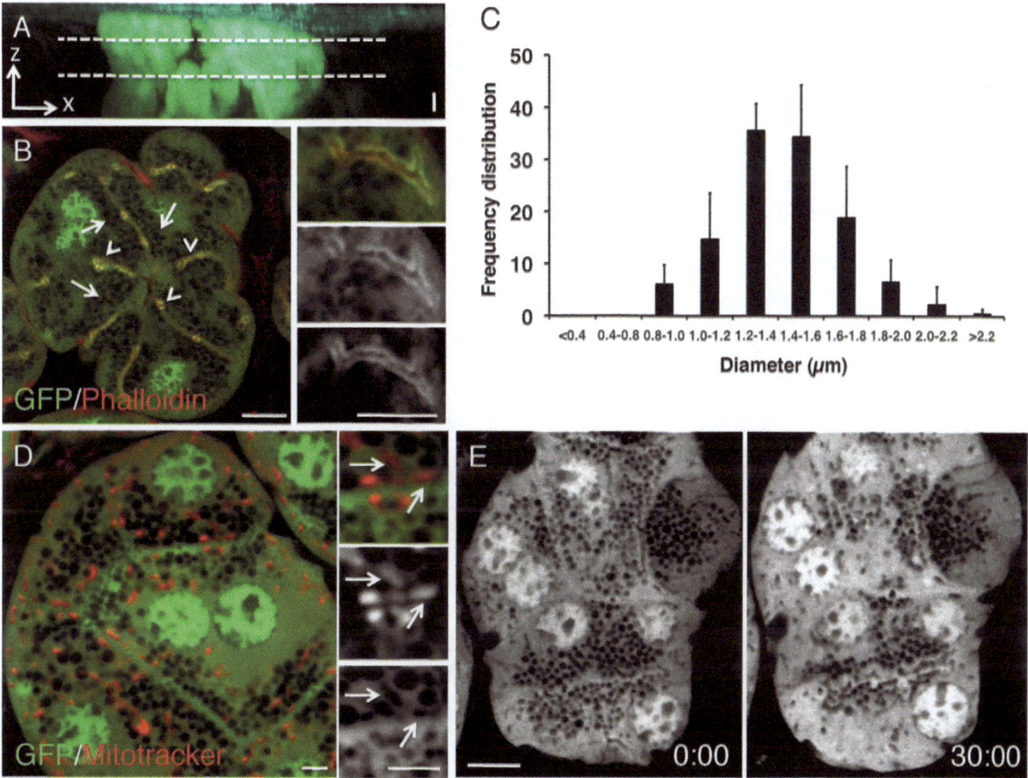

Fig. 2 Secretory granules visualized through the GFP mouse model. (**a**) The SGs of an anesthetized GFP mouse were exposed, and a Z-scan was performed by two-photon microscopy (excitation 930 nm, Plan-Apo 60x/1.2 NA water-immersion objective). The volume rendering was realized with Imaris (Bitplane), and the maximal projection of a side view (zx plane) is shown. Collagen fibers (cyan) appears in the first 2–5 μm from the surface of the tissue, and the two acini were fully reconstructed (*green*). Bar, 10 μm. (**b, c**) The SGs of a GFP mouse were excised and fixed with 2 % formaldehyde for 15 min. The glands were processed and labeled with Texas Red-phalloidin to reveal F-actin, as previously described [40]. The glands were imaged by confocal microscopy (excitation 488 and 561 nm, Plan-Apo 60x/1.4 NA oil objective). (**b**) Secretory granules appear as dark circular structures devoid of GFP (*arrows*) that are clustered around the F-actin-enriched canaliculi (*arrowheads*). Bar, 5 μm. *Insets*—High magnification of a canaliculus. Note the enrichment of both GFP (*middle panel*) and F-actin (*lower panel*) on the limiting membranes of the canaliculus. Bar, 5 μm. (**c**) For each acinus, the diameters of the secretory granules were measured manually, and their frequency distribution was calculated. Each bar represents the average frequency distribution for ten acini in the same animal (±SD). The measurements were repeated in four animals with similar results. (**d, e**) The SGs of anesthetized GFP mice were exposed and bathed for 20 min either with 500 nm of Far red-mitotracker (Invitrogen) (**d**) or with saline (**e**). (**d**) The SGs were imaged by confocal microscopy (excitation 488 and 647 nm). Mitochondria appear as elongated profiles (*arrows*). Bars, 3 μm. (**e**) The exposed SGs were imaged in time-lapse (excitation 488 nm, 1 frame/s), and 0.25 mg/kg of isoproterenol was injected SC to stimulate exocytosis. A snapshot was taken at time 0 and after 30 min from the injection. Note that the number of secretory granules is substantially reduced. Bar, 10 μm

This results in a sharp increase in the GFP levels around the limiting membranes of the secretory granules followed by a slow decline that mirrors their decrease in size (Fig. 3b, c). These "flashes of GFP" represent bona fide exocytic events since they are correlated with the

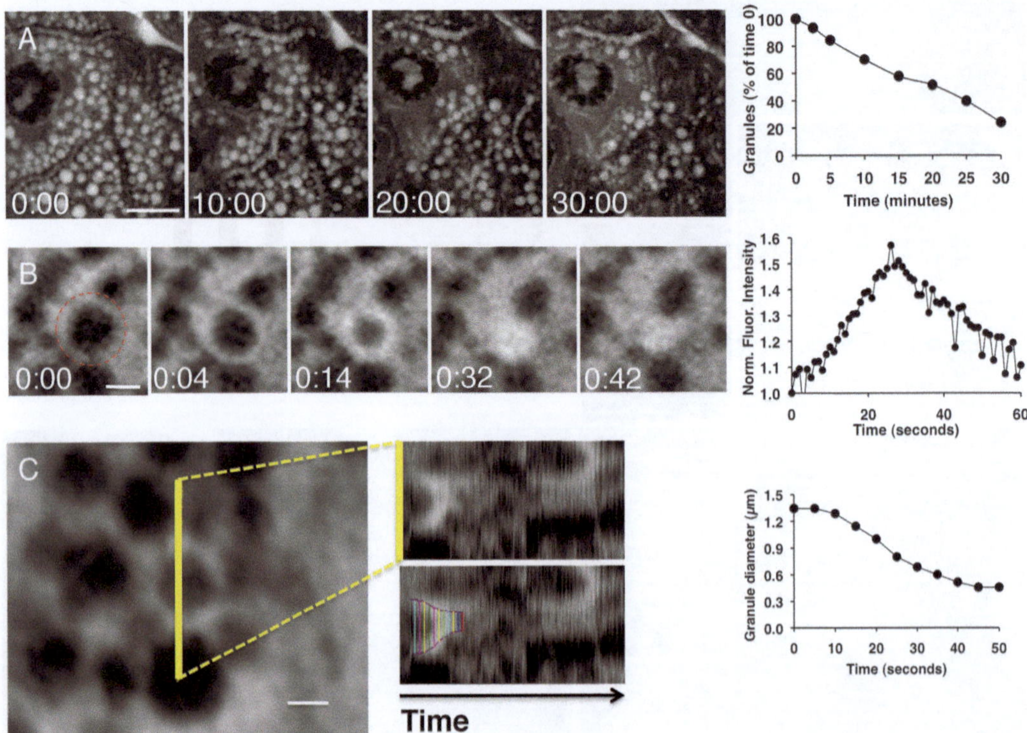

Fig. 3 Analysis of degranulation and lifetime of the secretory granules at the plasma membrane after fusion by using the GFP mouse. (**a–c**) The SGs of an anesthetized GFP mouse were exposed and imaged in time-lapse by confocal microscopy (excitation 488 nm, 1 frame/s). Isoproterenol (0.25 mg/Kg) was injected SC to stimulate exocytosis. Four movies, 10 min each, were acquired. (**a**) Frames at selected times were processed and used to score the number of secretory granules. Briefly, each frame was inverted by using Photoshop to highlight the secretory granules. Circular profiles larger than 800 nm were scored as secretory granules and reported as a function of time (graph on the right). Bar, 10 μm. (**b**, **c**) The movies were converted into time sequences by using Metamorph and analyzed to identify and select secretory granules close to the apical plasma membrane that exhibited an increase in the intensity of the GFP fluorescence around their limiting membranes. If needed, any shift in the xy plane was corrected by processing the time sequence with ImageJ (Stag-reg plug-ins). Importantly, time sequences with major shifts along the *z*-axis were not used for the analysis. A region of interest was drawn around the selected granule (**b**, *red broken circle*) or a line parallel to the apical pole (**c**, *yellow solid line*). The region measurement (**b**) or the kymograph (**c**) functions were used to record the integrated fluorescence intensity (IFI) or determine the diameter for each time frame, respectively (see Section 3). Representative curves are shown for a single granule. In a typical experiment 10–15 granules per acinus are evaluated and 3–4 animals per condition are used. Bars, 1 μm

decreasing number of secretory granules per acinar cell, as previously shown [13].

In order to characterize this process, we quantitatively evaluate three parameters: the extent of degranulation, the intensity of the GFP around individual granules, and their diameters over time. First, the time-lapse sequences are processed with ImageJ (StackReg plug-in) to correct for any slight shift in xy due to residual motion artifacts. In order to estimate the extent of degranulation, a minimum of 6 frames are selected: the first frame right after the injection of isoproterenol (time 0) and the others after 2.5, 5, 10, 20, and 30 min (Fig. 3a). The frames are processed with Photoshop, converted to grayscale mode, and inverted to highlight the secretory granules (Fig. 3a). All vesicular profiles with a diameter above 800 nm are scored as secretory granules, normalized to the number of granules at time 0, and reported as % of granules versus time (Fig. 3a, graph). This approach provides information on the kinetics of degranulation of the secretory granules and can be used to study several aspects of the mechanisms controlling regulated exocytosis. For example, we characterized the optimal doses of isoproterenol and showed that muscarinic stimulation does not elicit any exocytic events in vivo in the submandibular glands [13]. Moreover, this method can be coupled with the administration of pharmacological agents in order to dissect the molecular machinery that regulates this process.

Measuring both the GFP levels around the secretory granules and their diameter provides information on their lifetime after fusion with the plasma membrane. The GFP levels can be measured using Metamorph (Molecular Devices, Sunnyvale, CA). Specifically, a region of interest is drawn around each individual granule (Fig. 3b, red broken circle), and the integrated fluorescent intensity (IFI) is recorded for each time point by using the region measurement function. The value of IFI recorded 5 s before the sharp increase of the GFP levels is used as a baseline for the normalization. As for the measurement of the diameter of the secretory granules over time, we used the kymograph function of Metamorph. A line parallel to the apical plasma membrane is drawn across the diameter of an individual secretory granule, and a kymograph is generated, as shown in Fig. 3c. The diameters are then outlined manually in the kymograph at selected time points, measured, and reported as a function of time. Based on these measurements, we estimated that after fusion with the plasma membranes, the secretory granules last 40–60 s and undergo a gradual collapse [13]. An important remark is that although this process is relatively slow and can be fully documented by acquiring the time-lapses at a lower scan speed, we still recommend a slight oversampling to compensate for the small drifts in the *z*-axis that occasionally occur.

This mouse model provides invaluable information on regulated exocytosis in vivo, and it can be also used to study other exocrine glands, such as the pancreas, lacrimal glands, and mammary glands. However, it is not a viable option to study endocrine glands, such as endocrine pancreas and adrenal medulla, or neurons, since in these organs the size of the secretory vesicles are close to the limit of resolution of light microscopy and the exocytic events occur on the order of milliseconds.

3.4 The mTomato Mouse as a Model to Visualize the Apical Plasma Membrane During Regulated Exocytosis

To visualize the apical plasma membrane during regulated exocytosis, we used a mouse that expresses a membrane-targeted peptide fused with the fluorescent protein tandem-Tomato (mTomato mouse). This peptide is derived from the first 8 N-terminal amino acids of MARCKS (myristoylated alanine-rich kinase substrate), and it is myristoylated on the first methionine and palmitoylated at the cysteine in position 3 and 4 [55]. In SGs, the peptide is localized both at the basolateral and the apical plasma membrane, but it is not present on secretory granules or on any other intracellular organelle (Fig. 4a) [13]. Canaliculi can be easily identified when they are oriented in parallel (Fig. 4a, asterisk) or perpendicularly (4A, arrows) to the xy plane. This mouse model enables imaging the secretory granules after fusion with the plasma membrane since the fluorescent peptide diffuses into their limiting membranes (Fig. 4a, asterisks). Moreover, this mouse model permits us to visualize the morphological changes of the apical plasma membranes upon its integration with the membrane of each secretory granule, as shown by the large expansion in the diameter of the canaliculi (Fig. 4a, arrows). The advantage of this model versus the GFP mouse is the fact that only granules fused with the plasma membrane are visualized. This model was instrumental in confirming that in SGs in vivo regulated exocytosis occurs primarily via single fusion events rather than compound exocytosis [12, 13].

The kinetics of the gradual collapse of the secretory granules can be determined by measuring the fluorescence intensity of the td-Tomato peptide, as soon as it appears in the limiting membranes of the granules. Similarly to what has been shown for the GFP mouse, a region of interest is drawn around the secretory granules,

Fig. 4 (continued) stimulate exocytosis. Six movies, 10 min each, were acquired. (**a**) The fluorescent peptide is localized at the cell surface of the acini. The apical canaliculi can be easily identified for their peculiar morphology particularly when they are oriented perpendicularly to the xy plane. Indeed, they appear as small circular profiles that expand after stimulation of exocytosis (*arrows*). When the canaliculi are oriented in parallel to the xy plane, they are more difficult to identify and appear visible only after they enlarge or when secretory granules are observed after fusion (*asterisk*). Bars, 10 μm. (**b**, **c**) The movies were converted into time sequences by using Metamorph, corrected for shift in the xy plane (ImageJ, Stag-reg plug-ins), and analyzed to identify the apical pole and the sites where exocytosis occurs. (**b**) Three frames were processed to show the expansion of the apical plasma membrane (*arrows*) and a secretory granule after fusion (*arrowhead*). Bar, 5 μm. (**c**) A region of interest was drawn around the granules after the fluorescent peptide diffused from the apical plasma membrane (**b**, *dotted red line*) or a line parallel to the plasma membrane (**c**, *solid yellow line*). The region measurement (**b**) or the kymograph (**c**) functions were used to record the integrated fluorescence intensity (IFI) or determine the diameter for each time frame, respectively (see Section 3). Notably, the kinetics of collapse of the secretory granules were in agreement with the kinetics determined using the GFP mouse (Fig. 3). In a typical experiment 10–15 granules per acinus are evaluated, and 3–4 animals per condition are used. Bars, 1 μm. (**d**) Large vacuoles derived from the secretory granules fused at the plasma membrane were detected inside acinar cells. The diameter of the granules was determined as described above and reported as a function of time. Bar, 1 μm

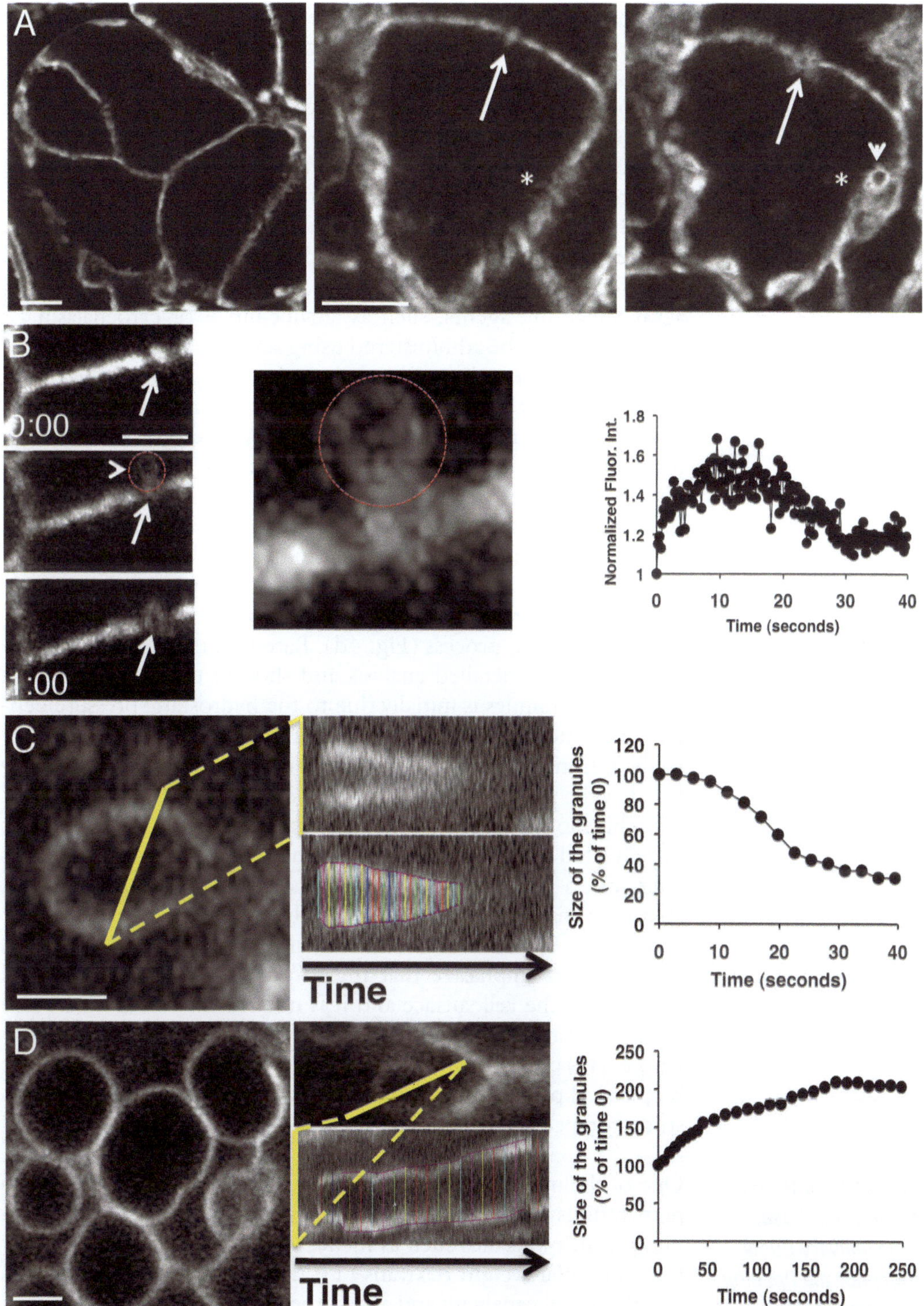

Fig. 4 Imaging the apical plasma membrane and the lifetime of the secretory granules by using the mTomato mouse. (**a–d**) mTomato mice were anesthetized, the SGs were exposed and either incubated with saline (**a–c**) or bathed in 10 μm cytochalasin D (**d**) as described in [13]. The SGs were imaged by using time-lapse confocal microscopy (excitation 561 nm, 200 ms/frame). After 20 min isoproterenol (0.25 mg/Kg) was injected SC to

and the IFI is recorded by using Metamorph (Fig. 4b, red broken circle). Alternatively, these kinetics can be studied by using kymographs to measure the diameter of the secretory granules (Fig. 3c). Notably, the kinetics of the gradual collapse determined with these methods are consistent with those determined using the GFP-mouse (Fig. 4c), indicating that the expression of these two reporter transgenes does not affect the dynamics of the process.

This approach is particularly useful to identify molecules that may regulate post-fusion events. For example, the role of the actin cytoskeleton in this process can be highlighted by administering actin-disrupting agents, such as latrunculin A or cytochalasin D. These drugs can be administered using several routes, such as bathing the glands on the microscope stage [13], retro-diffusion by injection through the salivary duct [40], or intra-organ injections. The optimal method may have to be determined by the investigator depending on the experimental conditions. The main effect of disrupting the actin cytoskeleton is to block the gradual collapse of the secretory granules, which instead progressively enlarge to form large vacuolar structures connected to the apical plasma membrane (Fig. 4d). The kymograph-based measurements of the granule diameters have revealed that the expansion of the secretory granules is a two-step process (Fig. 4d). Based on this finding, we performed a more detailed analysis and showed that the increase in size of the granules is initially due to the hydrostatic pressure generated by the secretion of fluids inside the canaliculi and later due to the fusion of other secretory granules with those that are already fused at the plasma membrane [12, 13].

This mouse provides invaluable information on several aspects of regulated exocytosis, such as the fate of the individual granules after fusion with the plasma membrane, the organization of the apical plasma membrane, and some biophysical properties of the membranes such as membrane mobility (graph in Fig. 4b). It is important to emphasize that the td-Tomato peptide is localized primarily at the cell surface and it is not enriched in the secretory granules or any other intracellular organelles before or after stimulation. This prevents the use of this mouse model to investigate the biophysical properties of the secretory granules before fusion or to study compensatory endocytosis.

3.5 Retro-Diffusion of Fluorescent Dyes in the Salivary Ducts to Monitor the Opening of the Fusion Pore and Compensatory Endocytosis

One of the main methods to study regulated exocytosis in ex vivo preparations from exocrine glands is to bathe the tissue in small fluorescent molecules such as lucifer yellow, sulforhodamine B, or low molecular weight dextrans [11, 24]. These dyes fill the ducts and the apical canaliculi and access the lumen of the granules upon the opening of the fusion pore. Moreover, this approach has been used to determine the size of the fusion pore by using probes with a known stokes radius, such as large molecular weight dextrans [21, 56]. In order to adapt this method to live animals, we introduced fine polyethylene cannulae into the Wharton's duct of anesthetized rats and used them to deliver fluorescent molecules to the acinar canaliculi [18, 40]. Since very small molecules have been shown to rapidly diffuse through the tight junctions in the SG epithelium [57], we routinely use 10 kDa dextran that has a relatively large stokes radius, and it is almost completely retained in the ducts.

After the submandibular SGs are surgically exposed, the anesthetized animals are placed on the microscope stage, and the dyes are allowed to slowly diffuse by gravity into the large ducts and the small acinar canaliculi [13]. After 20 min the acinar canaliculi are clearly visible, and after the identification of the appropriate imaging area, isoproterenol (0.1–0.25 mg/kg) is injected SC. It is important to emphasize that in order to detect the opening of the fusion pore and to monitor the influx of the dye, the acquisition speed has to be set to a minimum of 200–300 ms/frame (or higher depending on the instrumentation). After approximately 1 min, the dye appears inside the lumen of the secretory granules at the apical plasma membrane. As observed before, the secretory granules undergo a gradual collapse. Both the IFI and the diameter of the secretory granules can be measured as described in Fig. 3 and 4 showing that the kinetics of the gradual collapse in live rats is very similar to that measured in mice using the transgenic models (Fig. 5b graph).

Notably, this approach can be also used to image compensatory endocytosis, as well. Indeed, after few minutes following stimulation of regulated exocytosis, several small vesicular structures derived from the apical plasma membrane [18] are observed in the cytoplasm of the acinar cells (Fig. 5c, arrowheads and insets). It is important to emphasize that since compensatory vesicles are smaller than secretory granules (approximately 100 nm) [18], the gain of the detectors has to be adjusted to detect smaller structures. This results in a slight saturation of the fluorescent signal in the canaliculi (Fig. 5c, left panel).

Unfortunately, a major limitation of this approach in live animals is the fact that the visualization of both the opening of the fusion pore and the formation of compensatory vesicles are restricted to the first few minutes from the onset of regulated exocytosis. This is because fluorescent dextrans are washed out from the canaliculi after 3–5 min following the injection of isoproterenol, as clearly observed by comparing the levels of dextran in the canaliculi before and after stimulation [18, 51]. Unfortunately, any attempt to diffuse or inject additional amounts of dyes results in the disruption of the epithelial integrity. Nonetheless, this approach provides a very powerful tool to characterize the machinery regulating both processes. Indeed, the retro-diffusion of the fluorescent dyes can be coupled to the use of pharmacological agents. Moreover, cannulae can be inserted in mice providing the opportunity to extend this method to the transgenic mice described above.

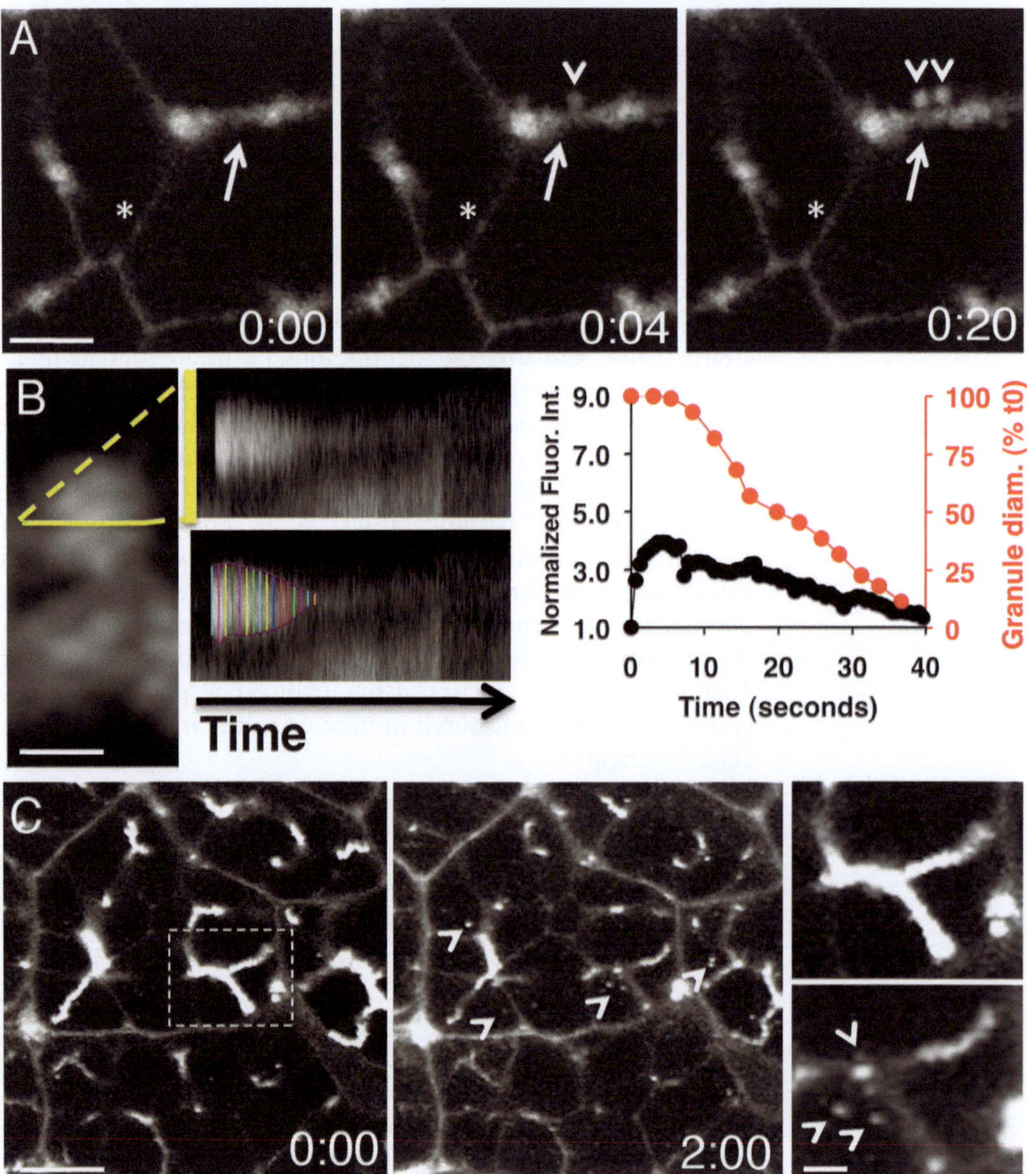

Fig. 5 Imaging the opening of the fusion pore and compensatory endocytosis by retro-diffusion of a fluorescent dye in the salivary ducts of live rats. (**a–c**) Sprague–Dawley rats were anesthetized, and a fine polyethylene cannula was inserted into the Wharton's duct. The SGs were exposed, and 10 μg/ml 10 kDa Texas Red dextran was retro-diffused by gravity as described in the Section 3. After 20 min the SGs were imaged by time-lapse confocal microscopy (excitation 561 nm, 200 ms/frame), and 0.25 mg/kg of isoproterenol was injected SC to stimulate exocytosis. (**a**) The acinar canaliculi are highlighted (*arrows*), and after the opening of the fusion pore, the fluorescent dextran accesses the lumen of the secretory granules (*arrowheads*). A small amount of dextran reaches the basolateral membrane (*asterisk*). Bar, 10 μm. (**b**) The IFI and diameter of the granules were determined as described in legends to Figs. 3 and 4. (**c**) After 2 min from the injection of isoproterenol, several small vesicles containing Texas Red dextran are observed in the cytoplasm of the acinar cells (*arrowheads* and *insets*). Note the reduction of the amount of dextran in the canaliculi, due to the secretion of fluids that wash out the dye. Bars, 10 μm and 1 μm (*insets*)

4 Conclusions

IVM has provided the means to investigate the dynamics of intracellular structures in live animals. Specifically, the combination of IVM with the ability to express or ablate genes in live animals has made possible addressing old and new questions in cell biology at a molecular level. Regulated exocytosis is a clear example of the full potential of this approach, as shown by the use of a series of selected transgenic mice (described here) and various tools, such as pharmacological agents and the transient expression of genes [18].

As for the mouse models, the GFP and mTomato strains have been instrumental in defining the general characteristics of regulated exocytosis in SGs in vivo such as regulation, rate of degranulation, modality of exocytosis (full collapse vs. compound exocytosis), and in providing a glimpse of the biophysical properties of the plasma membrane in vivo. We envision that these mice will be extremely valuable to characterize this process in other exocrine organs such as the pancreas, lacrimal glands, and mammary glands. In addition, the generation of new strains expressing either fluorescently tagged reporters for component of the actin cytoskeleton (e.g., Lifeact and myosin II) or Ca^{++} and cAMP sensors (e.g., Chameleon) will provide additional information on the integration between cell signaling and the cytoskeleton during regulated exocytosis. Moreover, further investigation may be possible by crossing these reporter mice with selected knockout animals.

The use of pharmacological agents and the expression of selected genes are also two very powerful tools. In this respect, the SGs represent an ideal organ due to the easy access through the oral cavity. Indeed, SGs have been used as a target organ to selectively deliver drugs, to express transgenes via non-viral- and viral-mediated approaches, and to acutely downregulate proteins via siRNA and shRNA. However, the development of novel and more sophisticated vehicles to accomplish tissue-specific targeting of molecules and genes will enable expanding this approach to the other exocrine organs.

In conclusion, we foresee that in the next few years, IVM will contribute to unraveling several molecular aspects of regulated exocytosis in the exocrine glands of live animals. This approach has the potential to be extended to other secretory organs, membrane-trafficking steps, and more broadly to other cell biological processes as well.

Acknowledgments

This research was supported by the Intramural Research Program of the NIH, National Institute of Dental, and Craniofacial Research.

References

1. Burgoyne RD, Morgan A (2003) Secretory granule exocytosis. Physiol Rev 83:581–632
2. Porat-Shliom N, Milberg O, Masedunskas A, Weigert R (2013) Multiple roles for the actin cytoskeleton during regulated exocytosis. Cell Mol Life Sci 70(12):2099–2121
3. Sudhof TC, Rizo J (2011) Synaptic vesicle exocytosis. Cold Spring Harb Perspect Biol 3
4. De Matteis MA, Luini A (2008) Exiting the golgi complex. Nat Rev Mol Cell Biol 9: 273–284
5. Luini A, Mironov AA, Polishchuk EV, Polishchuk RS (2008) Morphogenesis of post-Golgi transport carriers. Histochem Cell Biol 129:153–161
6. Parekh AB, Putney JW Jr (2005) Store-operated calcium channels. Physiol Rev 85:757–810
7. Petersen OH (2003) Localization and regulation of Ca2+ entry and exit pathways in exocrine gland cells. Cell Calcium 33:337–344
8. Seino S, Shibasaki T (2005) Pka-dependent and pka-independent pathways for camp-regulated exocytosis. Physiol Rev 85: 1303–1342
9. Hou JC, Min L, Pessin JE (2009) Insulin granule biogenesis, trafficking and exocytosis. Vitam Horm 80:473–506
10. Malacombe M, Bader MF, Gasman S (2006) Exocytosis in neuroendocrine cells: new tasks for actin. Biochim Biophys Acta 1763: 1175–1183
11. Kasai H, Kishimoto T, Nemoto T, Hatakeyama H, Liu TT, Takahashi N (2006) Two-photon excitation imaging of exocytosis and endocytosis and determination of their spatial organization. Adv Drug Deliv Rev 58:850–877
12. Masedunskas A, Porat-Shliom N, Weigert R (2012) Linking differences in membrane tension with the requirement for a contractile actomyosin scaffold during exocytosis in salivary glands. Commun Integr Biol 5:84–87
13. Masedunskas A, Sramkova M, Parente L et al (2011) Role for the actomyosin complex in regulated exocytosis revealed by intravital microscopy. Proc Natl Acad Sci U S A 108: 13552–13557
14. Pickett JA, Edwardson JM (2006) Compound exocytosis: mechanisms and functional significance. Traffic 7:109–116
15. Thorn P, Parker I (2005) Two phases of zymogen granule lifetime in mouse pancreas: Ghost granules linger after exocytosis of contents. J Physiol 563:433–442
16. Khandelwal P, Ruiz WG, Apodaca G (2010) Compensatory endocytosis in bladder umbrella cells occurs through an integrin-regulated and rhoa- and dynamin-dependent pathway. EMBO J 29:1961–1975
17. Masedunskas A, Sramkova M, Weigert R (2011) Homeostasis of the apical plasma membrane during regulated exocytosis in the salivary glands of live rodents. Bioarchitecture 1:225–229
18. Sramkova M, Masedunskas A, Parente L, Molinolo A, Weigert R (2009) Expression of plasmid DNA in the salivary gland epithelium: novel approaches to study dynamic cellular processes in live animals. Am J Physiol Cell Physiol 297:C1347–C1357
19. Jerdeva GV, Wu K, Yarber FA et al (2005) Actin and non-muscle myosin ii facilitate apical exocytosis of tear proteins in rabbit lacrimal acinar epithelial cells. J Cell Sci 118:4797–4812
20. Nightingale TD, Cutler DF, Cramer LP (2012) Actin coats and rings promote regulated exocytosis. Trends Cell Biol 22(6):329–337
21. Larina O, Bhat P, Pickett JA et al (2007) Dynamic regulation of the large exocytotic fusion pore in pancreatic acinar cells. Mol Biol Cell 18:3502–3511
22. Nemoto T, Kojima T, Oshima A, Bito H, Kasai H (2004) Stabilization of exocytosis by dynamic f-actin coating of zymogen granules in pancreatic acini. J Biol Chem 279:37544–37550
23. Gorr SU, Venkatesh SG, Darling DS (2005) Parotid secretory granules: crossroads of secretory pathways and protein storage. J Dent Res 84:500–509
24. Kasai H, Hatakeyama H, Kishimoto T, Liu TT, Nemoto T, Takahashi N (2005) A new quantitative (two-photon extracellular polar-tracer imaging-based quantification (tepiq)) analysis for diameters of exocytic vesicles and its application to mouse pancreatic islets. J Physiol 568: 891–903
25. Castle AM, Huang AY, Castle JD (2002) The minor regulated pathway, a rapid component of salivary secretion, may provide docking/ fusion sites for granule exocytosis at the apical surface of acinar cells. J Cell Sci 115: 2963–2973
26. Castle JD (1998) Protein secretion by rat parotid acinar cells. Pathways and regulation. Ann N Y Acad Sci 842:115–124
27. Chen Y, Warner JD, Yule DI, Giovannucci DR (2005) Spatiotemporal analysis of exocytosis in mouse parotid acinar cells. Am J Physiol Cell Physiol 289:C1209–C1219

28. Warner JD, Peters CG, Saunders R et al (2008) Visualizing form and function in organotypic slices of the adult mouse parotid gland. Am J Physiol Gastrointest Liver Physiol 295: G629–G640
29. Proctor GB, Carpenter GH (2007) Regulation of salivary gland function by autonomic nerves. Auton Neurosci 133:3–18
30. Behrendorff N, Dolai S, Hong W, Gaisano HY, Thorn P (2011) Vesicle-associated membrane protein 8 (vamp8) is a snare (soluble n-ethylmaleimide-sensitive factor attachment protein receptor) selectively required for sequential granule-to-granule fusion. J Biol Chem 286:29627–29634
31. Nemoto T, Kimura R, Ito K et al (2001) Sequential-replenishment mechanism of exocytosis in pancreatic acini. Nat Cell Biol 3: 253–258
32. Segawa A, Riva A (1996) Dynamics of salivary secretion studied by confocal laser and scanning electron microscopy. Eur J Morphol 34: 215–219
33. Segawa A, Terakawa S, Yamashina S, Hopkins CR (1991) Exocytosis in living salivary glands: direct visualization by video-enhanced microscopy and confocal laser microscopy. Eur J Cell Biol 54:322–330
34. Pickett JA, Thorn P, Edwardson JM (2005) The plasma membrane q-snare syntaxin 2 enters the zymogen granule membrane during exocytosis in the pancreatic acinar cell. J Biol Chem 280:1506–1511
35. Thorn P, Fogarty KE, Parker I (2004) Zymogen granule exocytosis is characterized by long fusion pore openings and preservation of vesicle lipid identity. Proc Natl Acad Sci U S A 101:6774–6779
36. Fernandez NA, Liang T, Gaisano HY (2011) Live pancreatic acinar imaging of exocytosis using syncollin–phluorin. Am J Physiol Cell Physiol 300:C1513–C1523
37. Masedunskas A, Weigert R (2008) Internalization of fluorescent dextrans in the submandibular salivary glands of live animals: a study combining intravital two-photon microscopy and second harmonic generation. SPIE, In, pp 68601V–68612V
38. Peter B, Van Waarde MA, Vissink A, s-Gravenmade EJ, Konings AW (1995) Degranulation of rat salivary glands following treatment with receptor-selective agonists. Clin Exp Pharmacol Physiol 22:330–336
39. Masedunskas A, Sramkova M, Parente L, Weigert R (2013) Intravital microscopy to image membrane traffickingin live rats Cell imaging techniques. Methods Mol Biol 931: 153–167
40. Masedunskas A, Weigert R (2008) Intravital two-photon microscopy for studying the uptake and trafficking of fluorescently conjugated molecules in live rodents. Traffic 9:1801–1810
41. Zipfel WR, Williams RM, Webb WW (2003) Nonlinear magic: multiphoton microscopy in the biosciences. Nat Biotechnol 21:1369–1377
42. Weigert R, Sramkova M, Parente L, Amornphimoltham P, Masedunskas A (2010) Intravital microscopy: a novel tool to study cell biology in living animals. Histochem Cell Biol 133:481–491
43. Svoboda K, Yasuda R (2006) Principles of two-photon excitation microscopy and its applications to neuroscience. Neuron 50:823–839
44. Cahalan MD, Parker I (2008) Choreography of cell motility and interaction dynamics imaged by two-photon microscopy in lymphoid organs. Annu Rev Immunol 26:585–626
45. Germain RN, Castellino F, Chieppa M et al (2005) An extended vision for dynamic high-resolution intravital immune imaging. Semin Immunol 17:431–441
46. Alexander S, Koehl GE, Hirschberg M, Geissler EK, Friedl P (2008) Dynamic imaging of cancer growth and invasion: a modified skin-fold chamber model. Histochem Cell Biol 130: 1147–1154
47. Amornphimoltham P, Masedunskas A, Weigert R (2011) Intravital microscopy as a tool to study drug delivery in preclinical studies. Adv Drug Deliv Rev 63:119–128
48. Ritsma L, Ponsioen B, van Rheenen J (2012) Intravital imaging of cell signaling in mice. Intravital 1:2–10
49. Dunn KW, Sandoval RM, Kelly KJ et al (2002) Functional studies of the kidney of living animals using multicolor two-photon microscopy. Am J Physiol Cell Physiol 283:C905–C916
50. Sandoval RM, Kennedy MD, Low PS, Molitoris BA (2004) Uptake and trafficking of fluorescent conjugates of folic acid in intact kidney determined using intravital two-photon microscopy. Am J Physiol Cell Physiol 287: C517–C526
51. Masedunskas A, Milberg O, Porat-Shliom N et al (2012) Intravital microscopy: a practical guide on imaging intracellular structures in live animals. Bioarchitecture 2
52. Masedunskas A, Porat-Shliom N, Weigert R (2012) Regulated exocytosis: novel insights from intravital microscopy. Traffic 13:627–634

53. Sramkova M, Masedunskas A, Weigert R (2012) Plasmid DNA is internalized from the apical plasma membrane of the salivary gland epithelium in live animals. Histochem Cell Biol 138(2):201–213
54. Hadjantonakis AK, Gertsenstein M, Ikawa M, Okabe M, Nagy A (1998) Generating green fluorescent mice by germline transmission of green fluorescent es cells. Mech Dev 76: 79–90
55. Muzumdar MD, Tasic B, Miyamichi K, Li L, Luo L (2007) A global double-fluorescent cre reporter mouse. Genesis 45:593–605
56. Bhat P, Thorn P (2009) Myosin 2 maintains an open exocytic fusion pore in secretory epithelial cells. Mol Biol Cell 20:1795–1803
57. Mazariegos MR, Tice LW, Hand AR (1984) Alteration of tight junctional permeability in the rat parotid gland after isoproterenol stimulation. J Cell Biol 98:1865–1877

Chapter 9

Rapid Analysis of Synaptic Vesicle Endocytosis in Synaptosomes

James A. Daniel and Phillip J. Robinson

Abstract

Synaptic vesicle endocytosis (SVE) is a critical mechanism by which synaptic transmission is regulated and maintained at presynaptic boutons. Small molecules that target SVE may prove to be useful not only for basic research into synaptic function but also in treating diseases that involve pathological neurotransmission. However, there are relatively few small molecules targeting SVE that have known molecular targets, and the development of these compounds has been a slow process. The lack of development on this front has been, in part, due to the laborious nature of traditional methods to study SVE, which are unsuited to the screening of compounds. This chapter describes a recently developed high-throughput approach to the analysis of SVE, which uses synaptosomes and high-content imaging. Rat forebrain synaptosomes are attached to glass microplates, and depolarization-induced SVE is measured by the uptake of the fluorescent styryl dye FM4-64. The synaptosomes may be used fresh or frozen and stored in liquid nitrogen, with the latter option greatly expanding the utility of the assay, by reducing the animal numbers required and allowing experiments to be done "on demand." The assay provides a rapid approach to screening small molecules for effects on SVE that will hopefully see an increase in the number of available small molecules that can target this important process at synapses.

Key words High-throughput screening, Endocytosis, Neurons, Synaptic, Vesicles, Styryl dyes, Synaptosomes

1 Introduction

1.1 Synaptic Vesicle Endocytosis

Intercellular communication in the nervous system occurs largely at specialized secretory structures called synapses. A typical synapse is comprised of a presynaptic bouton and a postsynaptic spine. The arrival of an action potential depolarizes the presynaptic bouton, which triggers voltage-dependent calcium influx and the release, by exocytosis, of neurotransmitter from synaptic vesicles (SVs). The neurotransmitter diffuses across the synaptic cleft that separates the presynaptic and postsynaptic structures, activating receptors on the postsynaptic spine. In this way, the electrical signal propagated along the axon is transformed into a chemical signal that can be passed on to the postsynaptic neuron. In the intact organism it is

Peter Thorn (ed.), *Exocytosis Methods*, Neuromethods, vol. 83,
DOI 10.1007/978-1-62703-676-4_9, © Springer Science+Business Media New York 2014

rare that a neuron fires a single action potential in isolation. More typically, a neuron fires action potentials repeatedly, resulting in multiple rounds of depolarization and hence the fusion of many SVs.

This creates at least three potential problems for the neuron: the fusion of many SVs results in the deposition of a large amount of membrane to the plasma membrane, which may cause an expansion and deformation of the presynaptic bouton; the fusion of these SVs also results in the deposition of SV proteins into the plasma membrane, which may interfere with subsequent fusion events and neuronal function; and assuming that the pool of SVs available for fusion is finite, with repeated activity the presynaptic bouton will eventually run out of SVs, causing a loss of neurotransmission.

In the intact organism, however, these scenarios do not seem to occur, even in systems that have extremely high rates of firing and SV fusion, such as mossy fiber synapses in the cerebellum (over 200 Hz) [1] and inner ear hair cells (which can fire at over 1,000 Hz and can replenish released SVs at a rate of 1,400 SV/s) [2]. This is because neurons possess the capacity to regenerate SVs locally, adjacent to the site(s) of SV fusion [3]. This process is referred to as synaptic vesicle endocytosis (SVE). Pharmacological or genetic blocks of SVE result in defects such as the three described above [4–7], the functional outcome of which is a failure of synaptic transmission. This failure is activity dependent [4, 7], suggesting that the loss of synaptic transmission is due to the gradual depletion of releasable SVs and, moreover, demonstrating the importance of SVE in maintaining synaptic transmission.

1.2 Small Molecule Development

Given its central role in maintaining neurotransmission, SVE has been of interest to researchers for the last several decades. The proteins that facilitate and regulate SVE have been extensively studied [8–10], generally using genetic strategies of perturbing protein function in cells and examining the effects on SVE (e.g., see [11, 12]). While this has led to a great increase in our understanding of the molecular mechanisms by which SVE is regulated, genetic approaches have significant limitations. In particular, within the machinery-regulating SVE, there appears to be considerable redundancy, in that when a single protein involved in SVE is genetically removed the functional outcome may often be minimal, presumably because of compensation within the system or the existence of multiple pathways that regulate endocytosis. This makes determining the function of single proteins by genetic approaches frequently problematic.

An alternative approach to genetic "knockout" or "knockdown" methods is the use of small molecules that target a protein involved in SVE. These compounds are, ideally, cell-permeable, reversible compounds that act in an acute manner to specifically inhibit the function of a known target protein. An ideal compound is also capable of being used in animal models, thereby allowing the same molecular tool to be employed in the full range of studies all the way from in vitro biochemistry (binding studies, enzymatic assays) through to in vivo research (disease models, behavioral studies). Ultimately, small molecules have the potential to be used as drugs for the acute treatment of diseases, which provides a significant clinical advantage for this approach compared to genetic approaches. Given that SVE is highly important for maintaining and shaping synaptic transmission, it seems like an ideal target for small molecule development, with scope for use in basic synaptic research and the treatment of neurological diseases. However, there are currently very few small molecules that specifically target SVE. The available inhibitors that have specific, identified molecular targets act on either dynamin [13–19] or clathrin [20], both of which are essential proteins in SVE. The lack of small molecule SVE inhibitors lies in part with the technical difficulty associated with development of the inhibitors. The approach to small molecule development is typically to begin with a large number of compounds with potential efficacy for the protein of interest and screen these compounds in vitro for an effect on the protein. In the case of dynamin, which is a GTPase, it is sufficient to screen a compounds library for inhibition or stimulation of GTPase activity in vitro. However, to screen the resultant compounds for a cellular effect, a cell-based assay must be developed. Screening is a laborious process, due to the large number of compounds that must be tested, so assays must be as simple and rapid as possible.

1.3 Traditional Approaches to the Study of SVE and the Need for High-Throughput Approaches

While a range of cell-based methods have been developed to study SVE, these are laborious and time intensive, rendering them unsuitable for small molecule screening. The approaches can be broken broadly into three categories, described below.

1.3.1 Membrane Capacitance Measurements

The electrical capacitance of the cell membrane is a function of its surface area. Thus, in neurons with very large presynaptic boutons and an abundance of SVs, exocytosis and endocytosis at synapses can be detected directly as an increase or decrease in membrane capacitance, respectively [21]. This is generally performed using a glass microelectrode in a cell-attached recording configuration. While this technique is powerful, it is only possible to record from very large presynaptic structures, such as the neuromuscular junction or inner ear hair cells in the periphery or some large, atypical structures in the brain, such as the Calyx of Held in the auditory brainstem. This is because most presynaptic boutons are smaller than the recording electrode tip. Capacitance is performed in acute preparations of tissue taken from animals. Cells are selected one at a time and recordings performed from a few individual terminals per day, at best. Thus, capacitance measurements do not provide a representative sample of SVE from a large number or diversity of presynaptic terminals, and these approaches are unsuitable for high-throughput screening of small molecules.

1.3.2 Externally Applied Fluorophores

The majority of SVE assays rely on the uptake by endocytosis of an externally applied fluorophore, which is diluted in the extracellular buffer and applied to neurons. Some are taken into vesicles as they undergo endocytosis, such as fluid-phase fluorescent compounds [22, 23] and fluorescent quantum dots [24, 25]. These methods can also be used to specifically study different modes of SVE by varying the size of the compound/bead to which the fluorophore is attached, based on the reasoning that smaller endocytic structures, such as single SVs, have relatively small fission pores and thus will exclude the entry of large fluorophore conjugates, such as high-molecular weight dextran. The styryl dyes, most notably FM1-43, bind to the plasma membrane directly and label the vesicles as they bud off from the plasma membrane and undergo internalization [26, 27]. Fluorescently labelled antibodies, directed against the lumenal portion of a SV protein, such as synaptophysin, can also be externally applied [28]. The antibodies bind to the proteins after fusion of SVs with the plasma membrane and are then taken up during SVE.

These protocols generally allow the study of single synaptic terminals by fluorescence microscopy, although it is possible to utilize FM dyes with synaptosomes in suspension using a cuvette-based fluorimeter [18, 29–31]. These methods generally utilize cultured neurons, although FM1-43 has been used to study SVE in tissue preparations [32–35], which are not suited to high-throughput applications, as discussed above. FM1-43 has been extensively used to study SVE in primary neuronal culture [27, 36–39]. Cultured neurons are prepared from neonatal tissue and require approximately 2 weeks to reach synaptic maturity before they can be used for experiments. With fluorescence microscopy, it is only possible to image around 10–40 synapses per microscopic field in real time. This places a limit of the number of synapses that can be studied per experiment. Each experiment takes, on average, an hour to complete and it is often difficult to find an appropriate microscopic field to image, due to cellular autofluorescence (primarily in the green/yellow emission spectrum) and accumulation of dye in cell bodies/dendrites which does not represent specific SVE (particularly using FM1-43). The equipment used is often relatively complex, requiring perfusion systems, imaging chambers, temperature regulation, and isolated current sources for electrical stimulation. The complex morphology of cultured neurons makes automated imaging approaches, such as those used in high-content screening, inappropriate, since many of the images acquired will either contain no synapses or will also contain extraneous labelled structures such as neuronal cell bodies. The fact that synapses must be manually selected and imaged over many hours to complete even a simple series of experiments means that these methods are unsuited to the rapid screening of small molecules.

1.3.3 Genetically Encoded Fluorescent Reporters of SVE

To date, all genetically encoded methods for studying SVE have been based around a pH-sensitive mutant of GFP, called pHluorin [40]. Generally, this approach requires pHluorin to be expressed as a fusion protein in cultured neurons. Transgenic mice have also been developed that express pHluorin fusion proteins in neurons [41]. The pHluorin protein is fused to the lumenal domain of a SV protein, such as VGLUT1 [42], synaptobrevin (referred to as synaptopHluorin) [43], or synaptophysin (referred to as SypHy) [44], targeting the protein to SVs and facilitating the detection of SVE events. pHluorin fluorescence is sensitive to pH, with maximal fluorescence at pH = 7.4. At the relatively acidic pH of the SV lumen, the fluorescence of a pHluorin fusion protein is suppressed. Upon fusion of the SV with the plasma membrane, the pHluorin construct is exposed to the extracellular buffer, which has a pH of 7.3–7.4. This causes a sudden increase in fluorescence, which can be used a measure of exocytosis. When the SV undergoes SVE, the vesicle begins to re-acidify and thus the fluorescence of the pHluorin protein declines. Assuming a constant rate of SV re-acidification, the rate of decline of fluorescence is an indirect measurement of the rate of SVE.

While pHluorin constructs have facilitated the study of SVE without using exogenous fluorophores, these methods once again require brain slices or, more commonly, cultured neurons that have been transfected or virally transduced to express the construct. These methods require molecular biology techniques and cell culture. Once transfection/transduction has been carried out, the cells will often only express the transgene transiently. The cells then require live cell microscopy to be carried out, which carries the same limitations as outlined for exogenous fluorophores listed above.

1.4 A High-Content Screening Approach to SVE

Outside of the field of SVE research, there has been a recent increase in the development of small molecule inhibitors with known protein targets. This has been facilitated by improvements in technologies that allow high-throughput in vitro assays, including cell-based assays. Liquid handling, imaging, and data analysis are increasingly able to be automated, greatly increasing the throughput of research and allowing techniques that were previously limited by time and labor factors. The use of high-throughput enzymatic assays has led to a boom in the number of available small molecules that inhibit endocytosis and cellular trafficking, although in most cases the molecular targets of these drugs are not known. Two exceptions are a number of drugs that specifically inhibit dynamin, a GTPase, and mechanoprotein involved in membrane scission, and the pitstop™ compounds, which inhibit clathrin, a structural protein. Until recently it was not possible to rapidly screen compounds for efficacy as SVE inhibitors at presynaptic boutons. This gap in available technology led to the development of the first SVE assay based on a high-content imaging platform. This chapter outlines this screening method and its application in assessing the efficacy of potential modulators of SVE.

1.5 Synaptosomes as a Model for Studying SVE

Synaptosomes are isolated presynaptic boutons, prepared from homogenized brain tissue. They have been used to study presynaptic function for over 50 years [45, 46], proving highly important in understanding the mechanisms that regulate neurotransmitter secretion. For the isolation of synaptosomes from rat brain, a crude preparation of homogenized cerebral tissue is separated on a discontinuous Percoll gradient by centrifugation, as described in detail previously [47]. Studies have demonstrated that synaptosomes retain physiological characteristics of intact synaptic boutons in the brain, including maintenance of a membrane potential, functional ion channels, respiratory capacity, and, most importantly, response to depolarizing stimuli, allowing the study of exocytosis and SVE [47–52].

For the high-content screening platform for quantifying SVE, we use isolated forebrain synaptosomes from adult rats. The synaptosomes are attached to a 96-well plate, depolarized and SVE is monitored by the uptake of FM4-64 (Fig. 1). Synaptosomes were selected for development of the high content-based SVE assay because they allow the analysis of SVE without interference from many of the experimental bottlenecks previously mentioned, such as the contribution of extraneous fluorescence by cell bodies and the need for cell culture. In addition, synaptosomes that have been fluorescently labelled appear as isolated fluorescent spots, much like the fluorescent beads that are often used to calibrate microscopes. These synaptosome images are highly suited to automated focusing and imaging, which is essential when using a high-content imaging system. The quantification of this data is also simple—since the vast majority of fluorescently labelled objects labelled in the sample are synaptosomes, it is a simple matter of quantifying the total fluorescence of each image.

Another highly advantageous aspect of synaptosomes is the fact that they retain viability after being frozen, stored in liquid nitrogen, and thawed [53, 54]. The methods for freezing and thawing are broadly similar to those used for the long-term storage of mammalian cells. We have stored frozen synaptosomes for up to 6 months with no significant loss of viability or functionality. This property of synaptosomes removes the need to prepare fresh samples whenever an experiment needs to be performed and allows the assay to be done "on demand." When fresh samples are prepared, it is often difficult to use all of the synaptosomes in a single set of experiments because of the large amount of synaptosomes

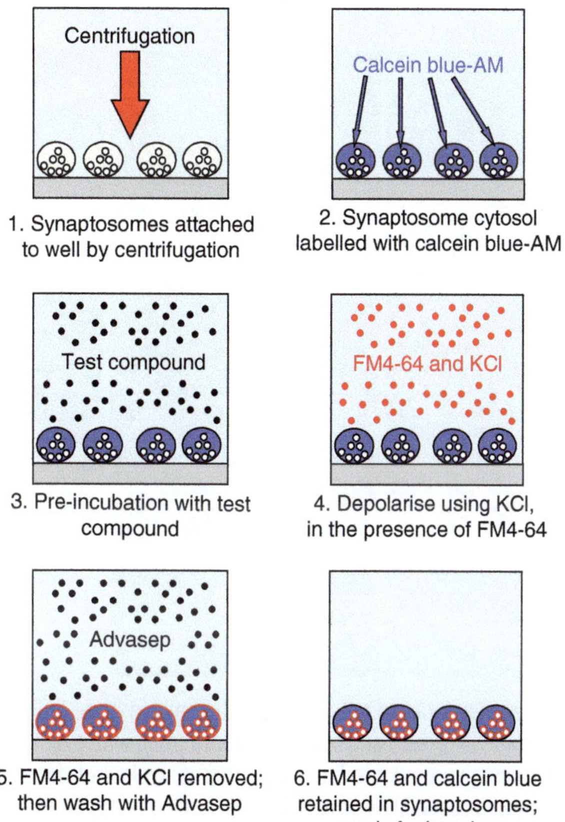

Fig. 1 Summary of the SVE assay. (*1*) Fresh or freeze/thawed synaptosomes in iso-osmotic sucrose are attached by centrifugation to the glass surface of a PEI-coated microplate well. (*2*) Synaptosomes are transferred to isotonic HBK containing the cell-permeable fluorophore, calcein blue-AM. This penetrates the cell, whereupon the fluorophore is trapped in the cytoplasm by cleavage of the AM ester, thus fluorescently labelling synaptosomes. (*3*) Synaptosomes are incubated in the compound of interest, typically a small drug-like molecule. (*4*) Synaptosomes are depolarized by the addition of 40 mM KCl to the well in the presence of FM4-64. FM4-64 labels the plasma membrane while depolarization stimulates SVE, which results in the formation of new SVs from the plasma membrane. These newly formed SVs are also labelled with FM4-64. (*5*) FM4-64 in the plasma membrane is removed by washing the synaptosomes in HBK containing Advasep. (*6*) The synaptosomes contain FM4-64-labelled SVs, the fluorescence of which can be visualized using a high-content screening system in order to quantify SVE

obtained from even a single rat brain. The ability to freeze synaptosomes greatly expands the efficiency of the assay in terms of animals required. We estimate that synaptosomes from a single rat brain can be used to screen up to 160 different compounds (assuming a single concentration of each compound is used, as is typical in a primary screen of a compound library).

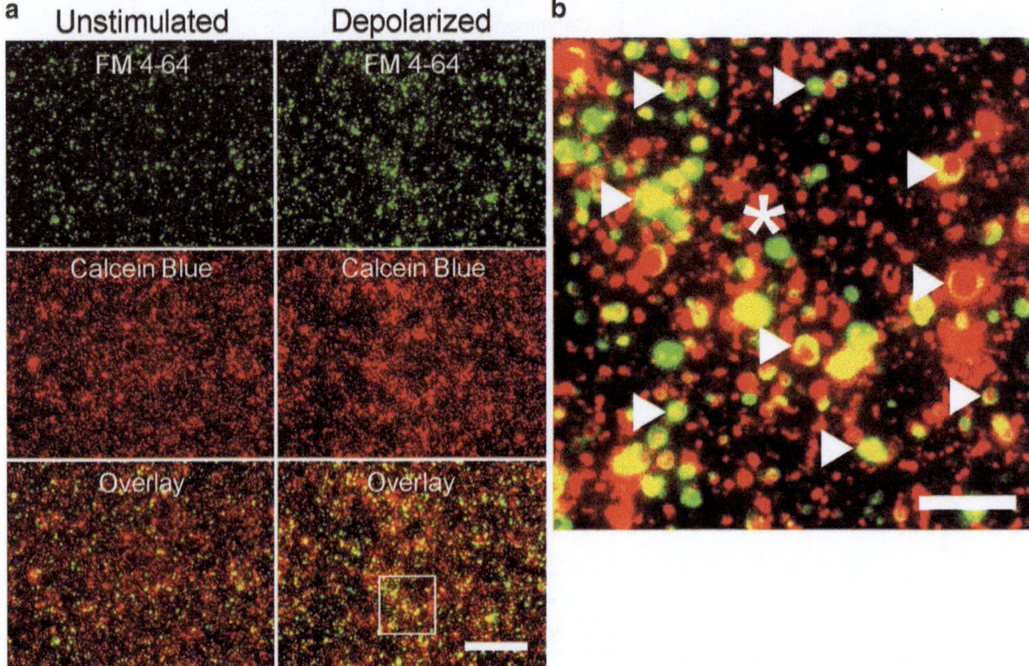

Fig. 2 Synaptosomes labelled with FM4-64 and calcein blue. (**a**) Images show synaptosomes that have been labelled with FM4-64 (*green*) and calcein blue-AM (*red*), with overlays of the fluorescence channels also shown. Both fluorophores give punctate labelling patterns, typical of fluorescently labelled synaptosomes, which are small, round structures. While some FM4-64 labelling is apparent in unstimulated synaptosomes (*left panels*), depolarization increases the amount of FM4-64 labelling markedly (*right panels*), without any apparent change in calcein blue labelling. (**b**) When the *inset* region of an overlaid image is viewed in greater detail, it is apparent that most of the FM4-64-labelled structures are also filled with calcein blue (*yellow*). This suggests that they are viable synaptosomes (*arrowheads*). Note the presence of many structures that are completely filled with calcein blue, but not FM4-64, which forms a halo or crescent moon shape, presumably due to the presence of large mitochondria in many synaptosomes. Some large FM-positive structures are apparent that do not have calcein blue labelling (*asterisk*). These may be nonviable membranous structures such as myelin particles, which often contaminate synaptosome preparations. Abundant calcein blue-positive puncta are also present that do not co-localize for FM4-64, suggestive of viable cellular debris that is incapable of SVE, which can also contaminate synaptosomal preparations. Scale bars = 40 μm in (**a**) and 10 μm in (**b**)

1.6 FM Dyes to Label SVE

The SVE assay uses uptake of the styryl dye FM4-64 to measure endocytosis (Fig. 2). Styryl dyes are derived from the parent compound FM1-43 [55]. The dyes possess a fluorescent head group and a hydrophobic tail. The tail facilitates binding to plasma membranes, which greatly increases the fluorescence of the molecule, such that the dyes have very low fluorescence when unbound. The assay is also effective using the styryl dye FM1-43, although the relatively broad yellow/green fluorescence of FM1-43 overlaps with many standard red fluorophores. By contrast, FM4-64 fluorescence emission can be used in combination with green fluorophores such as FITC, expanding the assay's versatility. For example, FITC-conjugated transferrin or dextran uptake may also be simultaneously monitored in synaptosomes labelled with FM4-64.

2 Materials

2.1 Reagents

Reagents can be obtained from major chemical manufacturers, except where noted.

Animals: Adult Sprague-Dawley rats are used in this protocol and must be used in accordance with the applicable animal ethics guidelines.

Reagents for preparing synaptosomes: See Dunkley et al. [47].

Sucrose/EDTA/Tris Buffer (SET): Contains 0.32 M sucrose, 1 mM EDTA, 5 mM Tris, pH 7.4. To prepare, dissolve 109.5 g sucrose, 372 mg EDTA·$2H_2O$, and 606 mg Tris in 1,000 ml of water. Adjust pH to 7.4 with HCl, filter sterilize, and store for up to 6 months at 4 °C.

10 % DMSO in SET: Add 9 ml of cold SET to 1 ml of tissue culture-grade DMSO and mix. Use immediately as this stock cannot be stored.

250 mM DTT: Dissolve 385.6 mg of DTT in 10 ml of SET. Store for up to 6 months at −20 °C in aliquots. Defrost only once and do not refreeze.

HEPES/Tris Buffer: Contains 87 mM HEPES. To prepare, dissolve 10.5 g of HEPES in 500 ml of water and adjust the pH to 7.4 by adding drops of saturated Tris base. Saturated tris base is prepared by adding 40 g of Tris to 100 ml water. Filter HEPES/Tris and store up to 6 months at 4 °C.

HEPES-Buffered Krebs-Like Buffer (HBK): The final composition of this buffer is 143 mM NaCl, 4.7 mM KCl, 1.3 mM $MgSO_4$, 1.2 mM $CaCl_2$, 20 mM HEPES, 0.1 mM NaH_2PO_4, 10 mM D-glucose, and pH 7.4. Begin by preparing stock solutions: 308 mM NaCl (9 g NaCl dissolved in 500 ml), 308 mM KCl (11.5 g in 500 ml), 154 mM $MgSO_4 \cdot 7H_2O$ (1.9 g in 50 ml), 1 M CaCl (7.36 g in 50 ml), and 100 mM Na_2HPO_4 (0.7 g in 50 ml). In a clean measuring cylinder, add stocks in the following order: 121 ml NaCl (308 mM), 4 ml KCl (308 mM), 2 ml $MgSO_4 \cdot 7H_2O$, 312 µl $CaCl_2$ (1 M), 60 ml HEPES/Tris (87 mM), pH adjusted to 7.4, and finally 260 µl Na_2HPO_4 (100 mM). Make up to 260 ml with water and add 0.48 g D(+)-glucose. Dissolve the glucose, filter, and store for up to 1 month at 4 °C.

1:15 PEI stock: Draw up 3.3 ml of 50 % PEI stock solution (Sigma-Aldrich cat. no. P3143) in a serological pipette. Dispense into a 50 ml tube containing 46.7 ml of water. The viscous PEI needs to be repeatedly pipetted back and forth to dissolve it in the water and wash all of the PEI out of the pipette barrel. To assist in this process, sit the pipette tip immersed in the tube for 5 min at a time and then repeat the pipetting. The 1:15 stock solution can be stored at 4 °C for up to a year.

10 mM Advasep-4: Chemical name: β-CD-sulfobutyl ether 4 (Advasep-4, CyDex cat. no. AR-0A4-005). Prepare the stock by dissolving 200.5 mg of Advasep-4 in 10 ml of HBK. Store indefinitely at −20 °C. Note that while Advasep-4 is only available from this supplier, Advasep-7 is a comparable, more widely available cyclodextrin compound.

5 mM Calcein blue-AM: Add 429 μl DMSO to vial containing 1 mg of calcein blue-AM to dissolve. The stock is stored, protected from light, in aliquots at −20 °C for up to 12 months. Can be frozen and thawed multiple times. Calcein blue-AM is available from Invitrogen (cat no. C-1429).

1 mM FM4-64: Reconstitute one 100 μg vial of FM4-64 in 167 μl of water. Store at −20 °C, protected from light, for up to 1 month. FM4-64 can degrade over time once reconstituted, so do not store for more than 1 month. FM4-64 can only be purchased from Invitrogen (cat. no. T-13320).

2.2 Equipment

- Equipment for preparing synaptosomes, as described in Dunkley et al. [47].
- Software for quantification of images (e.g., MetaXpress v.3.1 and Acuity v.2.0 from Molecular Devices).
- High-content screening system (e.g., ImageXpress Micro Widefield HCS System, Molecular Devices) with 20× fluor objective lens and fluorescence filters suitable for imaging calcein blue and FM4-64. For FM4-64, a custom filter set is ideal, e.g., Semrock filters: Ex FF01 500/24-25, Em FF01 675/67-25, and dichroic FF560 Di01-25 × 36.
- Gas tank containing carbogen and fit with a gas dispersion tube (e.g., Kimble Chase cat. no. 956500-0023).
- Glass bottom 96-well microplates (e.g., Matrical cat. no. MGB096-1-2-LG).
- Heating block for microplate (e.g., Thermomixer R, Eppendorf cat. no. 002670158, plus exchangeable thermoblock for MTPs and deep well plates, Eppendorf cat. no. 022670565).
- Manual 200 μl multichannel pipette (e.g., Pipetman Neo Multichannel, Gilson cat. no. F14404) and an electronic

multichannel pipette (e.g., Ovation 5–250 μl multichannel pipette, Vistalab cat. no. 1060-0250).

- Wide orifice pipette tips for dispensing synaptosomes, 250 and 1,000 μl volumes (e.g., Finntip wide orifice tips, Thermo Scientific cat. no. 9405-020 for 250 μl and 9405-050 for 1,000 μl).

3 Methods

3.1 Preparation of Synaptosomes

3.1.1 Method of Preparation

In the work described here, we used the method of Dunkley et al. [47] for preparation of rat cortical synaptosomes. This method is faster than other techniques and yields a sample that is highly enriched for viable synaptosomes. Briefly, whole forebrain (cerebrum) from adult male rats is homogenized in iso-osmotic sucrose and centrifuged at low speed to remove cellular debris. The supernatant containing the synaptosomes is centrifuged through a discontinuous gradient of Percoll. The gradient separates the homogenate into fractions based on density, with fractions 3 and 4 being the most highly enriched for synaptosomes. These fractions are pooled and washed twice by centrifugation to remove residual Percoll.

3.1.2 Freezing Synaptosomes

While fresh synaptosomes can be used in the assay, one key advantage of the method is that it allows the use of synaptosomes that have been frozen and stored in liquid nitrogen. Synaptosomes are washed as described in step 10 of Dunkley et al. [47]. The synaptosomes are then resuspended in 25 ml of the same sucrose-based buffer (SET) used throughout the earlier stages of the protocol. The suspension is centrifuged again for 10 min at 20,000 × *g*. This pellet is resuspended in SET such that the total protein content is approximately 1–2 mg/ml. Three hundred microliter aliquots of this suspension are dispensed and each is mixed with an equal volume of SET containing 10 % dimethyl sulfoxide (DMSO). The aliquots are then transferred to a polystyrene container and placed at −80 °C overnight for slow freezing. The container itself is constructed from the tube rack in a pack of 15 ml BD Falcon tubes (BD cat. no. 352099, www.bdbiosciences.com). A 2 cm thick sheet of polystyrene is then used as a lid on the top of the rack and the container sealed with tape. Aliquots are transferred to liquid nitrogen the next day for long-term storage.

3.2 Preparation of Glass Microplates

The preparation of plates is essential for good attachment and imaging of synaptosomes in the SVE assay. After testing various attachment substrates, we found that polyethyleneimine (PEI, 1:15,000 dilution from a 50 % stock solution) was the most effective, ensuring the attachment of individual synaptosomes to the glass bottom of microplates for imaging. The coating of glass with PEI and attachment of synaptosomes by centrifugation was originally reported by the David Nicholls laboratory [56]. Plates are coated with 100 μl of synaptosome suspension in each well of a 96-well plate and incubating overnight at 37 °C. The PEI is then removed and the plates are dried without washing. The coated plates remain stable for a week at room temperature or several weeks if stored at 4 °C.

3.3 Thawing and Attachment of Synaptosomes

Synaptosomes are thawed for 2 min in a 37 °C water bath, then placed on ice and 1 ml of ice-cold SET is added to each aliquot. Residual DMSO is then removed by centrifugation for 2 min in a microfuge (13,000 rpm). The supernatant is discarded and the synaptosomes resuspended in 600 μl of SET containing 250 mM DTT. A protein estimate should be carried out at this point using whichever method is most convenient. The synaptosomes are then diluted in SET+DTT to a final concentration of 20 μg/ml. One hundred microliters of this suspension is added to each well of a pre-coated 96-well microplate. Synaptosomes are then attached to the plate by centrifugation at low speed (1,500 × *g*) for 30 min.

Deviation from the resuspension and attachment protocols outlined here can result in synaptosomal clumping. Synaptosomes tend to stay well separated in the absence of salt, which is why iso-osmotic SET buffer is used for freezing, resuspension, and attachment. The use of an isotonic buffer, with relatively high salt content, can cause synaptosomes to clump together, leading to poor imaging data.

3.4 Labelling of Synaptosomes with Calcein Blue-AM

All subsequent steps of the protocol should be carried out at 30 °C. Physiological buffers should be infused before use with an oxygen-rich gas mix, such as carbogen, to ensure that synaptosomes have adequate oxygen for respiration. Multichannel pipettes are ideal for liquid handling across multiple wells. Buffer addition/removal is performed from left to right across the plate. Electronic pipettes are ideal for injection of buffers as they minimize the repetitive strain of the liquid handling in the assay; however, manual pipettes allow finer control of the aspiration rate when removing buffer from the plate, which can help to reduce disturbance of the synaptosome layer.

Many compounds can cause synaptosomes or cultured cells to detach from glass culture surfaces. This typically only happens at relatively high concentrations of compounds (>30 μM), although some can cause an almost total detachment of synaptosomes at 10–20 μM. This is problematic, since a decreased number of synaptosomes will obviously decrease the number that can take up FM4-64, resulting in lower FM4-64 fluorescence overall, which may be falsely interpreted as inhibition of SVE. In order to prevent these false negative observations, a cytosolic fluorophore should be applied to synaptosomes prior to treatment with any compounds that are to be screened. Calcein blue-AM, which is membrane permeable due to the acetoxymethyl ester (AM) group on the molecule, is ideal for cytosolic labelling in this assay. In the cytosol, the AM group is removed by cytosolic esterases, trapping the fluorescent calcein blue in the synaptosome. When imaging, calcein blue labelling appears as small fluorescent puncta (Fig. 2). The retention of calcein blue also shows that synaptosomes that have been frozen and then thawed have retained intact membranes. Other fluorescent cell labels, such as carboxyfluorescein succinimidyl ester (CFSE), can also be used in this assay instead of calcein blue-AM.

To monitor synaptosome attachment and viability with calcein blue-AM, SET buffer is removed from the plate after centrifugation and replaced with HBK containing 1 μM calcein blue-AM. The synaptosomes are then incubated at 30 °C for 30 min to allow internalization of the dye and cleavage of the AM ester, leaving the cell-impermeable free calcein blue trapped in the cytoplasm.

3.5 Pharmacological Treatment

This high-content method was originally developed to screen relatively large numbers of small molecules for efficacy as inhibitors of SVE. It is at this step in the protocol, immediately prior to the SVE assay itself, when pharmacological interventions should be performed. Small molecules or cell-permeable peptides may be used. Incubation times will vary between compounds, and for screening purposes we typically use a preincubation time of 30 min in the presence of test compounds before conducting the SVE assay. Compounds are diluted in HBK. If compounds are dissolved in a solvent such as ethanol or DMSO, do not exceed 1 % solvent concentration in the final assay. The buffer volume used for this incubation should be 9× the volume that is to be used for FM4-64 injection (see below).

Control samples are treated with vehicle alone, such as DMSO, to eliminate the possibility of nonspecific effects due to the effect of the vehicle. Two kinds of control samples are typically included in screening studies. Unstimulated control samples are treated with vehicle and FM4-64 dye but are not depolarized. These samples give an indication of the baseline (unstimulated) level of FM4-64 labelling in the sample. In our experience, unstimulated uptake can be inhibited by treatment with the dynamin inhibitor Dyngo™-4a, suggesting that baseline SVE is at least partly dynamin dependent. These unstimulated controls are included to ensure that stimulation mediates a significant increase in FM dye labelling. This is done by comparing unstimulated controls with stimulated controls. Stimulated control samples are treated with vehicle alone and depolarized, such that they represent the "untreated" level of stimulated SVE. Deviation from the stimulated control represents a stimulation or inhibition of SVE. We typically normalize drug-treated samples against the stimulated control samples, such that each sample measurement is scaled to represent a proportion of the control value (known as control-mean scaling). This scales the control samples such that they have a mean value of "1," as shown in Fig. 3. Treatment with an inhibitor of SVE mediates a dose-dependent decrease in FM4-64

a

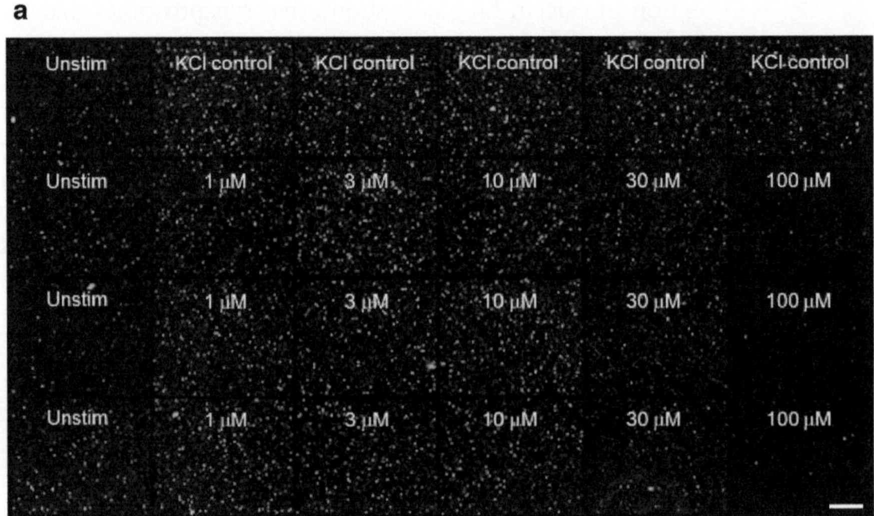

b

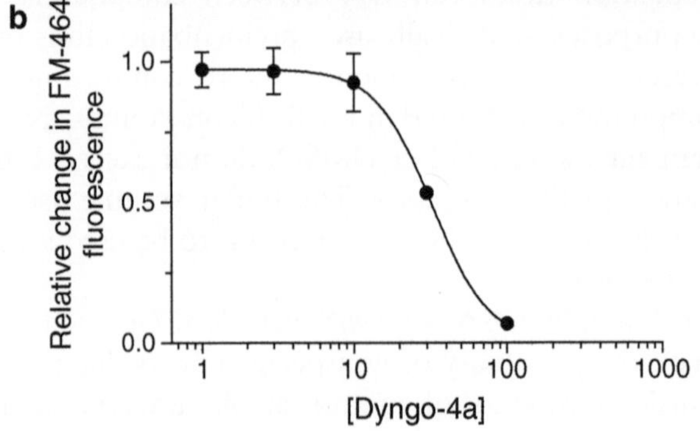

Fig. 3 Dose-dependent inhibition of FM4-64 labelling by the dynamin inhibitor, Dyngo™-4a. (**a**) A montage showing images from control and Dyngo-4a-treated synaptosomes, acquired using a high-content screening system. Synaptosomes were pretreated for 30 min either with 1 % DMSO (control wells) or the dynamin inhibitor, Dyngo-4a, at various concentrations (as shown). The *left column* of images is from unstimulated control samples, which were not depolarized. Wells marked with KCl control were depolarized by the injection of 40 mM KCl. The other samples were pretreated with the indicated concentration of Dyngo-4a and repeated in triplicate, as shown. The images show the relatively low level of FM4-64 labelling evident in unstimulated samples and also demonstrate the dose-dependent decrease in FM4-64 labelling in Dyngo-4a-treated samples compared to KCl control samples. Scale bar = 80 μm. (**b**) Fluorescence images were quantified and the results are summarized. The fluorescence intensity from Dyngo-4a-treated samples was normalized against depolarized, DMSO-treated control samples, thereby expressing the effects of Dyngo-4a treatment as a relative change in FM4-64 fluorescence. Dyngo-4a treatment results in a dose-dependent decrease in FM4-64 fluorescence, with a classical sigmoidal dose-response curve. Data represents three independent experiments. *Error bars* represent SEM

labelling (Fig. 3a). Figure 3b shows a dose-dependent decrease in FM4-64 labelling in the images acquired from this experiment using a high-content screening system.

It should also be noted that many compounds exhibit fluorescence at various wavelengths. Dose-dependent increases in synaptosomal fluorescence should be treated with caution since this could be the result of either a potentiation of FM4-64 uptake (increased SVE) or fluorescence of the compound itself. It is advisable in these instances to acquire images from synaptosomes that have been treated with the compound by not subjected to FM4-64 labelling, as this will allow the fluorescence of the compound alone to be visualized.

3.6 Depolarization and Uptake of FM4-64

Synaptosomes respond to changes in membrane potentials, as do synaptic boutons in the intact brain and can be depolarized by electrical or chemical means. Elevated K^+ is typically used in studies to induce neurotransmitter release, although 4-aminopyridine (4-AP) is also an effective depolarizing agent [57]. When used at relatively low concentrations (0.1–3 mM), 4-AP is considered to be a milder form of stimulation than elevated K^+. Synaptosomes also respond to a number of non-depolarizing secretogogues such as the calcium ionophore ionomycin. Depolarization triggers SV exocytosis through the influx Ca^{2+}, which binds to synaptotagmin and triggers vesicle fusion. Depolarization also triggers SVE (Fig. 2a), presumably to compensate for the loss of SVs induced by exocytosis. The specific mechanism linking depolarization and endocytosis is not well understood but may involve the SV protein synaptotagmin [58].

The depolarization step constitutes the SVE assay itself. Two stocks of concentrated FM4-64 dye are prepared, both at an FM4-64 concentration of 10 μM—one stock in water, which serves as a negative control, and one stock in KCl solution, typically a 400 mM KCl solution. To depolarize the synaptosomes, 1 volume of concentrated FM4-64 stock, prepared as above, is injected into a well, which contains 9 volumes of HBK. For example, for a buffer volume of 81 μl of HBK per well, add 9 μl of FM4-64 (in 400 mM KCl), giving a final volume of 90 μl and a final FM4-64 concentration of 1 μM and a final K^+ concentration of 44.7 mM. Depolarization is allowed to continue for the desired time (typically we use 2 min) before the buffer is removed and replaced with fresh HBK.

The concentration of KCl determines the strength of depolarization, so users may wish to vary the KCl concentration injected. Depolarization for 2 min across a range of KCl concentrations is shown in Fig. 4a. KCl additions above 40 mM do not mediate further increases in FM4-64 labelling. This is thought to be due to "clamping" of the voltage-gated calcium channels by the strong,

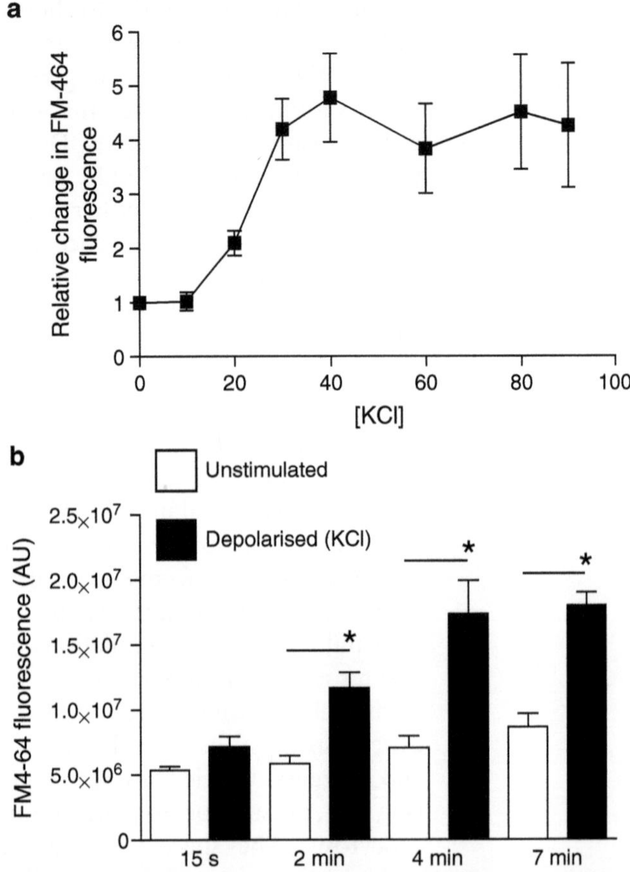

Fig. 4 SVE is dependent on the strength and time of depolarization. (**a**) Synaptosomes were depolarized for 2 min by the addition of KCl, in the presence of 1 μM FM4-64. Synaptosomes exhibited a dose-dependent increase in FM4-64 labelling with increasing (K^+), achieving maximal uptake of the dye at 40 mM KCl. (**b**) Synaptosomes were incubated with 1 μM FM4-64 and either left unstimulated (*white bars*) or depolarized by the addition of 40 mM KCl (*black bars*) for different periods of time. After 15 s, no significant difference in FM4-64 labelling was apparent between unstimulated and depolarized synaptosomes. However, when the incubation time was increased, depolarization resulted in a significant increase in FM4-64 labelling due to stimulation of SVE. Maximal FM4-64 labelling was apparent at 5 min, and further incubation did not result in further increase in FM4-64 fluorescence. *Asterisk* denotes $P<0.05$ according to an unpaired two-tailed *t*-test. FM4-64 fluorescence is expressed as arbitrary fluorescence units (AU). Experimental conditions for all data: nine images acquired per well, three wells per condition (i.e., triplicate wells), $n=3$ independent experiments. *Error bars* represent SEM

constant depolarization induced by KCl. Time of depolarization is also an important consideration. The amount of SVE is also dependent on the time of depolarization, as illustrated in Fig. 4b. After a very brief depolarization (15 s), FM4-64 uptake has not yet risen to a level that is significantly above the basal level of unstimulated synaptosomes. Labelling becomes stronger with longer depolarization times, reaching saturation at 5 min (with 40 mM KCl injection), at which point further time of depolarization does not increase labelling. Saturation presumably occurs because all SVs available for release have undergone recycling. Two minutes of depolarization gives submaximal FM4-64 labelling (Fig. 4b), allowing the observation of whether compounds being tested have either stimulatory or inhibitory effects on SVE.

Likewise, users may wish to examine different concentrations of FM4-64 in their own hands. An FM4-64 concentration should be used such that a depolarization-induced increase in FM4-64 labelling is easily detected. The rationale is that at very low concentrations of dye, the signal will be so faint that no difference will be detectable between the unstimulated control and the depolarized sample. At very high dye concentrations, the extraneous labelling of the plasma membrane and spontaneously recycling SVs will be so strong that the system will become saturated, once again diminishing the sensitivity of the system. An FM4-64 concentration of between 1 and 3 μM is optimal to detect stimulation-induced SVE (2 min depolarization with 40 mM KCl).

3.7 Washing

During labelling, FM4-64 binds to the plasma membrane. Any membrane that is internalized through SVE will remain labelled as SVs in the cytosol; however, there will also be a large amount of extraneous labelling of the plasma membrane, which will mask the fluorescence signal from the SVs unless it is removed. Replacing the buffer with dye-free HBK alone is insufficient to remove the extraneous plasma membrane FM4-64. We use Advasep-4, a cell-impermeable cyclodextrin that is able to extract styryl dyes from the plasma membrane. Advasep-7 is structurally similar and used more commonly; however, Advasep-4 is generally less expensive and equally effective.

To wash extraneous FM4-64 from the plasma membrane, 9 μl of 10 mM Advasep-4 is injected into each well (final volume = 90 μl per well). The contents of the well are mixed, either manually or using an orbital plate shaker (400 rpm) for 2 min. The Advasep-4/HBK is then removed and replaced with standard HBK. The plate is now ready for imaging and should be protected from light sources. The fluorescence is stable in synaptosomes for at least 2 h, but imaging immediately after the assay is optimal.

3.8 Imaging

One advantage of using a high-content system for this assay is that it allows automated acquisition of a large number of images. This increases the assay efficiency and reduces the potential operator bias. The assay has been developed using a Molecular Devices ImageXpress Micro, and the imaging conditions will need to be optimized depending on the system being used by each lab.

It is important to obtain a representative measure of fluorescent labelling from each well. We typically take nine equally spaced images per well at a magnification of 20×, resulting in 864 images across a 96-well microplate. Images close to the edges of the wells are not optimal, since these areas are more susceptible to disturbance through liquid handling. Acquisition of an entire plate (FM4-64 and calcein blue) takes approximately 45 min.

3.9 Quantitation of Fluorescence Data

Analysis of data, even when automated, is likely to be the most time-consuming step of the assay and time should be allocated accordingly.

Each image is first segmented by image thresholding to identify the structures of interest, in this case synaptosomes. Thresholding parameters are set within the analytical software, defining objects on the basis of their diameter and fluorescence intensity. The software then uses each fluorescence images to create a binary (black and white) image. This is automatically performed for all images across the plate. We use Molecular Devices MetaXpress software; however, many software packages can perform this kind of simple image segmentation. We typically set the object size limits at 1–4 μm and find that varying the size limits (e.g., to 1–2 μm) has no appreciable effect on the results. The fluorescence intensity limit that should be placed on the thresholding is determined by the fluorescence of the original images and so must be determined by the researcher.

After thresholding, quantitation of the binary image must be carried out. A number of different parameters can be calculated from the images, with the researcher having to choose between quantifying fluorescence on a per image basis (e.g., average number of objects per image, total fluorescence intensity per image) or a per object basis (e.g., average fluorescence intensity per object). Measuring the total fluorescence intensity per image is suitable, since almost all FM4-64-positive objects in the images are presumed to be synaptosomes.

3.10 Troubleshooting

Typically, problems in the assay will only be apparent at the imaging stage. Some common problems and ways to address them are listed below:

3.10.1 Faint FM4-64 Fluorescence

If the FM4-64 fluorescence appears faint, first ensure that the stock solution was reconstituted from the powder stock less than a month ago, since FM4-64 solutions are only stable for up to 1 month in the freezer. It is also possible that the FM4-64 concentration being used is too low. We find that a final assay concentration of 1 μM is optimal; however, concentrations up to 10 μM are routinely used in cell-based studies and should be tested if the fluorescence is too faint. Finally, it is possible that the samples have faded due to photobleaching. Ensure that the plates are protected from light immediately after assay completion and imaged as soon as possible thereafter.

3.10.2 Faint Calcein Blue Fluorescence

For optimal labelling, working calcein blue solutions (1 μM) should be used immediately. It is important to protect the samples from exposure to light once calcein blue labelling is completed. This is done by wrapping each plate in foil.

3.10.3 Depolarization Does Not Increase FM4-64 Labelling

If no difference can be detected between the FM4-64 labelling in the unstimulated and stimulated control samples, this suggests that the depolarization has not been effective. The time of depolarization and/or the strength of the stimulus (e.g., the concentration of KCl used to depolarize the synaptosomes) should be increased. We find that with an injection of 40 mM KCl, FM4-64 labelling is maximal after 5 min (Fig. 4b). Alternatively, it is possible that the Advasep wash has not been successful, in which case the wash time and/or Advasep concentration should be increased. Finally, it is possible that the synaptosomes are unresponsive to stimulation, in which case they are not viable. Calcein blue can also be used to monitor synaptosomal viability. If calcein blue labelling is weak and the synaptosomes are unresponsive to stimulation, it may be best to prepare new synaptosomes and rerun the assay.

3.10.4 Synaptosomes Appear as Clumps Rather Than Single Puncta

Clumped synaptosomes suggest a problem in either the PEI coating on the glass microplates or, more likely, the synaptosomes were not resuspended in SET. Synaptosomes should appear evenly distributed and punctate as in Fig. 2. Ensure that the synaptosomes are not transferred to physiological, isotonic buffer until after they have been attached to the glass by centrifugation.

3.10.5 Synaptosome Layer Is Not Uniform

If the synaptosomes are sparse, then it is possible that more synaptosomes need to be loaded per well. An increase in synaptosomal content to 3 μg per well is advisable. Also, ensure that the PEI-coated plates are not more than a week old if they have been stored at room temperature, since PEI will degrade over time. If the sparsely coated wells have been subjected to small molecule treatment, it is possible that the compound itself has induced synaptosome detachment from the glass. If the synaptosome layer is "patchy," with some images containing many more synaptosomes than others, this is likely due to mechanical disruption of some synaptosomes by liquid handling. Ensure that liquid handling is performed gently, injecting buffers down the sides of the wells rather than directly onto the glass surface. Also ensure that images are acquired from closer to the center of each well rather than the edges, which are subject to more mechanical disturbances.

4 Conclusions

This protocol describes a versatile high-throughput SVE assay utilizing synaptosomes. The use of styryl dyes to monitor SVE has become well established since their development almost 20 years ago [26]. The high-throughput SVE assay was based on established methods that utilize the styryl dye FM4-64. This assay has several advantages over traditional approaches for the purposes of screening compounds for SVE effects. This first is the assay speed. SVE is normally studied in cultured neurons, which require 1–3 weeks to mature. SVE is then indirectly examined at individual presynaptic terminals by live cell microscopy, either using styryl dye or by transfection with a genetically encoded reporter, such as synaptopHluorin. Fluorescence measurements are taken over time at single terminals across several experiments, and there is massive interference from neuronal cell bodies such that only well-isolated axons can be studied. This approach is unsuitable for screening compound libraries for SVE effects. However, using synaptosomes it is possible to screen multiple compounds at multiple concentrations in a single day. The 96-well layout makes liquid handling across large numbers of replicate samples possible. Semiautomated imaging and analysis with the HCS system further streamlines the assay.

The application of SVE for screening compounds as SVE inhibitors is demonstrated in this chapter using the dynamin inhibitor Dyngo-4a. However, the assay also has broader applicability to examining the molecular mechanisms that regulate SVE. Cell-permeable peptides or antibodies can modulate SVE [7, 23]. These peptides/antibodies could be screened using the SVE assay, allowing researchers to rapidly examine the role of specific proteins in SVE. This high-throughput approach to studying SVE will be useful both in studying the molecular machinery of SVE and the further development of compounds that modulate SVE.

A high-throughput SVE assay system was recently developed for use with cultured neurons [59]. The neurons are infected with adeno-associated virus (AAV) expressing the SVE reporter SypHy, electrically stimulated and imaged using a microlens array. While this assay system is powerful, it still requires a number of more laborious steps, including neuron culture and generation of concentrated AAV stocks. It also requires the use of a specific high-throughput screening platform, called the MANTRA system, which may not be available to the majority of researchers. However, the MANTRA system is accessible to genetic manipulations, meaning that genome screening using this system is a possibility that is not currently available using a synaptosome-based approach. The availability of both cell- and synaptosome-based screening platforms suggests that the future is bright for the development of synaptically active small molecules.

References

1. Saviane C, Silver RA (2006) Fast vesicle reloading and a large pool sustain high bandwidth transmission at a central synapse. Nature 439:983–987
2. Moser T, Beutner D (2000) Kinetics of exocytosis and endocytosis at the cochlear inner hair cell afferent synapse of the mouse. Proc Natl Acad Sci U S A 97:883–888
3. Smith SM, Renden R, Von Gersdorff H (2008) Synaptic vesicle endocytosis: fast and slow modes of membrane retrieval. Trends Neurosci 31:559–568
4. Newton AJ, Kirchhausen T, Murthy VN (2006) Inhibition of dynamin completely blocks compensatory synaptic vesicle endocytosis. Proc Natl Acad Sci U S A 103:17955–17960
5. Ceccarelli B, Hurlbut WP (1980) Ca2+-dependent recycling of synaptic vesicles at the frog neuromuscular junction. J Cell Biol 87:297–303
6. Watanabe O, Meldolesi J (1983) The effects of alpha-latrotoxin of black widow spider venom on synaptosome ultrastructure. A morphometric analysis correlating its effects on transmitter release. J Neurocytol 12:517–531
7. Anggono V, Smillie KJ, Graham ME, Valova VA, Cousin MA, Robinson PJ (2006) Syndapin I is the phosphorylation-regulated dynamin I partner in synaptic vesicle endocytosis. Nat Neurosci 9:752–760
8. Cousin MA, Robinson PJ (2001) The dephosphins: dephosphorylation by calcineurin triggers synaptic vesicle endocytosis. Trends Neurosci 24:659–665
9. Sudhof TC (2004) The synaptic vesicle cycle. Annu Rev Neurosci 27:509–547
10. McMahon HT, Boucrot E (2011) Molecular mechanism and physiological functions of clathrin-mediated endocytosis. Nat Rev Mol Cell Biol 12:517–533
11. Milosevic I, Giovedi S, Lou X, Raimondi A, Collesi C, Shen H, Paradise S, O'Toole E, Ferguson S, Cremona O, De Camilli P (2011) Recruitment of endophilin to clathrin-coated pit necks is required for efficient vesicle uncoating after fission. Neuron 72:587–601
12. Raimondi A, Ferguson SM, Lou X, Armbruster M, Paradise S, Giovedi S, Messa M, Kono N, Takasaki J, Cappello V, O'Toole E, Ryan TA, De Camilli P (2011) Overlapping role of dynamin isoforms in synaptic vesicle endocytosis. Neuron 70:1100–1114
13. Harper CB, Martin S, Nguyen TH, Daniels SJ, Lavidis NA, Popoff MR, Hadzic G, Mariana A, Chau N, McCluskey A, Robinson PJ, Meunier FA (2011) Dynamin inhibition blocks botulinum neurotoxin type-A endocytosis in neurons and delays botulism. J Biol Chem 286: 35966–35976
14. Hill TA, Mariana A, Gordon CP, Odell LR, McGeachie AB, Chau N, Daniel JA, Gorgani NN, Robinson PJ, McCluskey A (2010) Iminochromene inhibitors of dynamin I & II GTPase activity and endocytosis. J Med Chem 53:4094–4102
15. Hill TA, Gordon CP, McGeachie AB, Venn-Brown B, Odell LR, Chau N, Quan A, Mariana A, Sakoff JA, Chircop M, Robinson PJ, McCluskey A (2009) Inhibition of dynamin mediated endocytosis by the dynoles—synthesis and functional activity of a family of indoles. J Med Chem 52:3762–3773
16. Hill TA, Odell LR, Quan A, Abagyan R, Ferguson G, Robinson PJ, McCluskey A (2004) Long chain amines and long chain ammonium salts as novel inhibitors of dynamin GTPase activity. Bioorg Med Chem Lett 14:3275–3278
17. Odell LR, Howan D, Gordon CP, Robertson MJ, Chau N, Mariana A, Whiting AE, Abagyan R, Daniel JA, Gorgani NN, Robinson PJ, McCluskey A (2010) The pthaladyns: GTP competitive inhibitors of dynamin I and II GTPase derived from virtual screening. J Med Chem 53:5267–5280
18. Quan A, McGeachie AB, Keating DJ, van Dam EM, Rusak J, Chau N, Malladi CS, Chen C, McCluskey A, Cousin MA, Robinson PJ (2007) MiTMAB is a surface-active dynamin inhibitor that blocks endocytosis mediated by dynamin I or dynamin II. Mol Pharmacol 72:1425–1439
19. Robertson MJ, Hadzic G, Ambrus J, Hyde E, Yuri Pomè D, Chau N, Whiting AE, Robinson PJ, McCluskey A (2012) The Rhodadyn™ series, a new class of small molecule inhibitors of dynamin GTPase activity. ACS Med Chem Lett 3:352–356
20. von Kleist L, Haucke V (2011) At the crossroads of chemistry and cell biology: inhibiting membrane traffic by small molecules. Traffic 13:495–504
21. Li W, Zhang Y, Bouley R, Chen Y, Matsuzaki T, Nunes P, Hasler U, Brown D, Lu HA (2011) Simvastatin enhances aquaporin-2 surface expression and urinary concentration in vasopressin-deficient Brattleboro rats through modulation of Rho GTPase. Am J Physiol Renal Physiol 301:F309–F318
22. Clayton EL, Cousin MA (2009) Quantitative monitoring of activity-dependent bulk endocytosis

of synaptic vesicle membrane by fluorescent dextran imaging. J Neurosci Methods 185:76–81

23. Clayton EL, Anggono V, Smillie KJ, Chau N, Robinson PJ, Cousin MA (2009) The phospho-dependent dynamin-syndapin interaction triggers activity-dependent bulk endocytosis of synaptic vesicles. J Neurosci 29:7706–7717
24. Zhang Q, Cao YQ, Tsien RW (2007) Quantum dots provide an optical signal specific to full collapse fusion of synaptic vesicles. Proc Natl Acad Sci U S A 104:17843–17848
25. Zhang Q, Li Y, Tsien RW (2009) The dynamic control of kiss-and-run and vesicular reuse probed with single nanoparticles. Science 323:1448–1453
26. Betz WJ, Bewick GS (1992) Optical analysis of synaptic vesicle recycling at the frog neuromuscular junction. Science 255:200–203
27. Ryan TA, Reuter H, Wendland B, Schweizer FE, Tsien RW, Smith SJ (1993) The kinetics of synaptic vesicle recycling measured at single presynaptic boutons. Neuron 11:713–724
28. Hua Y, Sinha R, Thiel CS, Schmidt R, Huve J, Martens H, Hell SW, Egner A, Klingauf J (2011) A readily retrievable pool of synaptic vesicles. Nat Neurosci 14:833–839
29. Cousin MA, Robinson PJ (2000) Two mechanisms of synaptic vesicle recycling in rat brain nerve terminals. J Neurochem 75:1645–1653
30. Cousin MA, Robinson PJ (1998) Ba^{2+} does not support synaptic vesicle retrieval in rat isolated presynaptic nerve terminals. Neurosci Lett 253:1–4
31. Cousin MA, Robinson PJ (2000) Ca^{2+} inhibition of dynamin arrests synaptic vesicle recycling at the active zone. J Neurosci 20:949–957
32. Betz WJ, Ridge RM, Bewick GS (1993) Comparison of FM1-43 staining patterns and electrophysiological measures of transmitter release at the frog neuromuscular junction. J Physiol Paris 87:193–202
33. Betz WJ, Bewick GS (1993) Optical monitoring of transmitter release and synaptic vesicle recycling at the frog neuromuscular junction. J Physiol (Lond) 460:287–309
34. Rizzoli SO, Richards DA, Betz WJ (2003) Monitoring synaptic vesicle recycling in frog motor nerve terminals with FM dyes. J Neurocytol 32:539–549
35. Kay AR, Alfonso A, Alford S, Cline HT, Holgado AM, Sakmann B, Snitsarev VA, Stricker TP, Takahashi M, Wu LG (1999) Imaging synaptic activity in intact brain and slices with FM1-43 in C. elegans, lamprey, and rat. Neuron 24:809–817
36. Harata N, Ryan TA, Smith SJ, Buchanan J, Tsien RW (2001) Visualizing recycling synaptic vesicles in hippocampal neurons by FM 1-43 photoconversion. Proc Natl Acad Sci U S A 98:12748–12753
37. Ryan TA, Reuter H, Smith SJ (1997) Optical detection of a quantal presynaptic membrane turnover. Nature 388:478–482
38. Anggono V, Cousin MA, Robinson PJ (2008) Styryl dye-based synaptic vesicle recycling assay in cultured cerebellar granule neurons. Methods Mol Biol 457:333–345
39. Cousin MA, Nicholls DG (1997) Synaptic vesicle recycling in cultured cerebellar granule cells: role of vesicular acidification and refilling. J Neurochem 69:1927–1935
40. Sankaranarayanan S, De Angelis D, Rothman JE, Ryan TA (2000) The use of pHluorins for optical measurements of presynaptic activity. Biophys J 79:2199–2208
41. Li Z, Burrone J, Tyler WJ, Hartman KN, Albeanu DF, Murthy VN (2005) Synaptic vesicle recycling studied in transgenic mice expressing synaptopHluorin. Proc Natl Acad Sci U S A 102:6131–6136
42. Voglmaier SM, Kam K, Yang H, Fortin DL, Hua Z, Nicoll RA, Edwards RH (2006) Distinct endocytic pathways control the rate and extent of synaptic vesicle protein recycling. Neuron 51:71–84
43. Miesenbock G, De Angelis DA, Rothman JE (1998) Visualizing secretion and synaptic transmission with pH-sensitive green fluorescent proteins. Nature 394:192–195
44. Granseth B, Odermatt B, Royle SJ, Lagnado L (2006) Clathrin-mediated endocytosis is the dominant mechanism of vesicle retrieval at hippocampal synapses. Neuron 51:773–786
45. Gray EG, Whittaker VP (1962) Isolation of nerve endings from brain: an electron microscopic study of the cell fragments derived by homogenization and centrifugation. J Anat 96:79–88
46. De Robertis E, Pellegrino de Iraldi A, Rodriquez de Lores Arniaz G, Salganicoff L (1962) Cholinergic and noncholinergic nerve endings in rat brain. I. Isolation and subcellular distribution of acetylcholine and acetylcholinesterase. J Neurochem 9:23–35
47. Dunkley PR, Jarvie PA, Robinson PJ (2008) A rapid percoll gradient procedure for preparation of synaptosomes. Nat Protoc 3:1718–1728
48. Whittaker VP (1993) Thirty years of synaptosome research. J Neurocytol 22:735–742

49. Erecinska M, Nelson D, Silver IA (1996) Metabolic and energetic properties of isolated nerve ending particles (synaptosomes). Biochim Biophys Acta 1277:13–34
50. Nicholls DG (2003) Bioenergetics and transmitter release in the isolated nerve terminal. Neurochem Res 28:1433–1441
51. Raiteri L, Raiteri M (2000) Synaptosomes still viable after 25 years of superfusion. Neurochem Res 25:1265–1274
52. Ghijsen WE, Leenders AG, Lopes Da Silva FH (2003) Regulation of vesicle traffic and neurotransmitter release in isolated nerve terminals. Neurochem Res 28:1443–1452
53. Smillie KJ, Cousin MA (2012) Akt/PKB controls the activity-dependent bulk endocytosis of synaptic vesicles. Traffic 13:1004–1011
54. Nichols RA, Wu WCS, Haycock JW, Greengard P (1989) Introduction of impermeant molecules into synaptosomes using freeze/thaw permeabilization. J Neurochem 52:521–529
55. Cochilla AJ, Angleson JK, Betz WJ (1999) Monitoring secretory membrane with FM1-43 fluorescence. Annu Rev Neurosci 22:1–10
56. Choi SW, Gerencser AA, Nicholls DG (2009) Bioenergetic analysis of isolated cerebrocortical nerve terminals on a microgram scale: spare respiratory capacity and stochastic mitochondrial failure. J Neurochem 109:1179–1191
57. Tibbs GR, Barrie AP, Van Mieghem FJE, McMahon HT, Nicholls DG (1989) Repetitive action potentials in isolated nerve terminals in the presence of 4-aminopyridine: Effects on cytosolic free Ca^{2+} and glutamate release. J Neurochem 53:1693–1699
58. Yao J, Kwon SE, Gaffaney JD, Dunning FM, Chapman ER (2012) Uncoupling the roles of synaptotagmin I during endo- and exocytosis of synaptic vesicles. Nat Neurosci 15:243–249
59. Hempel CM, Sivula M, Levenson JM, Rose DM, Li B, Sirianni AC, Xia E, Ryan TA, Gerber DJ, Cottrell JR (2011) A system for performing high throughput assays of synaptic function. PLoS ONE 6:e25999

Chapter 10

Isolated Neurohypophysial Terminals: Model for Depolarization–Secretion Coupling

José R. Lemos, James McNally, Edward Custer, Adolfo Cuadra, Hector Marrero, and Dixon Woodbury

Abstract

Here we discuss the useful properties of a preparation of isolated neurosecretory nerve terminals obtained from mammalian neurohypophyses (posterior pituitaries). These nerve terminals release the two neuropeptides, oxytocin and vasopressin, which are easily assayed by radioimmunoassay and/or ELISA. Depolarization-induced exocytosis is dependent on the same parameters as those regulating release from the whole, intact neurohypophysis. The isolated nerve terminals can be identified and also imaged for intracellular Ca^{2+} and a functional synapse can be reconstituted using their purified neurosecretory granules. Furthermore, some nerve terminals are large enough to allow the use of patch-clamp techniques and the monitoring of exocytosis by capacitance and amperometric measurements. We also present evidence which show that the isolated neurohypophysial nerve terminals represent a powerful tool for studying the mechanisms underlying depolarization–secretion coupling (D-SC).

Key words Neurosecretion, Nerve terminals, Depolarization–secretion coupling, Release, Exocytosis, Capacitance, Amperometry, Bilayers, Reconstitution, Patch-clamp, Imaging, Calcium, Receptors, Channels, Identification, ELISA, Posterior pituitary

1 Introduction

The neurohypophysis consists almost entirely of neurosecretory nerve terminals and glial cells (pituicytes). The nerve terminals, whose cell bodies (magnocellular neurons) are located in the supraoptic and paraventricular nuclei of the hypothalamus [1], are much more numerous than the pituicytes. Each axon emerging from a hypothalamic magnocellular neuron gives rise to an average of 2,000 nerve terminals. The hypothalamic-neurohypophysial system (HNS) mostly secretes two neurohormones, vasopressin (AVP) and oxytocin (OT). Both are packaged into neurosecretory granules and exocytosed from the neurohypophysis where they can easily be detected by specific radioimmunoassays and/or by ELISAs in the perfusate of isolated nerve terminals [2]. As discussed below,

Peter Thorn (ed.), *Exocytosis Methods*, Neuromethods, vol. 83,
DOI 10.1007/978-1-62703-676-4_10, © Springer Science+Business Media New York 2014

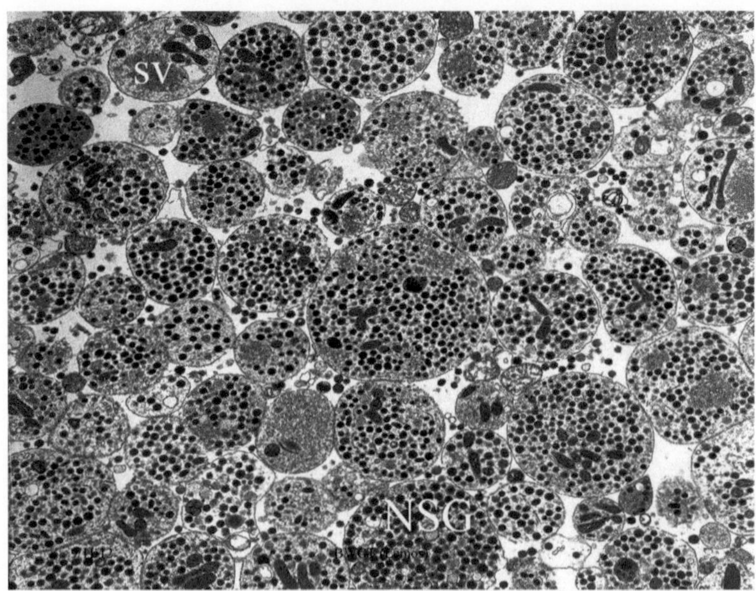

Fig. 1 EM of isolated NH terminals (NHT). Electron micrograph of isolated NHT after centrifugation (see 2.1.1). The NHTs contain only mitochondria, neurosecretory granules (NSG), and micro- or synaptic vesicles (SV). Largest NHT at center is ~5 μm in diameter (Courtesy of JJ Nordmann)

the preparation of isolated neurohypophysial (NH) terminals is extremely simple and, in contrast to classical brain synaptosomes, is more homogeneous, especially in regard to the products that they release.

The integrity and purity of the isolated terminals are essential for this preparation. Fortunately, in the intact NH, blood vessels, glia, and other cells are clearly distinguishable from the terminals. Electron microscopy (Fig. 1 and ref. [3]) and immunocytochemistry [2] have been used to characterize this preparation, ensuring that only nerve terminals are isolated with minimal contamination from the other components. Numerous studies have confirmed that the physiological functioning of isolated terminals is the same as that of the intact NH [4–7]. While most nerve terminals have a diameter of about 1 μm, some are large enough (>5 μm) to allow the use of the patch-clamp technique [8, 9]. In addition, we can confirm by immunoblotting [10] or specific ELISAs [11] after patch-clamp recordings, whether the terminal is vasopressinergic or oxytocinergic, allowing for characterization of the distinct physiological properties between these two terminal subclasses.

Here we summarize the methods that can be utilized with this preparation and data obtained with this preparation over the last few years [1, 7, 11–17]. We hope to demonstrate that this isolated nerve terminal preparation represents an effective tool (but *see* **Note 1a**) for studying the coupling between electrical and neurosecretory activities in these neurohypophysial terminals, as well as providing essential clues about mechanisms behind neurotransmission in general [1].

2 Materials and Methods

2.1 Preparation of HNS Terminals

2.1.1 Isolation of NH Terminals

Neurohypophyses can be obtained from various types of mammals: e.g., mice, rats, and cows. We have found that the *pars intermedia*, which adheres to the neurohypophysis, partially inhibits the release of AVP and OT [18], so we usually remove it before preparing the terminals. The neurohypophyses are rinsed with normal saline and, before homogenization, are dipped in the homogenization medium containing 270 mM sucrose, 10 mM HEPES, pH 7.0, and 0.01 mM EGTA. The control of the ionized calcium concentration in the homogenization medium is of extreme importance for the maintenance of their secretory mechanism (*see* **Note 1b**).

For hormone release assays (see below), after gentle homogenization (without the use of any enzymes), the sample is centrifuged at 600 × *g* for 5 min in order to eliminate any cell debris (pituicytes, capillaries, red blood cells) although for patch-clamping, we do not have to centrifuge [9]. We found that centrifugation at a lower speed (300 × *g*) will also eliminate the cell debris but that this will allow the larger nerve terminals to remain in the supernatant. The supernatant is then centrifuged at 3,400 × *g* for 20 min and the resulting pellet consists almost entirely of nerve terminals full of peptidergic neurosecretory granules (Fig. 1), as judged from their positive labeling with anti-neurophysin antibodies [2].

For electrophysiology (see below), the dissociated neurohypophysis is plated into a 35 mm polystyrene cell culture dish (Corning, Corning, NY). After a waiting period of 2–3 min in order for the terminals to settle down to the bottom of the culture dish, the preparation is perfused (at room temperature) with a low free Ca^{2+} Locke's saline solution, containing (in mM) NaCl 140; KCl 5; EGTA 2; $CaCl_2$ 1.9 (3 μM free Ca^{2+}); HEPES 10; glucose 10; $MgCl_2$ 1.2; and pH 7.25, 298–302 mOsmol/L, to insure removal of any floating debris. This low calcium saline solution is then slowly replaced with a higher free Ca^{2+} (2.2 mM) Normal Locke's solution following the initial rinse.

For fast agonist application in *electrophysiology* experiments, loosely plated terminals may be preferred. Loose plating allows the picking up of terminals, with the patch pipette, in order to positioning them before the orifice of a super-fusion apparatus. The following procedure promotes loose plating of terminals to the cell culture dish. Isolated terminals from the dissociated neurohypophysis, suspended in homogenizing buffer, are deposited onto the center of a culture dish and allowed to settle for 2–3 min. Then low-Ca^{2+} Locke's saline solution is slowly added to the droplet containing the terminals (about 100 μL added directly to it), followed by adding additional low-Ca^{2+} Locke's to the outer edges of the dish. This is continued so that the droplet is absorbed with the media added to the surrounding edge of the dish, which should take an additional 2–4 min. The low-calcium solution is then replaced, by perfusion, with a higher free Ca^{2+} (2.2 mM) Normal Locke's solution (perfusion should flow at approximately 2 mL/min). Terminals selected for patching should demonstrate movement upon super-fusion application of media while loosely fixed to the dish. This procedure promotes loose plating of the terminals and allows them to be lifted from the dish after obtaining a GΩ seal for electrophysiology and/or suctioning into pipette for identification (see below).

For fluorometric imaging, a secure attachment of the terminals to the substrate is preferred. This can be achieved by plating onto either glass coverslips, polystyrene, or glass bottom culture dishes and enhanced with polylysine (50 μg/mL, Sigma Aldrich, St. Louis, MO) coating their surface. In this procedure the suspension containing the dissociated neurohypophysis is mixed with an equal volume of Normal Locke's buffer. Prior to plating, a drop of Normal Locke's solution is added to the dish and the terminal containing suspension is deposited over the droplet of Normal Locke's. After 2–3 min of allowing the suspension to settle, Normal Locke's is slowly added to the edges of the dish until the media absorbs the droplet containing the suspension. After 30–60 min the terminals can be subject to imaging.

2.1.2 Identification

We have developed [11] a method that can reliably determine the AVP and/or OT content of individual rat neurohypophysial terminals (NHTs) ≥5 μm in diameter, the size used for electrophysiological recordings. A specific and sensitive Enzyme Linked ImmunoAssay (ELISA: Assay Designs, Inc., Ann Arbor, MI) is used to determine the content of AVP and/or OT in individual terminals isolated and collected as described below. Assay sensitivity, which is the smallest amount of measurable AVP or OT that is reliably not zero, is determined in assay buffer supplied by the manufacturer. This is accomplished by running replicates ($n = 8$) of different AVP or OT concentrations and comparing those to the same number of replicates containing no hormone. The sensitivity of the AVP and OT ELISA's in assay buffer is 0.25 and 1.0 pg per well, respectively, which equates to 2.5 pg/mL of AVP and 10 pg/mL of OT. To determine if terminal isolation, collection, storage, or preparation conditions affects the sensitivity of the AVP or OT ELISA, standard curves are generated in Normal Locke's solution, pipette solution, and 0.1 and 0.5 % Triton X-100, and standards are prepared, frozen, and stored overnight. None of the above conditions affects the AVP or OT standard curves so all standard curves for AVP or OT are prepared in the assay buffer supplied by the manufacturer of the ELISA.

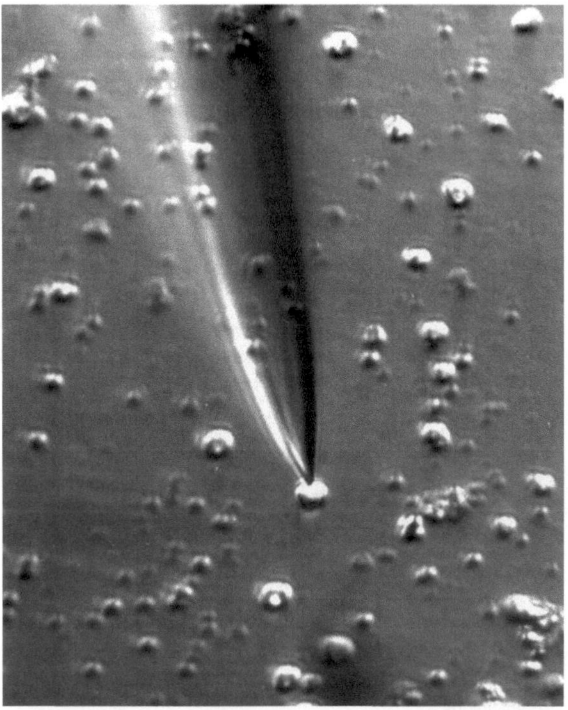

Fig. 2 Patch-clamping isolated NH terminals. An approximately 5 μm diameter terminal is sealed to a recording pipet. Under both phase contrast and Hoffman optics (seen here), NH terminals are readily distinguishable from cells of the *pars intermedia* and from pituicytes by their spherical shape and lack of nuclei, as well as their small size. RBCs are readily identified due to their concavity

Using an inverted microscope the NHTs are visually identified by their characteristic appearance, spherical shape, lack of nuclei, and relatively small size (5–12 μm in diameter). The size of each NHT is determined using a calibrated ocular micrometer (±0.25 μm). Thin borosilicate glass pipettes (Drummond Scientific Co., Broomall, PA) with a resistance of 3–6 MΩ containing Normal Locke's solution are placed against each individual NHT until a GΩ seal resistance is obtained (see Fig. 2). The content of each NHT is aspirated into a borosilicate glass pipette with the total volume less than 10 μL. Once the terminal is aspirated, the tip of the glass pipette is broken inside the microcentrifuge tube and immediately frozen at −20 °C until the NHT is processed and assayed for AVP and/or OT content [10].

The detergent Triton X-100 (0.5 %) at a volume of 120 μL is added to each microcentrifuge tube containing individual NHT in order to lyse the terminal and release AVP and/or OT into the assay buffer. When individual terminals are assayed for AVP or OT alone, the entire volume of the 0.5 % Triton X-100 buffer is assayed. When individual NHTs are assayed for both AVP and OT, the 0.5 % Triton X-100 buffer is divided (100 μL was assayed for OT and 20 μL is assayed for AVP) based on the greater sensitivity of the AVP-ELISA compared to the OT ELISA. A similar percentage of AVP-positive terminals ≥5 μm in diameter were identified immunocytochemically (73 %) and assaying individual NHT with a specific AVP-ELISA (72 %) when NHTs were assayed for AVP alone, whereas a higher percentage of OT-positive terminals ≥5 μm in diameter were identified immunocytochemically (43 %) compared to assaying individual NHT with a specific OT ELISA (26 %). Furthermore, the percentage of AVP-positive (60 %) and OT-positive (18 %) NHTs decreased when NHTs were assayed for both AVP and OT. Therefore, the best method to reliably identify AVP-positive NHTs is to assay the entire content of individual NHTs only for AVP, since this allows for the conclusion that AVP-negative terminals would contain only OT [11]. More recently, GFP-AVP and RFP-OT transgenic rats have become available [19].

2.2 Methods Applicable to HNS Terminals

2.2.1 Electrophysiology

Patch-clamp methods have already been applied to a number of secretory systems, such as chromaffin cells [20, 21], acinar cells [22], and pituitary tumor cells [23], for both macroscopic current and single channel recordings. We were the first to patch-clamp isolated nerve terminals from a crustacean neurosecretory system, the sinus gland [8]. It has also become possible to extend this analysis to rat and murine neurohypophysial nerve terminals [9, 12, 24]. Using these techniques we have been able to identify and characterize a number of voltage-gated channels (see Table 1) and ligand-gated receptors (see Table 2) in AVP vs. OT terminals.

Isolated NH nerve terminals (NHTs) are within the optimal size range (5–15 μm dia.) for application of the whole-cell recording (WCR) method. NHT readily form >2 GΩ seals with fire-polished electrodes, making feasible the application of patch-clamp techniques [25] to study the ionic channels of their membrane. The power of these techniques and their applicability to the analysis of Ca^{2+} channels, of particular importance for release (see above), is well documented [9, 20, 26, 27]; reviews: [28, 29].

Perforated Patch Recordings: Whenever possible we use the perforated-patch configuration for recording Ca^{2+} currents in order to avoid any rundown [30] of the currents. We use amphotericin B (at 180 μg/mL) to create a monovalent-selective perforated patch [31, 32]. This creates a low electrical resistance that exhibits only low permeability to Ca^{2+} and other divalents but allows the complete substitution of Cs^{+} (or other cations) for K^{+} in the terminal cytosol. This effectively blocks all outward K^{+} currents and allows the recording of uncontaminated inward Ca^{2+} currents. The advantage of the perforated vs. "(w)hole cell" configuration is that the terminal is relatively "intact" without substantial washout of its cytosolic components [30]. As a result we can record viable Ca^{2+} currents from neurohypophysial terminals for hours instead of

Table 1
Voltage-gated channel types in AVP vs. OT neurohypophysial terminals

Channel	Calcium					Potassium	Sodium				
Subtype (#) subunits	L-type [99] α1C&D	N-type [100] α1B & β3,4	N-type [100] α1B & β2a	Q-type [42] α1A	R-type [34] α1E	A-type [35] αShaker	BK-type [101] αsloβ4	BK-type [102] αsloβ1[a]	Mink [103]	Delayed [104] rectifier?	Na [38]
AVP	+	–	+	+	–	+	+	–	+	+	+
OT	+	+	–	–	+	+	+	–	+	+	+
AVP&OT	+	–?	+	+	+	+	+?	–	+?	+	+

Table is arranged horizontally by channel, subtype (subunits when known), and whether found (+) in AVP, OT, or double labeled (AVP&OT) NH terminals. Next to each channel are references (#) for each

[a]Indicates that this is found only in hypothalamic cell bodies

Table 2
Ligand-gated receptor types in AVP vs. OT neurohypophysial terminals

Receptor	Purine			Opioid		ACh		Glycine	Salusinβ	Relaxin	GABA	Dopamine
Subtype (#) subunits	P2X [40] 2,3&4	P2X [13] 7	A [33] 1	μ-Type [105]	κ-Type [106]	Nicotinic [100] α7	Nicotinic [100] α4β2	Homomeric [107] α1&2	MrgA [108] 1	LGR [109] 7/8	A [110]	D [111] 4
AVP	+	+	+	+	+	+	+	–	+	+	+	+
OT	–	+	+	+	+	+	+	+	+	+	–	–
AVP&OT	+?	+	+	+	+	+?	+?	+	+?	+?	?	?

Table is arranged in rows horizontally by receptor, sub-type (subunits when known), and whether found (+) in AVP, OT, or double labeled (AVP&OT) NH terminals. Next to each receptor are references (#) for each minutes [33, 34]. This more physiological preparation allows experiments that would have been otherwise impossible (*see* **Note 2**).

Channels: There are two ***main*** voltage-dependent outward currents in NHT (see Table 1). One is a fast, inactivating current that can be blocked with 4-aminopyridine [35] and the other is a non-inactivating current which can be blocked with tetrandrine [36, 37]. The pharmacology and kinetics of the initial transient outward current identify it as an A-current. The second, slower developing outward current, which is dependent upon intracellular Ca^{2+} concentrations, shows very slow inactivation [38]. This calcium-activated K^{+} (BK) current can be blocked by extracellular ATP [39]. These and two other K currents (see Table 1) are largely responsible for returning the resting membrane potential (RMP) to hyperpolarized (−80 mV) values in these NHTs [35, 38, 40].

Additionally, there are fast, voltage-dependent inward currents that are blocked by tetrodotoxin (TTX), as well as slow, inward currents that are not blocked by TTX but disappear in minimal Ca^{2+} saline [31]. These characteristics suggest that they are voltage-dependent Na^{+} and Ca^{2+} currents (see Table 1). There are two kinetic components to the slower inward Ca^{2+} current, which are differentially inactivated depending upon the holding potential [38, 41]. These kinetic components appear to be composed of three distinct subtypes of Ca^{2+} channels in each kind of peptidergic terminal: L, N, and Q in vasopressinergic terminals [42] vs. L, N, and R in oxytocinergic terminals [34].

Receptors: Many types of ligand-gated receptors have been characterized in the isolated AVP vs. OT NHT (see Table 2). For example, ATP has been shown to generate excitatory ionic currents in neurons from both the peripheral and central nervous system acting via the ionotropic purinergic receptors P2XR [43–46]. ATP is found to be co-released with AVP and OT and regulates hormone release in the NH [47]. A direct role for P2X2R was shown to mediate changes in $[Ca^{2+}]i$ and release of AVP in NHT [48]. Follow-up studies showed P2X2R and P2X3R generated ion currents in isolated NHT, by showing inhibition to the antagonists suramin and pyridoxal-phosphate-6-azophenyl-2′,4′-disulfonic acid (PPADS) [40].

Immunocytochemistry and patch-clamp studies in brain and cultured neurons have shown that different P2XR are trafficked to different subcellular locations [49, 50]. This appears to be the case in the HNS as well (see Table 2). Tissue punches of the supraoptic nucleus (SON), which along with the PVN contain the cell bodies that project to NH, further revealed the presence of mRNA transcripts for various P2XR subtypes including P2X2R, P2X3R, P2X4R, P2X6R, and P2X7R [51]. Immunofluorescence studies demonstrated differences in the subcellular distribution of P2XR at the NHT by showing the presence of P2X2R, P2X3R, P2X4R, and P2X7R immunoreactivity, but not the P2X6 [40]. Furthermore, OT containing terminal showed a further difference by revealing the presence of only P2X7R immunoreactivity [13, 40]. Stimulation of dissociated magnocellular neurons of the SON with ATP demonstrated different kinetic responses when compared with the nerve endings of the neurohypophysis [13].

The P2X7R has been implicated in regulating the release of neurotransmitters [52]. Studies using HNS explants [53] suggest that the P2X7R receptor is involved in the sustained release of OT and AVP. When ATP is coadministered with noradrenergic agonist phenylephrine, OT and AVP secretion was inhibited by the P2X7R-selective antagonist brilliant blue-G (BBG). Of interest, BBG showed no effect when the explant was stimulated by ATP, alone [54]. The requirement of the phenylephrine implicates the release of ATP by surrounding glia, as adrenergic receptors are reported to be poorly expressed in hypothalamic neurons [55, 56], and further suggests that glial cells may be required to generate the levels of ATP sufficient to stimulate the P2X7R. Our laboratory has been characterizing purinergic currents, in isolated NHT, with properties that are consistent with P2X7R and have further observed that it plays a role in neuropeptide release [57].

2.2.2 Imaging

Calcium Imaging

The isolated NHT preparation is readily adaptable to a number of fluorescence imaging techniques to monitor both the influx of external Ca^{2+} and mobilization of Ca^{2+} from intracellular stores (see Fig. 3). Utilization of the Ca^{2+}-indicator fura-2 [58–60] has

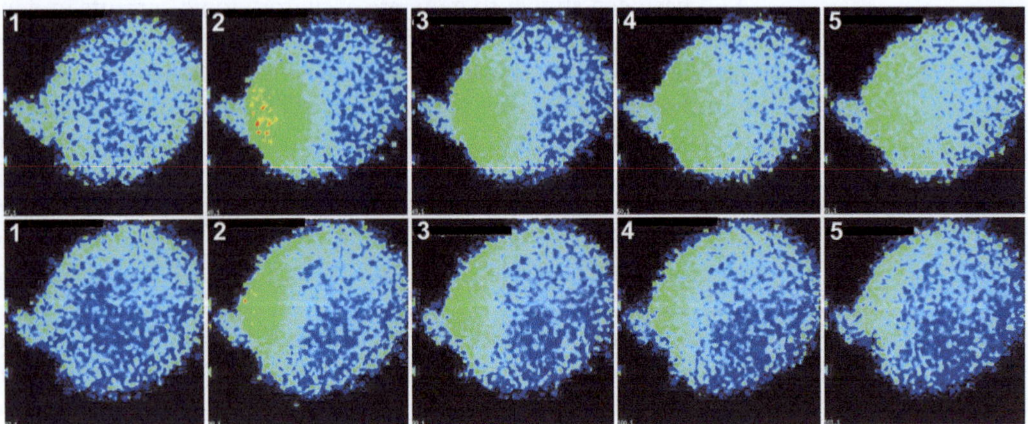

Fig. 3 Imaging calcium syntillas in NHT. The two sets of images (*upper* and *lower*) each show an individual calcium syntilla recorded from whole-cell patched NHT (H.P. = −60 mV) loaded with the calcium indicator dye, Fluo-3. Each image (*1–5*) represents a single frame from high-speed fluorescence records (at 50 Hz) starting immediately prior to syntilla appearance, and then the following four frames. In these fluorescent images, cooler colors (*blue*) represent lower calcium levels, while warmer colors (*red*) represent higher calcium levels. These recordings were performed in the absence of extracellular calcium. Thus, these events are indicative of calcium release from intracellular stores

allowed characterization of bulk $[Ca^{2+}]i$ in these terminals. Recently, a number of groups have reported highly localized Ca^{2+} release events, similar to Ca^{2+} sparks in the muscle, in a number of neuronal preparations [61]. Specifically, our group, in conjuction with the UMass Biomedical Imaging Group, has utilized high-speed Ca^{2+} imaging techniques to characterize such Ca^{2+} release events in murine NHT [12]. These events, termed Ca^{2+} syntillas, were observed to emanate from a ryanodine-sensitive intracellular Ca^{2+} pool and displayed an increase in frequency with depolarization in the absence of Ca^{2+} influx. Further, the voltage dependence of these events was observed to be mediated by dihydropyridine receptors [62]. These findings suggest that voltage-dependent Ca^{2+} syntillas may represent an additional factor involved in DSC.

Fluorescence imaging (Fig. 3) of these intracellular Ca^{2+} transients was performed [12, 15, 62, 63] using fluo-3 indicator dye (*see* **Note 3**). NHTs, isolated as defined above, were whole-cell patched using a pipette solution of (mM) 135 KCl, 2 $MgCl_2$, 30 HEPES, 4 Mg-ATP, 0.3 Na-GTP, 0.05 K_5-fluo3, and 15 DA (pH 7.2). High-speed Ca^{2+} imaging was performed using a custom built inverted wide field microscope equipped with a Nikon 100× (1.3 NA) oil immersion objective and high-resolution CCD camera, providing 133 nm of specimen per image pixel. Imaging was performed at a rate of 50 Hz (10 ms exposure) for 4 s at a time. The total amount of Ca^{2+} (250–400k) released per syntilla was quantified by measuring its "signal mass" as previously described [12, 64, 65].

2.2.3 Secretion Assays

False Transmitter Amperometry

Despite our knowledge of the nature of Ca^{2+} syntillas elucidated from the calcium imaging studies discussed above, defining the physiological role these events play in the NHT has proven to be more difficult. If such localized Ca^{2+} release events occur in the precise location of the final exocytotic event(s), then they may be directly involved in triggering exocytosis. However, directly addressing this hypothesis was difficult as no method capable of vizualing individual release events in these CNS terminal was available. To get around this issue, we adapted an amperometric method for studying vesicle fusion to this system which relies on loading the secretory granules with the false-transmitter dopamine [66]. Thus, allowing, for the first time, the recording of individual exocytotic events from peptidergic NHT (see Fig. 4). Interestingly, simultaneous use of this technique along with the high-speed Ca^{2+} imaging described above has enabled us to establish that spontaneous neuropeptide release and Ca^{2+} syntillas do *not* display any observable temporal or spatial correlation [15], confirming similar findings in chromaffin cells [63]. While these results provide strong evidence against syntillas playing a direct role in eliciting spontaneous release, they do not rule out indirect modulatory effects of syntillas on secretion.

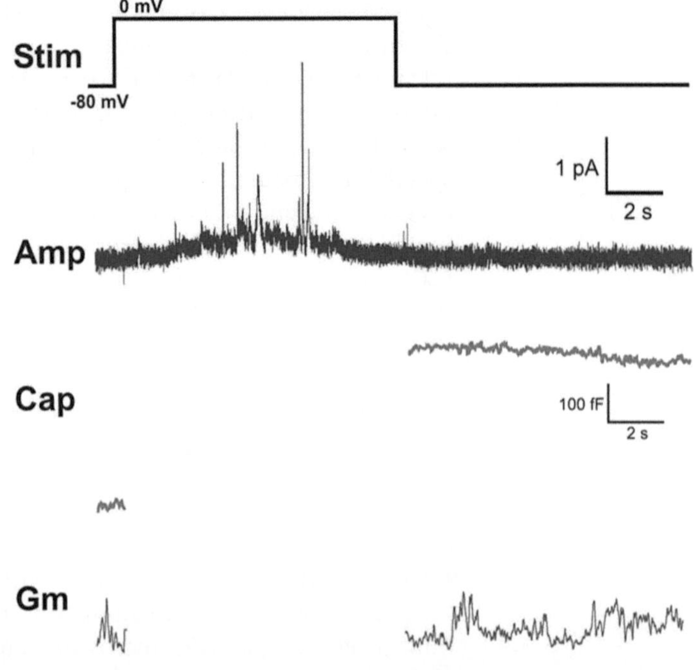

Fig. 4 Amperometric and capacitance recordings can be utilized to monitor exocytotic activity from dissociated NHT. The traces shown in this figure were recorded simultaneously from a whole-cell patched NHT following false-transmitter loading. The terminal was stimulated with a 10 s depolarizing step pulse (−80 to 0 mV). During this stimulation, amperometric spikes (upward deflections) are clearly evident in the amperometric trace (Amp). Similarly, the capacitance record (Cap) shows a robust increase elicited by the depolarizing step. Both the increase in capacitance and appearance of amperometric spikes are indicative of evoked exocytotic activity. Also shown is the membrane conductance recording (Gm) which is not changed by the depolarizing step, highlighting that the observed change in capacitance was not an artifact

Loading NHT with the false transmitter was accomplished through the inclusion of the electrochemically detectable transmitter dopamine (DA; 15 mM) in the pipette solution. NHTs were patched in the whole-cell configuration and voltage-clamped at a holding potential of either −80 mV (capacitance recordings; Fig. 4) or −60 mV (calcium imaging experiments; Fig. 3). DA was allowed to dialyze into the NHT, where it can then traffic into acidic compartments, including the large dense core vesicles (LDCV) [66]. This process generally required about 5 min for sufficient loading. During the course of DA loading, a 5 μm diameter carbon fiber electrode (CFE) voltage-clamped at 700 mV was placed in close apposition to the plasma membrane of the terminal. When DA-loaded LDCVs undergo exocytosis, DA is co-released from the NHT along with native neuropeptides. When this DA contacts the CFE, it is rapidly oxidized, which can be visualized as a transient spike in current flow through the electrode (*see* **Note 4**). Successful loading was determined by the appearance of spontaneous amperometric events at a resting membrane potential of −80 mV [15].

Analysis of amperometric spikes was performed using an Igor pro (Wavemetrics) macro, quanta analysis, written by Eugene Mosharov of David Sulzer's lab (available at www.sulzerlab.org). Amperometric events were detected using a peak threshold cutoff of 2.5 times the root mean square of trace noise. Calculation of the total amperometric charge was performed using Origin 7 (Originlab Corp.). Amperometric charge was calculated by integrating baseline subtracted amperometric current traces.

Capacitance

For experiments examining stimulus-induced capacitance changes (see Fig. 4), it was found that special care should be taken in the preparation and handling of the terminals in order to obtain reliable and reproducible results [67]. The differences, from the general procedures given above, are based on the minimization of mechanical and/or chemical perturbations that would render the terminals capacitance-wise unresponsive to electrical stimulation (*see* **Note 5a**).

Stimulus-induced capacitance changes were measured using the piecewise method, as configured for either the EPC-7 amplifier (List-Medical, Germany, customized for capacitance and resistance dithering, software from Wavemetrics lock-in emulation program integrated in Igor) or the EPC-9 (lock-in emulation program: 800–1,200 Hz sine, 10 mV PP max., 15–25-point integrations). The elicited change in capacitance was recorded as the average between 250 and 750 ms following cessation of stimulus. This was done in order to avoid possible interference of "tail artifacts" [68] that last for up to 90–100 ms after the end of the stimulus while still capturing the capacitance maxima with minimal endocytosis interference. In many cases for isolated terminals, the artifact was still significant (30 % of initial values) at 100–120 ms after cessation of stimulus. The occurrence of decaying positive capacitance, as well as negative capacitance, was observed at various degrees in all samples tested [67]. These are interpreted as the result of endocytotic activity (reduction in surface area) and are expected to occur as part of the normal physiological cycle of membrane processing. However, no correlation was observed between the type of stimulus protocol used and the magnitude (and/or occurrence) of endocytosis (*see* **Note 5b**).

Release

AVP and OT release can be easily studied from a purified preparation of isolated NHT (*see* **Note 1b**). All of the following procedures are done inside of an incubator with the temperature maintained at 37 °C. Following centrifugation at 3,400 × *g*, the pellet containing the purified NHTs is resuspended in up to 1 mL of Normal Locke's saline by gently mixing the pellet and supernatant with a 200 μL pipette tip. The resuspended solution containing the purified NHTs is then equally aliquoted into 1.7 mL microcentrifuge tubes and brought up to a final volume of 0.8 mL with Normal Locke's saline. The solution containing the NHTs is then loaded onto Acrodisc LC 13 mm syringe filters with 0.45 μm PVDF membranes (Pall Corporation, Ann Arbor, MI) with a 1 mL Sub-Q 26 gauge 5/8 in. latex free tuberculin syringe (Becton, Dickson and Company, Franklin Lakes, NJ). The loading is accomplished by first adding 100 μL of Normal Locke's saline into the syringe followed by the solution containing the NHTs and then an additional 100 μL of Normal Locke's saline, resulting in the NHTs sandwiched in between the saline buffer. This insures that the saline buffer is perfused onto the filters prior to and after the NHTs have been loaded. In addition, a prefilter (Acrodisc 0.45 μm 25 mm Syringe filter; Pall Corporation, Ann Arbor, MI) is placed prior to the 13 mm Acrodisc filter containing the NHTs to prevent any blockage or air bubbles from coming in contact with the NHTs. Once the NHTs have been loaded onto the filters, they are perfused by means of a peristaltic pump at 50 μL/min for a period of 30 min. The flow rate is then gradually increased to the desired flow rate for the experiment and allowed to stabilize for at least 20 min at the final flow rate prior to the beginning of the sampling period. The flow rate utilized is entirely dependent on the number of NHTs loaded onto the filters which determines the amount of AVP and/OT in the perfusate. Timed perfusate samples are collected by means of a fraction collector and stored at −80 °C until assayed for AVP and/or OT by an ELISA (Enzo Life Sciences, Plymouth Meeting, PA). The analysis of rapid changes in the amount of AVP and/or OT secreted is limited more by the rate at which one can collect the samples than by the sensitivity of the ELISA/radioimmunoassay. This is demonstrated in Fig. 5 where the pulsatile release of AVP in response to repetitive depolarizations with high KCl is clearly determined in samples collected every 5 s but not in samples collected at 30 or 60 s. We have also studied the time course of the high K-induced AVP release by measuring the amount of hormone secreted during sequential 15 s periods. Whereas the basal release from the nerve terminals isolated from one rat NH was about 100 pg AVP/15 s, the amount released at the peak of the K-evoked secretion was 300 pg/15 s [69].

2.2.4 Reconstitution of a Functional Synapse

Exocytosis is the process of cellular secretion in which substances contained in intracellular vesicles/neurosecretory granules are discharged from the cell by fusion of the vesicular membrane with the outer cell membrane. A comprehensive understanding of factors involved in exocytosis is of paramount importance in the study of synaptic function (e.g., D-SC). Extensive work has been performed

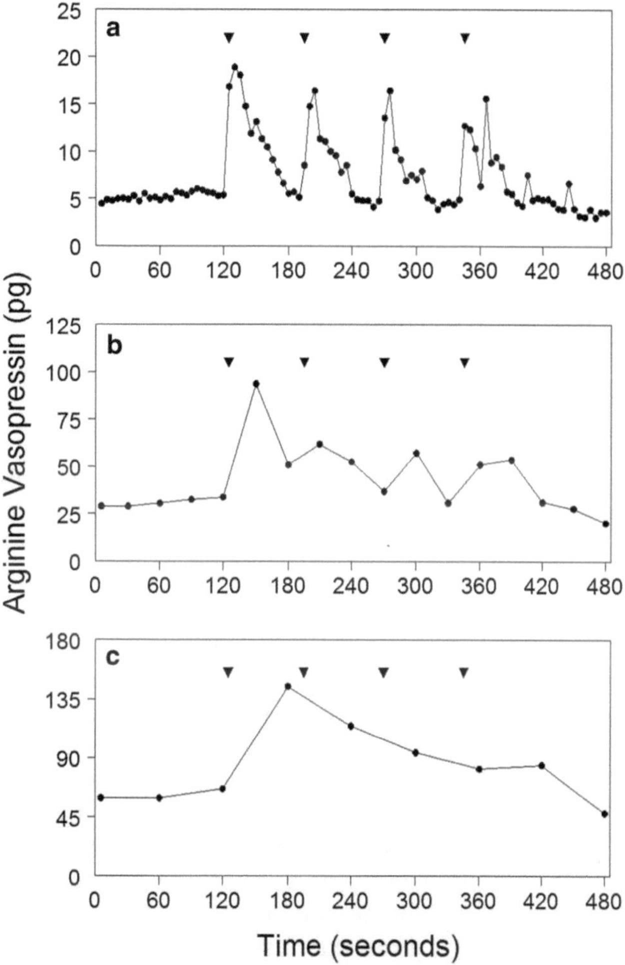

Fig. 5 Effect of sampling rate on the pattern of neuropeptide release from neurohypophysial terminals: The precision with which to detect transient release of AVP from neurohypophysial terminals repetitively stimulated with high (50 mM) potassium is dependent on the sampling frequency. When NHTs are stimulated with high potassium chloride (50 mM) in a manner simulating an endogenous burst, samples collected every 5 s (panel **a**) clearly detect transient increases in AVP release in response to each high potassium stimulation. When the sampling frequency is decreased to 30 (panel **b**) or 60 s (panel **c**), the transient increase in AVP in response to each high potassium stimulation becomes less defined

in this area, yielding a vast array of protein and lipid components that may be involved in this process. However, how these components are able to come together to elicit this function remains controversial. While molecular biology is a powerful tool, genetic manipulations, such as deletion or enhancement of proteins thought to be involved in fusion and release, can only tell us so much about what role these factors play in this process. Truly defining their role necessitates a simple *reconstituted* system using defined components.

Bilayer/Model system: For exocytosis to occur in the synapse, fusion of the vesicular membrane and the presynaptic membrane must occur. Thus, the first step in the in vitro reconstitution of this process necessitated the development of a relatively simple model system that can be used to detectably follow vesicle fusion. The artificial bilayer (black lipid membrane or BLM) has proven to be more than sufficient for the modeling of the presynaptic membrane (Fig. 6a). Artificial bilayers, first described by Mueller [70], are ideal in that their lipid composition can be controlled, as well as the environment on either side of the bilayer, making the system extremely adaptable. In addition, proteins and other factors can be reconstituted into the bilayer, adding to the system's versatility.

Initial work performed with this model system utilized artificial vesicles loaded with fluorescent dye as a means to detect fusion. Using a fluorescence microscope, observation of the bilayer showed that artificial vesicles loaded with fluorescent dye fuse to the bilayer and in so doing release their fluorescent contents [71, 72]. This simple system lacked many of the important biological components involved in fusion, and the amount of vesicle fusion observed was very low, requiring the addition of an osmotic gradient across the bilayer to drive the fusion. Fusion was increased only when the vesicle (*cis*) side of the bilayer was made hyperosmotic, a condition which should induce vesicles to shrink. Interestingly, addition of an ion channel to the vesicle membrane resulted in a large increase in number of vesicle fusions [72]. This observation was quantitatively explained by increased entry of the osmotic agent through the vesicular channel. For vesicles in contact with the bilayer membrane, the increased vesicle osmolarity causes water to flow across both the bilayer and vesicle membranes inducing these vesicles to swell and eventually fuse with the bilayer [72, 73].

Reconstitution/Fusion Assays: With a fusion assay (*see* **Note 6a**) in hand [74, 75], the planar bilayer system allows for the "ground up" reconstitution of mechanisms required for fusion, through incorporation of the cellular machinery (e.g., SNARE proteins [76]) into the model system. With present recombinant technology, large amounts of SNARE proteins, and other proteins believed to be involved in exocytosis, can be produced in *E. coli* and subsequently purified (*see* **Note 6b**). These recombinant proteins can be reconstituted into both artificial vesicles and bilayers. This allows individual as well as different subsets of proteins to be studied in a simplified and controlled environment.

Three separate techniques have been successfully used to reconstitute proteins into the planar bilayer. The first technique is to wet a pipette tip with a solution containing the purified protein and simply "brushing" the protein into a preformed bilayer (*see* **Note 6c**). Alternatively, purified protein can be added directly to the lipid solution used to form the bilayer or added to the solution surrounding the bilayer and allowed to spontaneously insert.

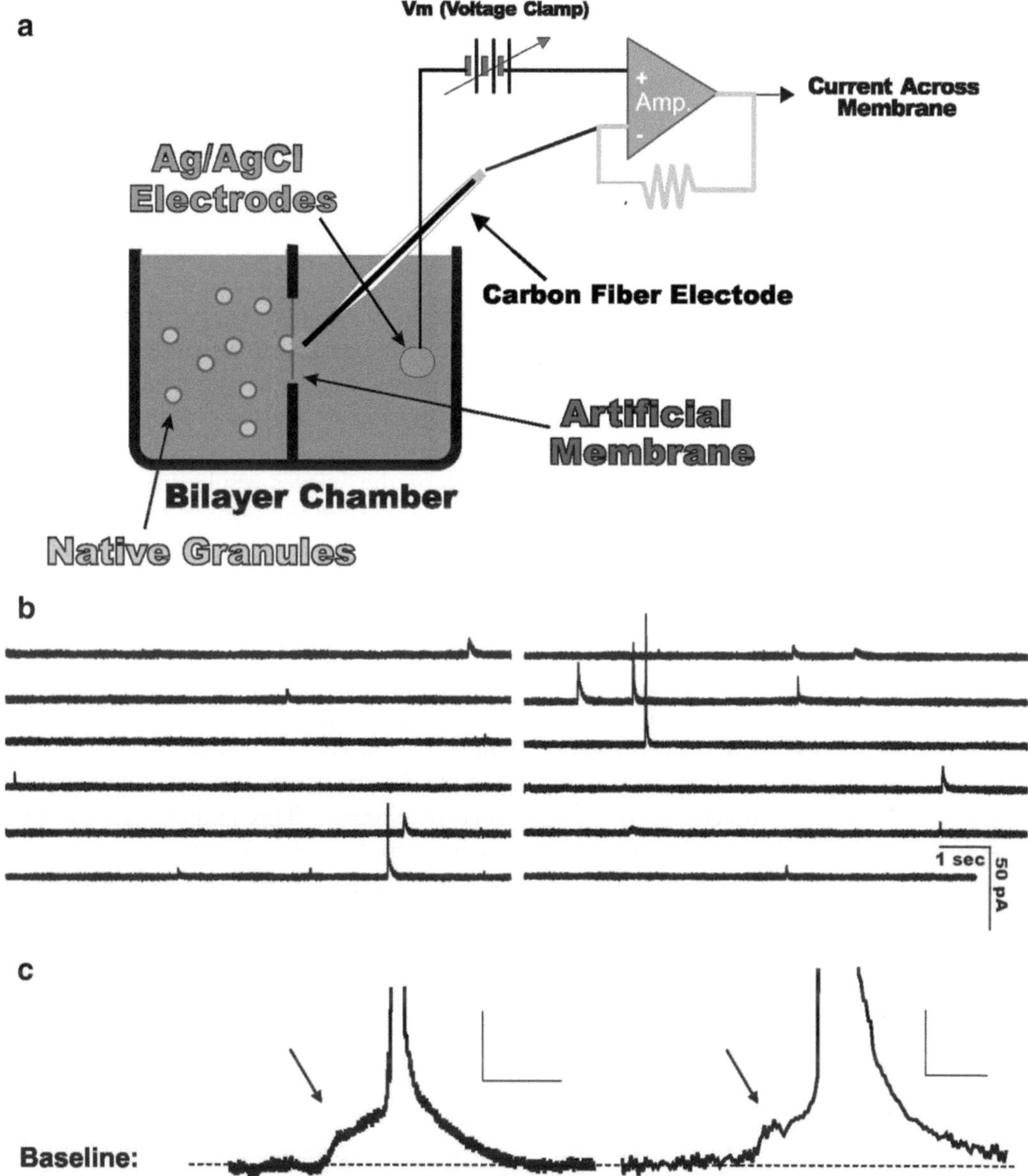

Fig. 6 Amperometric detection with planar lipid bilayer setup. (**a**) Amperometry can be used as a means of detecting vesicular content release across the planar bilayer. This is accomplished by placing a carbon fiber electrode (CFE, 30 μm diameter) in close apposition to the *trans* side of the planar bilayer. Native granules or artificial vesicles containing an oxidizable transmitter, such as dopamine (DA), can then be added on the opposite (*cis*) side. The CFE is held at a constant voltage known to oxidize the transmitter, and the current across the CFE monitored. When vesicles fuse with the bilayer the oxidizable transmitter will be released into the *trans* side of the bilayer and come in contact with the CFE. When the transmitter contacts the CFE it will be oxidized, resulting in a current spike [98]. (**b**) The spontaneous fusion of native chromaffin granules with a bilayer containing syntaxin can be detected as amperometric spikes (upward deflections in the current trace). Section **b** shows all of the spontaneous amperometric events seen during a 5-min recording. Without syntaxin in the bilayer such spontaneous amperometric events were not seen. (**c**) Several amperometric spikes appear to possess a pre-spike "foot," which are indicative of fusion pore formation and of slow release of dopamine prior to full vesicular fusion. The presence of feet indicates that all of the components required for fusion pore formation are present in our simplified system. The scale bars in **c** represent 200 ms on the *x*-axis and 0.5 pA on the *y*-axis

Artificial vesicles: can be prepared in several different ways. For a more native (and less defined) system, secretory vesicles isolated from cells can be used. In this case vesicles must be tagged with ion channels so that fusion can be detected. We routinely make such "modified secretory vesicles" by adding to them nystatin and ergosterol to form functional nystatin channels (*see* **Note 6a**). For a fully defined system, artificial vesicles (liposomes) can be made in the presence of purified SNARE proteins and, again, an ion channel as a fusion tag.

Through the reconstitution of essential components, spontaneous fusion of vesicles to a planar bilayer can be observed [16, 77, 78]. Initial work using this model system has been aimed at determining the minimal configuration of proteins required to drive fusion. It has been determined that simply the addition of the SNARE protein syntaxin-1 to the bilayer results in the ability of modified secretory vesicles to fuse with the bilayer. As noted above, even without proteins fusion would be expected following the addition of an osmotic gradient (hyperosmotic on the vesicle side), but after adding syntaxin-1 spontaneous fusion is observed in the absence of such a driving force. Modified neurosecretory granules from isolated NHT, presumably containing native SNAREs in their membrane, are unable to spontaneously fuse to a protein-free bilayer [16]. Furthermore, protein-free vesicles, containing only nystatin, ergosterol, and purified lipids, do **not** spontaneously fuse to a bilayer regardless of the presence of syntaxin. This finding has also been extended to large dense core granules isolated from bovine chromaffin cells [79], in addition to small clear synaptic vesicles isolated from the *Torpedo* electric organ [77, 78]. The fact that this result is the same for both large dense core granules and small clear synaptic vesicles illustrates the remarkable conservation of the process of SNARE driven fusion. Additionally, it attests to the ability of the single protein, syntaxin, to interact with proteins on the vesicle membrane to overcome the energy barrier to fuse vesicles to the bilayer membrane. Variations of this result have been observed by other workers [80, 81] but are in *contrast* to reconstitution studies measuring fusion of one population of liposomes to another [82, 83], where the additional SNARE protein, SNAP-25, is needed. These differences suggest that different combinations of SNAREs are needed with different membrane environments perhaps due to changes in lipid composition or membrane property [84].

In order to complete our model synapse, the artificial bilayer system requires a postsynaptic element through which vesicle content release can be detected. While use of the nystatin/ergosterol technique provides an elegant means of detecting membrane fusion between vesicles and an artificial bilayer, it provides no information about vesicular content release. It is for this reason that amperometric detection has been incorporated into our artificial synapse (*see* Fig. 6).

Amperometry is a means of electrochemical detection that can be used to follow the release of oxidizable agents (see above). By placing a CFE in close apposition to our artificial bilayer, as shown in Fig. 6a, oxidizable agents released through the membrane will come in contact with the electrode. If the electrode is held at a potential capable of oxidizing this agent, this contact will result in current flow through the electrode that can be easily monitored. This technique is far superior to the fluorescence technique described above in that it requires much less equipment and provides more telling data about the kinetics of vesicular content release (*see* **Note 6d**).

As discussed earlier, using the nystatin/ergosterol technique, it has been shown that modified chromaffin granules are able to undergo spontaneous fusion with an artificial bilayer containing the SNARE protein syntaxin. Native chromaffin granules contain the oxidizable neurotransmitter dopamine. This makes these granules compatible with this amperometric technique. As shown in Fig. 6b, using amperometry to detect release, unmodified native chromaffin granules have been found to spontaneously fuse to a bilayer containing only syntaxin [79].

Several important points can been determined from this result. First, it provides a validation of results obtained using the nystatin/ergosterol technique. A major criticism of the nystatin/ergosterol technique was that adding artificial components, such as nystatin and ergosterol, to granules might serve to influence their ability to fuse with the bilayer. Since amperometric detection of release provides the same result using completely native granules, the additional components used in the nystatin/ergosterol technique play no significant role. The ability to detect spontaneous release amperometrically shows that the interaction of syntaxin with native membrane proteins does in fact result in full membrane fusion that allows for passage of vesicular contents to the opposite side of the bilayer (see Fig. 6a).

Another important finding is the presence of small current steps that precede some of the amperometric spikes observed [15]. These currents steps (Fig. 6c) appear similar in appearance to pre-spike "foot" signals first observed in chromaffin cells [85, 86]. This pre-spike foot signal is attributed to the slow release of oxidizable transmitter through a fusion pore prior to pore widening and complete collapse of the vesicle into the plasma membrane [87]. The appearance of pre-spike feet preceding amperometric spikes seen in our extremely simple model system implies that all of the components required to form a fusion pore are present. In addition, while it is known that atypical SNARE motif combinations [77, 88, 89], like those forming in our model system, arise in vitro, it has not been determined if these less stable SNARE combinations are physiologically relevant. These amperometric results would argue

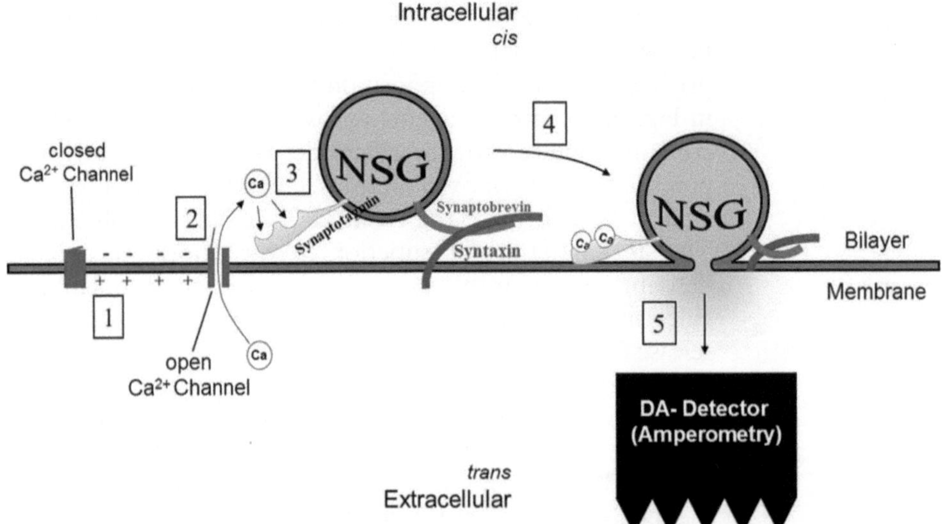

Fig. 7 Model of a reconstituted synapse using a planar lipid bilayer. In order to reconstitute calcium-dependent release in this system, voltage-gated calcium channels can be incorporated into the bilayer (*1*). Upon depolarization, the calcium channels will open and allow calcium (Ca^{2+}) to move across the bilayer (*2*). The incorporation of components necessary for fusion, such as syntaxin and possibly SNAP-25, into our model system will allow calcium (*3*) to stimulate the fusion of a secretory granule (NSG), containing both synaptotagmin and synaptobrevin (*4*). If the secretory vesicle contains an oxidizable transmitter (e.g., DA), vesicle fusion and transmitter release can be detected with an amperometric electrode (*5*), representing the postsynaptic side of our model synapse

that, at least in this case, one atypical SNARE combination is in fact relevant, given that pore formation and vesicular content release can be seen. The successful addition of an amperometric electrode, as the postsynaptic element of our model system, allows us to monitor the kinetics of reconstituted release with a temporal capability far greater than that allowed using other fusion assays. Additionally, amperometric detection of release should aid in the determination of the physiological relevance of non-cognate SNARE complexes seen in earlier in vitro work.

Future work will involve elucidation of the factors necessary for the conformation of *calcium* dependence on fusion and release in this system (Fig. 7). Overall, this system provides a powerful tool that can be used to study the numerous proteins and environmental parameters thought to play a role in exocytosis. By combining information obtained using in vitro techniques such as this with information obtained in vivo, a clearer picture of the machinery involved in neurotransmitter release will emerge.

3 Notes

1. *Preparation and release*
   (a) One of the preparation's major drawback is that the isolated NHT appear to have no genetic machinery [88, 90] and, thus, molecular biological tools cannot be ***directly*** applied (although one can take advantage of transgenic animals [19, 91, 92]). Another shortcoming of the preparation is that once isolated, mammalian terminals are electrophysiologically viable for only 3–8 h and cannot be cultured.
   (b) The presence of EGTA may be the key to obtaining ***isolated*** NHT which can release hormones as a result of depolarization. Indeed, although isolated neurosecretory nerve terminals have already been described more than 40 years ago [93], it is only more recently that a preparation has been obtained from which the hormones are released by exocytosis. For example, we thought that nerve terminals isolated from bovine neurohypophyses did not release AVP in response to high concentrations of potassium because secretion had ***already*** occurred at the slaughterhouse during hemorrhaging of the animals. That is, we thought that the mechanism leading to exocytosis had probably been inactivated. However, subsequent experiments [94] have shown that isolated bovine nerve terminals do respond to depolarization if prepared in a medium containing EGTA. In fact, these findings are not too surprising. First, during homogenization the cytoplasm of some of the nerve terminals is exposed to the external medium as judged by the presence of Lucifer Yellow in such terminals following homogenization in a medium containing this fluorescent marker [2]. Secondly, we found, using the Ca^{2+}-indicator fura-2, that a HEPES-buffered sucrose medium without EGTA contains 1–10 μM free calcium [58], i.e., a cytoplasmic concentration which is large enough to induce hormone release [94].
2. *Perforated patching*
   The disadvantages of perforated patches are mainly that one loses "control" of the internal milieu of the terminal and cannot, therefore, introduce large or membrane impermeable probes/agents directly into the terminal, as with WCR. The wait time for perforation is another disadvantage. Two things are important in order to successfully obtain a GΩ seal: (1) front-fill pipette with pipette solution, before backfilling with Amphotericin containing pipette solution and (2) approach

NHT rapidly, before allowing the two phases of pipette solution to mix. Generally, a perforation period of 20 min up to 1 h can be expected when performed properly. The perforation period will vary with the distance of the front-filled solution from the pipette tip orifice and the degree of taper of the tip itself. The narrower the taper, the longer the perforation period will take. However if the taper is not sufficiently narrow and the front-fill is too shallow, the DMSO from the backfill solution (used to dissolve amphotericin) will reach the orifice too quickly and prevent obtaining a GΩ seal. One also has to be more careful about the access resistance (we reject any >10 MΩ), since it will always be higher than with the conventional "(w)hole"-cell configuration, especially when measuring capacitance (see Fig. 4).

3. *Ca^{2+} imaging*
   The main limiting factor for the analysis of intracellular Ca^{2+} fluctuations, such as syntillas, lies in the speed of the equipment used to observe these events. First, the CCD camera must be able to acquire images at a high enough rate (>25 Hz) to visualize these millisecond scale events. Further, the Kd of the Ca^{2+} fluorophore used must be low enough to allow for the proper temporal resolution to observe these rapid events (see Fig. 3). Thus, more common Ca^{2+} indicators such as Fura-2 cannot be used to visualize syntillas. It is also important to maintain terminals in a Ca^{2+}-free buffer when recording syntillas to ensure that external calcium entry does not obscure intracellular Ca^{2+} release events. Similarly, as the frequency of syntillas is voltage dependent, it is important to maintain whole-cell patched terminals at a holding potential where syntillas occur at a rate sufficient to ensure that they will often occur during imaging, but slow enough that do not greatly overlap, resulting in a global increase in intracellular Ca^{2+}.

4. *False transmitter loading and amperometry*
   The principle benefit derived from using the false-transmitter loading technique with isolated NHTs is that it provides the ability monitor secretory activity from these terminals at the quantal level and with high temporal resolution (see Fig. 4). Increasingly, specialized electrodes are being developed that promise to provide the ability to electrochemically monitor the release of transmitters not generally assayable by amperometric detection. However, the electrodes required for such detection generally must be specially coated and are both expensive and notoriously difficult to work with. Thus, the false-transmitter loading technique provides the added benefit of allowing for the use of standard non-oxidizable electrodes (carbon fiber, gold, platinum, etc.), which do not require such treatment.

Despite these benefits, the technique is limited to an extent by the small amplitude of the amperometric events observed. This issue is an inherent problem with false-transmitter loading. In systems which natively package an oxidizable transmitter into their secretory vesicles (i.e., chromaffin cells), these transmitters are at very high concentrations (1–2 M), which one could not hope to attain through false-transmitter loading [15]. To circumvent such issues, it is essential to allow enough time following transition into whole-cell configuration to allow for sufficient false-transmitter loading to occur. Further, it is of the utmost importance to perform these experiments on equipment in which assiduous efforts have been made to minimize electrical interference (ground loops, line noise, vibrations, etc.). With low noise levels, a sufficient signal-to-noise ratio can be obtained allowing detection and analysis of quantal release events. Finally, in regard to coupling this technique with Ca^{2+} imaging, exposing the CFE to the fluorescent light required for Fluo3 analysis of Ca^{2+} can result in the emergence of a photoelectric current that may interfere with amperometric detection. To minimize this issue, it is necessary to use an iris in line with the path of the fluorescent light to focus the light specifically upon the patched terminal and limit exposure of the CFE.

5. *Capacitance*
   (a) The special procedures required for consistent capacitance response are (1) slow and monitored (using microscope) trituration of the neurohypophysis when isolating the terminals and done in solutions containing no more than 5 μM EGTA, (2) lower $CaCl_2$ and $MgCl_2$ in both wash (0.2 mM and 0.1 mM, respectively, after depositing terminals) and in "Normal" solutions (1.2–1.4 mM and 0.3 mM, respectively), (3) lower KCl in all potassium-containing solutions (to 2.5 mM), (4) avoiding mechanical movements once patches are established (lifting patched terminals was a particular problem), and (5) slower perforation times (usually achieved by filling the pipette tip for longer times and/or using low concentrations of perforating agents).
   (b) The induced capacitance response was found to be dependent on both the ionic contents of the solutions used and the stimulus patterns used to evoke such responses. Specifically, shifts in ionic concentrations expected to arise following physiological stimulation of NHT in situ render optimal capacitance changes [67]. Further, use of stimulus patterns derived from actual action potentials trains recorded from the terminals themselves increased capacitive responses [67].

6. *Reconstituted synapse*
   (a) The addition of ion channels to the vesicle membrane has provided our model (see Fig. 7) with a more simplistic and elegant means to monitor vesicular fusion. Upon fusion, vesicular channels are incorporated into the planar bilayer, allowing for individual fusion events to be visualized by monitoring the bilayer conductance. The use of ion channel incorporation to follow vesicle fusion is often problematic due to the fact that the step-increase in membrane conductance resulting from ion channel incorporation is permanent and will eventually result in amplifier saturation and bilayer instability. In order to avoid this permanent increase in membrane conductance, the antibiotic nystatin can be used as a marker channel. Low concentrations of nystatin are capable of forming ion channels in membranes that contain the sterol, ergosterol [74, 95–97]. Adding both nystatin and ergosterol to the vesicle membrane results in the presence of active nystatin channels in the vesicles. When vesicles containing nystatin channels fuse with the planar bilayer, the active nystatin channel are transferred into the planar bilayer, yielding an increase in membrane conductance. Because the planar bilayer is sterol free, the ergosterol associated with active nystatin channels dissipates, causing the channel to close resulting in a decay of the original conductance increase. This transient increase in bilayer conductance from fusion of nystatin and ergosterol containing vesicles has been shown previously [74, 75].

       In addition to using this nystatin/ergosterol technique to measure fusion, it also provides one more way (*see* **Note 6c** below) to add proteins to the artificial bilayer and thereby determined if these proteins play a role in membrane fusion. Since fusion events can be recorded over a long period of time, effects on the rate of fusion can also be assessed.
   (b) Techniques for producing recombinant proteins from *E. coli* are wildly used and need little mention here except that for an unknown reason, the SNARE protein SNAP-25 does not produce well from bacteria. Best results occur when bacteria are induced to produce the recombinant SNAP-25 while still at a lower density. Growing bacteria to only an optical density at 620 nm (OD_{620}) of 0.6–0.8 leads to higher purity protein. At higher ODs, other protein bands appear in SDS-PAGE gels. These bands presumably are fragments or aggregates of SNAP-25.
   (c) It seems surprising that protein can be reconstituted into a planar bilayer simply by brushing the preformed membrane with a pipet tip dipped in a solution containing purified protein. Normally, planar bilayers are formed by

dipping the tip of a pipet into a decane solution containing lipids and then brushing a bubble formed on the end of the pipet over a hole made in a plastic partition. However, after a membrane has been formed, it is possible to re-brush the membrane with a bubble formed on the end of a new pipet tip. Following re-brushing, the formation of a new membrane can be observed electrically as the sudden drop and then rise in membrane capacitance. If the new pipet is first dipped in an aqueous solution containing the protein, then some protein is picked up by the pipet and delivered to the bilayer membrane during brushing. The success of the technique was discovered accidently when a protein-containing membrane was formed following re-brushing with a pipet that had been dipped in the protein solution instead of the decane/lipid solution (in this first case the protein was an ion channel, so the successful reconstitution was detected from the channel conductance). This technique is quick and easy but provides little means to control the amount of protein incorporated into the bilayer [78].

(d) A key advantage of amperometric detection is the ability to detect fusion from unmodified native secretory vesicles. However, positioning the detection electrode close to the planar bilayer membrane is not an easy process. Initially, our planar bilayer apparatus needed to be modified to allow for the incorporation of a manipulator capable of positioning the CFE in close proximity to the *trans* side (side opposite to where vesicles are added) of the artificial bilayer (see Fig. 6a). Experimental CFE placement was achieved as follows: First, the plastic partition containing the aperture upon which the artificial bilayer will be formed was placed in the recording apparatus. Then the CFE was placed in a micromanipulator and positioned to insure that the CFE tip could travel cleanly through this aperture and not contact the partition. A dissecting scope mounted above the bilayer apparatus was used to monitor this process. Larger sized apertures (~200 μm) were generally used for these experiments to make this easier. Following rough CFE placement, the electrode was retracted through the aperture, solutions were added to the chambers, and a bilayer formed as described above. After successful bilayer formation, the micromanipulator was then used to move the CFE tip into close apposition with the bilayer. This was accomplished by monitoring the bilayer capacitance as the CFE tip was moved into position. Contact between the bilayer and the CFE tip was observed as an increase in capacitance. Once observed the tip was backed off slightly, and the experimental procedure was begun.

Acknowledgements

Many thanks to the father of the "kikis," Jean J. Nordmann, for developing and teaching us this wonderful preparation. We also wish to acknowledge seminal work in developing this preparation by Vincent Coccia, Govindan Dayanithi, Valerie DeCrescenzo, Alex Dopico, Thomas Knott, Joseph Lockhart, Karen Ocorr, Sonia Ortiz-Miranda, Mark Savage, Edward Stuenkel, Peter Thorn, Steven Treistman, Cristina Velazquez, Gang Wang, Xiaoming Wang, and Yong Yin. Supported by NIH grants: NS29470, DA10487 and NS063192 to J.R.L.

References

1. Lemos JR (2012) Magnocellular neurons in *eLS*. John Wiley & Sons, Ltd: Chichester. DOI: 10.1002/9780470015902.a0000176.pub2
2. Nordmann JJ, Dayanithi G, Lemos JR (1987) Isolated neurosecretory nerve endings as a tool for studying the mechanism of stimulus–secretion coupling. Biosci Rep 7:411–426
3. Morris JF, Pow DV (1993) New anatomical insights into the inputs from hypothalamic magnocellular neurons. In: North WG, Moses AM, Share L (eds) The neurohypophysis: a window on brain function, Annals of the New York Academy of Science. The New York Academy of Sciences, New York, pp 16–33
4. Knott TK, Dayanithi G, Coccia V, Custer EE, Lemos JR, Treistman SN (2000) Tolerance to acute ethanol inhibition of peptide hormone release in the isolated neurohypophysis. Alcohol Clin Exp Res 24:1077–1083
5. Wang XM, Lemos JR, Dayanithi G, Nordmann JJ, Treistman SN (1991) Ethanol reduces vasopressin release by inhibiting calcium currents in nerve terminals. Brain Res 551:338–341
6. Wang XM, Dayanithi G, Lemos JR, Nordmann JJ, Treistman SN (1991) Calcium currents and peptide release from neurohypophysial terminals are inhibited by ethanol. J Pharmacol Exp Ther 259:705–711
7. Marrero HG, Lemos JR (2003) Loose-patch clamp currents from the hypothalamo-neurohypophysial system of the rat. Pflugers Arch 446:702–713
8. Lemos JR, Nordmann JJ, Cooke IM, Stuenkel EL (1986) Single channels and ionic currents in peptidergic nerve terminals. Nature 319:410–412
9. Lemos JR, Nordmann JJ (1986) Ionic channels and hormone release from peptidergic nerve terminals. J Exp Biol 128:53–72
10. Wang XM, Treistman SN, Lemos JR (1991) Direct identification of individual vasopressin-containing nerve terminals of the rat neurohypophysis after 'whole-cell' patch-clamp recordings. Neurosci Lett 124:125–128
11. Custer EE, Ortiz-Miranda S, Knott TK et al (2007) Identification of the neuropeptide content of individual rat neurohypophysial terminals. J Neurosci Methods 163:226–234
12. De Crescenzo V, ZhuGe R, Velazquez-Marrero C et al (2004) Ca^{2+} syntillas, miniature ca^{2+} release events in terminals of hypothalamic neurons, are increased in frequency by depolarization in the absence of ca^{2+} influx. J Neurosci 24:1226–1235
13. Knott TK, Hussy N, Cuadra AE et al (2012) ATP appears to act via different receptors in terminals vs. somata of the hypothalamic neurohypophysial system. J Neuroendocrinol 24:681–689
14. Knott TK, Marrero HG, Fenton RA, Custer EE, Dobson JG Jr, Lemos JR (2007) Endogenous adenosine inhibits cns terminal Ca(2+) currents and exocytosis. J Cell Physiol 210:309–314
15. McNally JM, De Crescenzo V, Fogarty KE, Walsh JV, Lemos JR (2009) Individual calcium syntillas do not trigger spontaneous exocytosis from nerve terminals of the neurohypophysis. J Neurosci 29:14120–14126
16. McNally JM, Woodbury DJ, Lemos JR (2004) Syntaxin 1a drives fusion of large dense-core neurosecretory granules into a planar lipid bilayer. Cell Biochem Biophys 41:11–24
17. Yin Y, Dayanithi G, Lemos JR (2002) Ca(2+)-regulated, neurosecretory granule channel involved in release from neurohypophysial terminals. J Physiol 539:409–418
18. Nordmann JJ, Dayanithi G, Cazalis M (1986) Do opioid peptides modulate, at the level of the

nerve endings, the release of neurohypophysial hormones? Exp Brain Res 61:560–566

19. Ueta Y, Dayanithi G, Fujihara H (2011) Hypothalamic vasopressin response to stress and various physiological stimuli: visualization in transgenic animal models. Horm Behav 59:221–226
20. Fenwick EM, Marty A, Neher E (1982) A patch-clamp study of bovine chromaffin cells and of their sensitivity to acetylcholine. J Physiol 331:577–597
21. Kidokoro Y (1985) Electrophysiology of adrenal chromaffin cells. In: Poisner AM, Trifaro JM (eds) The electrophysiology of the secretory cell. Elsevier, New York
22. Maruyama Y, Peterson OH (1982) Single-channel currents in isolated patches of plasma membrane from basal surface of pancreatic acini. Nature 299:159–161
23. Hagiwara S, Byerly L (1983) The calcium channel. Trends Neurosci 6:189–193
24. Mason WT, Dyball RE (1986) Single ion channel activity in peptidergic nerve terminals of the isolated rat neurohypophysis related to stimulation of neural stalk axons. Brain Res 383:279–286
25. Hamill OP, Marty A, Neher E, Sakmann B, Sigworth FJ (1981) Improved patch-clamp techniques for high-resolution current recording from cells and cell-free membrane patches. Pflugers Arch 391:85–100
26. Carbone E, Lux HD (1984) A low voltage-activated, fully inactivating Ca channel in vertebrate sensory neurones. Nature 310:501–502
27. Fox AP, Nowycky MC, Tsien RW (1987) Kinetic and pharmacological properties distinguishing three types of calcium currents in chick sensory neurons. J Physiol 394:149–172
28. Dunlap K, Luebke JI, Turner TJ (1995) Exocytotic Ca^{2+} channels in mammalian central neurons. Trends Neurosci 18:89–98
29. Kawamoto EM, Vivar C, Camandola S (2012) Physiology and pathology of calcium signaling in the brain. Front Pharmacol 3:61
30. Byerly L, Yazejian B (1986) Intracellular factors for the maintenance of calcium currents in perfused neurones from the snail, Lymnea stagnalis. J Physiol 370:631–650
31. Wang X, Treistman SN, Wilson A, Nordmann JJ, Lemos JR (1993) Calcium channels and peptide release from neurosecretory terminals. News Physiol Sci 8:64–68
32. Rae J, Cooper K, Gates P, Watsky M (1991) Low access resistance perforated patch recordings using amphotericin b. J Neurosci Methods 37:15–26
33. Wang G, Dayanithi G, Custer EE, Lemos JR (2002) Adenosine inhibition via a(1) receptor of N-type Ca^{2+} current and peptide release from isolated neurohypophysial terminals of the rat. J Physiol 540:791–802
34. Wang G, Dayanithi G, Newcomb R, Lemos JR (1999) An R-type Ca^{2+} current in neurohypophysial terminals preferentially regulates oxytocin secretion. J Neurosci 19:9235–9241
35. Thorn PJ, Wang XM, Lemos JR (1991) A fast, transient K^+ current in neurohypophysial nerve terminals of the rat. J Physiol 432:313–326
36. Wang G, Lemos JR (1992) Tetrandrine blocks a slow, large-conductance, Ca(2+)-activated potassium channel besides inhibiting a non-inactivating Ca^{2+} current in isolated nerve terminals of the rat neurohypophysis. Pflugers Arch 421:558–565
37. Wang G, Jiang MX, Coyne MD, Lemos JR (1993) Comparison of effects of tetrandrine on ionic channels of isolated rat neurohypophysial terminals and Y1 mouse adrenocortical tumor cells. Zhongguo Yao Li Xue Bao 14:101–106
38. Wang X, Treistman SN, Lemos JR (1992) Two types of high-threshold calcium currents inhibited by ω-conotoxin in nerve terminals of rat neurohypophysis. J Physiol 445:181–199
39. Lemos JR, Wang G (2000) Excitatory versus inhibitory modulation by ATP of neurohypophysial terminal activity in the rat. Exp Physiol 85:67s–74s
40. Knott TK, Velazquez-Marrero C, Lemos JR (2005) ATP elicits inward currents in isolated vasopressinergic neurohypophysial terminals via p2x2 and p2x3 receptors. Pflugers Arch 450:381–389
41. Treistman SN, Bayley H, Lemos JR, Wang XM, Nordmann JJ, Grant AJ (1991) Effects of ethanol on calcium channels, potassium channels, and vasopressin release. Ann N Y Acad Sci 625:249–263
42. Wang G, Dayanithi G, Kim S et al (1997) Role of Q-type Ca^{2+} channels in vasopressin secretion from neurohypophysial terminals of the rat. J Physiol 502(Pt 2):351–363
43. Edwards FA, Gibb AJ, Colquhoun D (1992) ATP receptor-mediated synaptic currents in the central nervous system. Nature 359:144–147
44. Silinsky EM, Gerzanich V, Vanner SM (1992) ATP mediates excitatory synaptic transmission in mammalian neurones. Br J Pharmacol 106:762–763
45. Ueno S, Harata N, Inoue K, Akaike N (1992) ATP-gated current in dissociated rat nucleus solitarii neurons. J Neurophysiol 68:778–785
46. Zimmermann H (1994) Signalling via ATP in the nervous system. Trends Neurosci 17: 420–426
47. Sperlagh B, Mergl Z, Juranyi Z, Vizi ES, Makara GB (1999) Local regulation of vaso-

pressin and oxytocin secretion by extracellular ATP in the isolated posterior lobe of the rat hypophysis. J Endocrinol 160:343–350
48. Troadec JD, Thirion S, Nicaise G, Lemos JR, Dayanithi G (1998) ATP-evoked increases in $[Ca2+]_i$ and peptide release from rat isolated neurohypophysial terminals via a P2X2 purinoceptor. J Physiol 511(Pt 1):89–103
49. Bobanovic LK, Royle SJ, Murrell-Lagnado RD (2002) P2X receptor trafficking in neurons is subunit specific. J Neurosci 22:4814–4824
50. Rubio ME, Soto F (2001) Distinct localization of P2X receptors at excitatory postsynaptic specializations. J Neurosci 21:641–653
51. Shibuya I, Tanaka K, Hattori Y et al (1999) Evidence that multiple P2X purinoceptors are functionally expressed in rat supraoptic neurones. J Physiol 514(Pt 2):351–367
52. Sperlágh B, Vizi ES, Wirkner K, Illes P (2006) P2X7 receptors in the nervous system. Prog Neurobiol 78:327–346
53. Sladek CD, Song Z (2012) Diverse roles of G-protein coupled receptors in the regulation of neurohypophyseal hormone secretion. J Neuroendocrinol 24:554–565
54. Gomes DA, Song Z, Stevens W, Sladek CD (2009) Sustained stimulation of vasopressin and oxytocin release by ATP and phenylephrine requires recruitment of desensitization-resistant P2X purinergic receptors. Am J Physiol Regul Integr Comp Physiol 297:R940–R949
55. Gordon GR, Baimoukhametova DV, Hewitt SA, Rajapaksha WR, Fisher TE, Bains JS (2005) Norepinephrine triggers release of glial ATP to increase postsynaptic efficacy. Nat Neurosci 8:1078–1086
56. Sawyer CH, Clifton DK (1980) Aminergic innervation of the hypothalamus. Fed Proc 39:2889–2895
57. Cuadra AE, Knott T, Custer E, Lemos JR (2013) P2X7 Receptor-Mediated Currents in Rat Hypothalamic Neurohypophysial System (HNS) Terminals. J Cell Physiol in press
58. Brethes D, Dayanithi G, Letellier L, Nordmann JJ (1987) Depolarization-induced Ca^{2+} increase in isolated neurosecretory nerve terminals measured with fura-2. Proc Natl Acad Sci USA 84:1439–1443
59. Stuenkel EL, Nordmann JJ (1993) Intracellular calcium and vasopressin release of rat isolated neurohypophysial nerve endings. J Physiol 468:335–355
60. Sasaki N, Dayanithi G, Shibuya I (2005) Ca^{2+} clearance mechanisms in neurohypophysial terminals of the rat. Cell Calcium 37:45–56
61. Berridge MJ (2006) Calcium microdomains: organization and function. Cell Calcium 40:405–412
62. De Crescenzo V, Fogarty KE, Zhuge R et al (2006) Dihydropyridine receptors and type 1 ryanodine receptors constitute the molecular machinery for voltage-induced Ca^{2+} release in nerve terminals. J Neurosci 26: 7565–7574
63. ZhuGe R, DeCrescenzo V, Sorrentino V et al (2006) Syntillas release Ca^{2+} at a site different from the microdomain where exocytosis occurs in mouse chromaffin cells. Biophys J 90:2027–2037
64. Sun XP, Callamaras N, Marchant JS, Parker I (1998) A continuum of insp3-mediated elementary Ca^{2+} signalling events in Xenopus oocytes. J Physiol 509(Pt 1):67–80
65. ZhuGe R, Fogarty KE, Tuft RA, Lifshitz LM, Sayar K, Walsh JV Jr (2000) Dynamics of signaling between Ca(2+) sparks and Ca(2+)-activated K(+) channels studied with a novel image-based method for direct intracellular measurement of ryanodine receptor Ca(2+) current. J Gen Physiol 116:845–864
66. Kim KT, Koh DS, Hille B (2000) Loading of oxidizable transmitters into secretory vesicles permits carbon-fiber amperometry. J Neurosci 20:RC101
67. Marrero HG, Lemos JR (2010) Ionic conditions modulate stimulus-induced capacitance changes in isolated neurohypophysial terminals of the rat. J Physiol 588:287–300
68. Giovannucci DR, Stuenkel EL (1997) Regulation of secretory granule recruitment and exocytosis at rat nerohypophysial nerve endings. J Physiol 498:735–751
69. Cazalis M, Dayanithi G, Nordmann JJ (1987) Hormone release from isolated nerve endings of the rat neurohypophysis. J Physiol 390:55–70
70. Mueller P, Rudin DO, Tien HT, Wescott WC (1962) Reconstitution of cell membrane structure in vitro and its transformation into an excitable system. Nature 194:979–980
71. Woodbury DJ, Hall JE (1988) Role of channels in the fusion of vesicles with a planar bilayer. Biophys J 54:1053–1063
72. Woodbury DJ, Hall JE (1988) Vesicle-membrane fusion. Observation of simultaneous membrane incorporation and content release. Biophys J 54:345–349
73. Cohen FS, Niles WD, Akabas MH (1989) Fusion of phospholipid vesicles with a planar membrane depends on the membrane permeability of the solute used to create the osmotic pressure. J Gen Physiol 93:201–210
74. Woodbury DJ (1999) Nystatin/ergosterol method for reconstituting ion channels into planar lipid bilayers. Methods Enzymol 294:319–339
75. Woodbury DJ, Miller C (1990) Nystatin-induced liposome fusion. A versatile approach

to ion channel reconstitution into planar bilayers. Biophys J 58:833–839

76. Jahn R, Fasshauer D (2012) Molecular machines governing exocytosis of synaptic vesicles. Nature 490:201–207
77. Rognlien KT, Woodbury DJ (2003) Reconstituting snare proteins into blms. In: Tein HT, Ottova-Leitmannova A (eds) Planar lipid bilayers (BLM) and their applications. Elsevier Science, Amsterdam, pp 479–488
78. Woodbury DJ, Rognlien K (2000) The t-snare syntaxin is sufficient for spontaneous fusion of synaptic vesicles to planar membranes. Cell Biol Int 24:809–818
79. McNally JM, Woodbury DJ, Lemos JR (2003) Syntaxin 1a is sufficient for spontaneous fusion of both neurohypophysial and chromaffin dense core granules to a planar lipid bilayer. In: 43rd annual meeting of the American Society for Cell Biology, San Francisco
80. Liu Q, Chen B, Yankova M et al (2005) Presynaptic ryanodine receptors are required for normal quantal size at the Caenorhabditis elegans neuromuscular junction. J Neurosci 25:6745–6754
81. Fix M, Melia TJ, Jaiswal JK et al (2004) Imaging single membrane fusion events mediated by snare proteins. Proc Natl Acad Sci USA 101:7311–7316
82. Weber T, Zemelman BV, McNew JA et al (1998) SNAREpins: minimal machinery for membrane fusion. Cell 92:759–772
83. Tucker WC, Weber T, Chapman ER (2004) Reconstitution of Ca^{2+}-regulated membrane fusion by synaptotagmin and snares. Science 304:435–438
84. Lee DE, LeW MG, Woodbury DJ (2012) Vesicle fusion to planar membranes is enhanced by cholesterol and low temperature. Chem Phys Lipids 166:45–54
85. Chow RH, von Ruden L, Neher E (1992) Delay in vesicle fusion revealed by electrochemical monitoring of single secretory events in adrenal chromaffin cells. Nature 356:60–63
86. Zhou Z, Misler S, Chow RH (1996) Rapid fluctuations in transmitter release from single vesicles in bovine adrenal chromaffin cells. Biophys J 70:1543–1552
87. Chow RH, Rüden L (1995) Electrochemical detection of secretion from single cells. In: Sakmann BaN E (ed) Single channel recording. Plenum, New York, pp 245–275
88. Bowen ME, Weninger K, Brunger AT, Chu S (2004) Single molecule observation of liposome-bilayer fusion thermally induced by soluble N-ethyl maleimide sensitive-factor attachment protein receptors (snares). Biophys J 87:3569–3584
89. Liu TT, Tucker WC, Bhalla A, Chapman ER, Weisshaar JC (2005) SNARE-driven, 25-millisecond vesicle fusion in vitro. Biophys J 89:2458–2472
90. Bredt DS, Hwang PM, Snyder SH (1990) Localization of nitric oxide synthase indicating a neural role for nitric oxide. Nature 347:768–770
91. Custer EE, Knott TK, Cuadra AE, Ortiz-Miranda S, Lemos JR (2012) P2X purinergic receptor knockout mice reveal endogenous ATP modulation of both avp and ot release from the intact neurohypophysis. J Neuroendocrinol 24:674–680
92. Zhang BJ, Kusano K, Zerfas P, Iacangelo A, Young WS III, Gainer H (2002) Targeting of green fluorescent protein to secretory granules in oxytocin magnocellular neurons and its secretion from neurohypophysial nerve terminals in transgenic mice. Endocrinology 143:1036–1046
93. LaBella FS, Sanwal M (1965) Isolation of nerve endings from the posterior pituitary gland. Electron microscopy of fractions obtained by centrifugation. J Cell Biol 25(Suppl):179–193
94. Lee CJ, Dayanithi G, Nordmann JJ, Lemos JR (1992) Possible role during exocytosis of a ca(2+)-activated channel in neurohypophysial granules. Neuron 8:335–342
95. Andreoli TE, Monahan M (1968) The interaction of polyene antibiotics with thin lipid membranes. J Gen Physiol 52:300–325
96. Cass A, Finkelstein A, Krespi V (1970) The ion permeability induced in thin lipid membranes by the polyene antibiotics nystatin and amphotericin b. J Gen Physiol 56:100–124
97. Woodbury DJ (1990) Vesicle-membrane fusion detected by simultaneous electrical and optical measurements. In: Proceedings of the twelfth annual international conference of the IEEE Engineering in Medicine and Biology Society 12(4):1747–1748
98. Woodbury DJ, McNally JM, Lemos JR (2007) SNARE-induced fusion of liposomes to a planar bilayer. In: Leitmannova Liu A (ed) Planar lipid bilayers. Elsevier Press, London
99. Lemos JR, Nowycky MC (1989) Two types of calcium channels coexist in peptide-releasing vertebrate nerve terminals. Neuron 2:1419–1426
100. Ortiz-Miranda S, Dayanithi G, Velásquez-Marrero C, Custer E, Treistman SN, Lemos JR (2010) Differential modulation of N-type calcium channels by μ-opioid receptors in oxytocinergic vs. vasopressinergic neurohypophysial terminals. J Cell Physiol 225:276–288
101. Wang G, Thorn PJ, Lemos JR (1992) A novel large-conductance Ca(2+)-activated potassium

channel and current in nerve terminals of the rat neurohypophysis. J Physiol 457:47–74

102. Wynne PM, Puig SI, Martin GE, Treistman SN (2009) Compartmentalized β subunit distribution determines characteristics and ethanol sensitivity of somatic, dendritic, and terminal large-conductance calcium-activated potassium channels in the rat central nervous system. J Pharmacol Exp Ther 329:978–986
103. Kilic G, Stolpe A, Lindau M (1996) A slowly activating voltage-dependent K+current in rat pituitary nerve terminals. J Physiol 497:711–725
104. Wilke RA, Ahern GP, Jackson MB (1998) Membrane excitability in the neurohypophysis. Adv Exp Med Biol 449:193–200
105. Ortiz-Miranda S, Dayanithi G, Custer E, Treistman SN, Lemos JR (2005) Micro-opioid receptor preferentially inhibits oxytocin release from neurohypophysial terminals by blocking R-type Ca^{2+} channels. J Neuroendocrinol 17:583–590
106. Rusin KI, Giovannucci DR, Stuenkel EL, Moises HC (1997) K-opioid receptor activation modulates Ca^{2+} currents and secretion in isolated neuroendocrine nerve terminals. J Neurosci 17(17):6565–6574
107. Hussy N, Bres V, Rochette M et al (2001) Osmoregulation of vasopressin secretion via activation of neurohypophysial nerve terminals glycine receptors by glial taurine. J Neurosci 21:7110–7116
108. Saito T, Dayanithi G, Saito J et al (2008) Chronic osmotic stimuli increase salusin-β-like immunoreactivity in the rat hypothalamo-neurohypophyseal system: possible involvement of salusin-β on [Ca2+]i increase and neurohypophyseal hormone release from the axon terminals. J Neuroendocrinol 20:207–219
109. Dayanithi G, Cazalis M, Nordmann JJ (1987) Relaxin affects the release of oxytocin and vasopressin from the neurohypophysis. Nature 325:813–816
110. Zhang SJ, Jackson MB (1995) GABAa receptor activation and the excitability of nerve terminals in the rat posterior pituitary. J Physiol 483(Pt 3):583–595
111. Wilke RA, Hsu S-F, Jackson MB (1998) Dopamine D4 receptor mediated inhibition of potassium current in neurohypophysial nerve terminals. J Pharmacol Exp Ther 284: 542–548

Chapter 11

The Sea Urchin Egg and Cortical Vesicles as Model Systems to Dissect the Fast, Ca^{2+}-Triggered Steps of Regulated Exocytosis

Prabhodh S. Abbineni, Elise P. Wright, Tatiana P. Rogasevskaia, Murray C. Killingsworth, Chandra S. Malladi, and Jens R. Coorssen

Abstract

Exocytosis is a fundamental process utilized by all eukaryotic organisms; this elegantly efficient process mediates such diverse functions as fertilization, synaptic transmission, and wound healing. Membrane fusion, the defining step of this process, has been well conserved through evolution. However, the mechanisms defining the priming, docking, and merger of two apposed native bilayer membranes have not been fully elucidated. Sea urchin cortical vesicles are locked at a stage just prior to Ca^{2+}-triggered membrane fusion and are thus an ideal system for fully defining the mechanisms underlying this process. Here we describe detailed methods to isolate these native secretory vesicles, monitor the fusion process, assess the minimal essential biochemical components, and identify their ultrastructural interactions that define the triggered exocytotic pathway.

Key words Calcium, Docking, Exocytosis, Membrane fusion, Priming, Secretory vesicle proteome/lipidome

1 Introduction

Exocytosis is a fundamental and essential cellular process enabled by a complex series of overlapping, integrated molecular interactions by which vesicles in different intermediate states/pools progress to fusion and content release. Effective dissection of this pathway requires the analysis of a given step (and perhaps those directly linked with it) without having multiple interconnected steps influence functional assessments or the identification of critical components [1–4]. Considering this demand for tightly coupled functional and molecular assays to most rigorously define underlying mechanism(s), there are few model systems that enable the direct quantitative assessment of individual stages of exocytosis [4, 5]. Detailed analysis of fast, Ca^{2+}-triggered fusion is thus

Peter Thorn (ed.), *Exocytosis Methods*, Neuromethods, vol. 83,
DOI 10.1007/978-1-62703-676-4_11, © Springer Science+Business Media New York 2014

ideally carried out in isolation. This enables the clear definition of components essential to the fundamental fusion mechanism (FFM) as well as those providing support or modulatory roles that define the phenotype of a given physiological fusion machine (PFM) [6–16].

Cortical vesicles (CV) isolated from unfertilized sea urchin eggs are stage-specific (i.e., fusion-ready), preserve native function (fusion) in response to the endogenous trigger (Ca^{2+}), and are amenable to a range of biochemical manipulations that enable direct assessment of mechanism [1–6, 17, 18]. Analyses indicate that Ca^{2+}-triggered CV-CV and CV-plasma membrane (PM) fusion (exocytosis) proceed via the same mechanism [2, 7, 19–21]; Ca^{2+}-activity curves for exocytosis from neurons and other secretory cells also overlap those for CV fusion (i.e., these are translationally invariant), indicating a common fusion mechanism [1–4, 21, 22]. Indeed, Ca^{2+} sensitivities and graded release responses comparable to those of urchin CV are seen in mammalian central neurons [23, 24]. Molecular components of fundamental, conserved mechanisms are most often also highly conserved, or the mechanism ceases to be. Among other proteins, there is thus a high degree of SNARE (*s*oluble *N*-ethylmaleimide-sensitive factor *a*ttachment *re*ceptor protein) conservation in mammalian synaptic vesicles relative to urchin CV, as well as similar lipid profiles [2, 25–27]. CV are isolated through simple subcellular fractionation; eggs are first sheared to yield cell surface complexes (CSC)—sheets of PM with docked CV (Fig. 2). Incubating CSC in chaotropic buffers dissociates CV from the PM and these are then isolated via differential centrifugation. Thus, considering the simplicity of isolation and the ease of direct functional/molecular assessments, CV have been described as an "ideal" stage-specific system with which to dissect the native pathway of fast, Ca^{2+}-triggered fusion [4, 5, 28]: (a) With ~15,000 fully docked/fusion-ready CV per egg (Fig. 5), and as a gravid female produces grams of eggs, there is ample material for the tightly coupled functional and molecular assays essential to dissect mechanism and identify even low abundance components; a mammalian central synapse has, at best, a few release ready vesicles that are impossible to isolate in that state. (b) Fusion-ready CV are easily isolated to extremely high purity; this is not the case for mammalian vesicles (Section 3.6 and Fig. 1d). (c) Mammalian vesicles de-prime within ~2–4 min of isolation, requiring cytosol to reverse this, yielding a heterogeneous population; CV remain fusion-active in buffer, in the absence of all cytosolic components. (d) CV are fully primed, whether docked to the PM or isolated from it, "locked" at a step downstream of SNARE interactions but still awaiting Ca^{2+} triggering to initiate fusion [2]. (e) Isolated CV retain the minimal essential machinery for docking, Ca^{2+}-sensing/triggering and

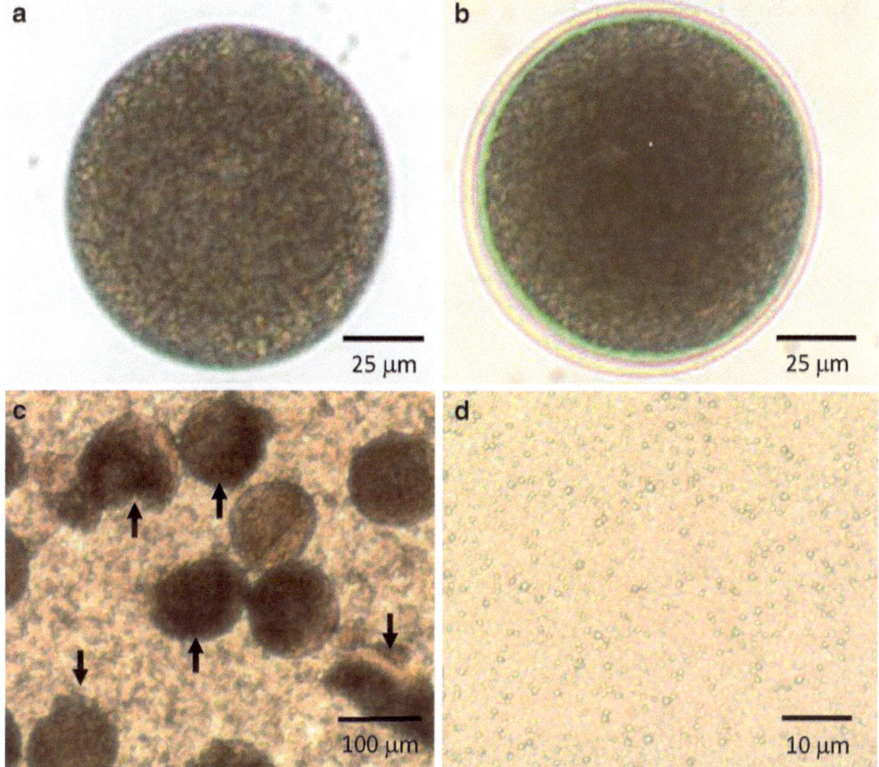

Fig. 1 (**a**) Freshly spawned egg collected in ASW. (**b**) Following fertilization the raised envelope resulting from the cortical reaction is clearly visible; this confirms the viability of Ca^{2+}-triggered CV fusion. (**c**) Eggs following initial disruption in a Potter-Elvehjem homogenizer; mostly ghosts (i.e., broken eggs) with cytosol leaking out are visible (indicated with *arrows*). (**d**) Final purified CV isolate used in experiments

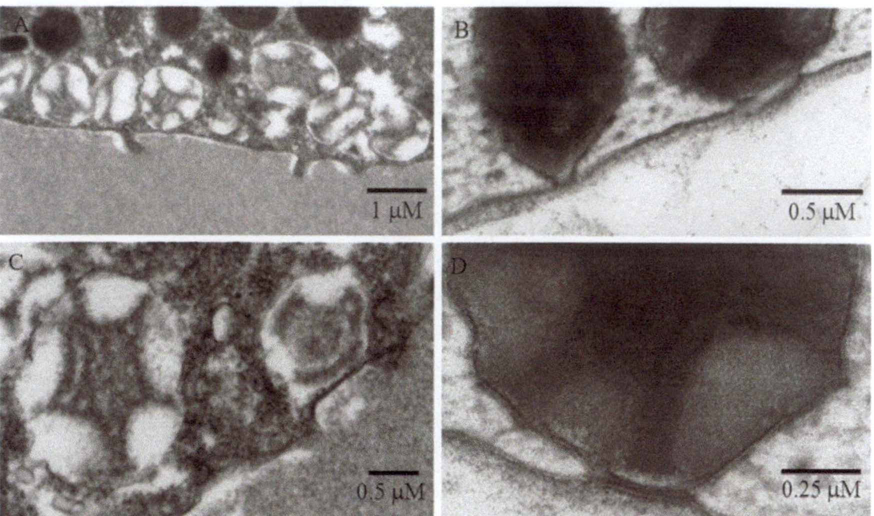

Fig. 2 (**a**) CV that have been chemically fixed for EM. The contents of the CV appear extracted, enlarged, and aggregated, and some CV have undocked from the PM; this is more clearly seen in the close-up in panel **c**. (**b**) An image of CV prepared by rapid freezing and freeze substitution; note that these CV retain their content and structure. (**d**) The rapidly frozen preparation shows the lipid bilayer of both the CV-limiting membrane and cell PM; with this preparation ultrastructural details of the native docking/fusion site are amenable to analysis. In addition, coat material is visible on the outer surface of both the CV and PM, as are likely cytoskeletal tethers aiding in the attachment of the CV to the PM

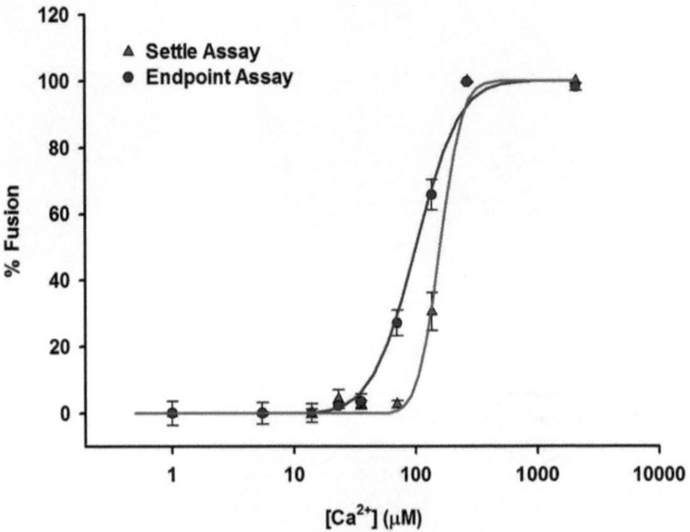

Fig. 3 Combined endpoint and settle assay fusion curves. There is a slight rightward shift in Ca^{2+} sensitivity when CV are allowed to settle into contact

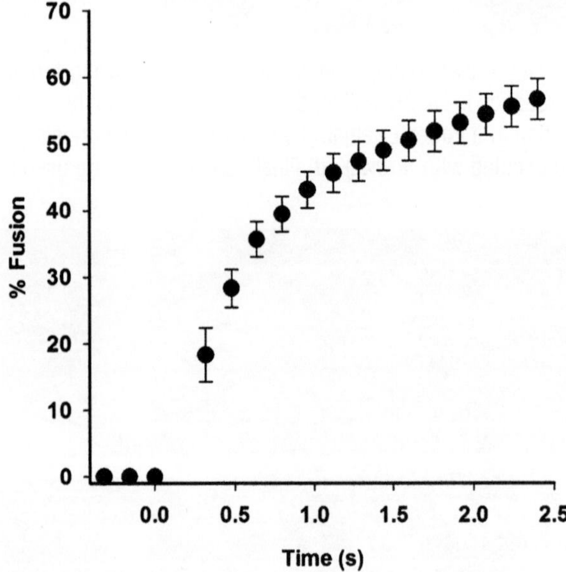

Fig. 4 A standard kinetic assay output measuring the rate of CV fusion

fusion (i.e., the FFM and at least some components of the PFM) [1, 2, 6, 8, 29], and these can be easily assessed (Figs. 3 and 4). (f) CV-CV and CV-PM docking, fusion, and molecular analyses are of reasonably high throughput. Working with purified CV "concentrates" the native docking/fusion machinery, enabling unrestricted access and more sensitive/focused molecular analyses as there is no PM. Thus, the power of the approach lies in the ease with which the composition of purified CV can be manipulated and quantified to identify molecular components essential for docking, priming, and Ca^{2+}-triggered fusion. No other such minimal, functional system is available [30].

2 Materials

2.1 Harvesting and Manipulation of Whole Eggs, CSC Preparation, and CV Isolation

Gravid sea urchins (we have worked with *Strongylocentrotus purpuratus, Lytechinus pictus, Strongylocentrotus franciscanus, Strongylocentrotus drobachiensis, Heliocidaris erythrogramma, and Centrostephanus rodgersii)*, 1 × 250 mL beaker/urchin, iris scissors, 0.5 M KCl, sea water (artificial or native), ice.

Artificial sea water (ASW): 435 mM NaCl, 15 mM $MgSO_4{\cdot}7H_2O$, 40 mM $MgCl_2$, 11 mM $CaCl_2$, 10 mM KCl, 10 mM HEPES (free acid), 1 mM EDTA, pH adjusted to 8.0 with NaOH.

Ca^{2+} free artificial sea water (CFASW): 445 mM NaCl, 50 mM $MgCl_2$, 10 mM KCl, 2.5 mM $NaHCO_3$, 2 mM EGTA, 1 mM EDTA, pH adjusted to 8.0 with NaOH.

Intracellular media (IM) 2× concentration: 420 mM glutamate (free acid), 1 M glycine, 20 mM NaCl, 20 mM PIPES (free acid), 2.2 mM $MgCl_2$, 10 mM EGTA, pH adjusted to 6.7 with KOH. Note: spike in 2 mM DTT into 1× solution prior to use.

Protease inhibitors (PI): 1 mM benzamidine HCl, 0.4 μg mL^{-1} aprotinin, 0.4 μg mL^{-1} leupeptin, 2 μg mL^{-1} AEBSF, and 0.4 μg mL^{-1} pepstatin (NB: dissolve pepstatin in dimethyl sulfoxide then add to the PI stock).

EGTA/EDTA buffer: 0.5 M EGTA, 0.5 M EDTA, pH adjusted to 8.5 with NaOH.

KEA buffer: 450 mM KCl, 50 mM NH_4Cl, 1 mM benzamidine HCl, 5 mM EGTA, pH adjusted to 9.1 with KOH.

PKME buffer: 50 mM PIPES (free acid), 425 mM KCl, 10 mM $MgCl_2$, 1 mM benzamidine HCl, 5 mM EGTA, pH adjusted to 8.0 with KOH.

Baseline intracellular media (BIM) 2× concentration: 420 mM glutamate (free acid), 1 M glycine, 20 mM NaCl, 20 mM PIPES (free acid), 0.1 mM $CaCl_2$ (Note: usually purchased as 1 M stock solution), 2 mM $MgCl_2$ (Note: usually purchased as 1 M stock solution), 2 mM EGTA, pH adjusted to 6.7 with KOH. Note: spike in 2 mM DTT and 2.5 mM magnesium adenosine triphosphate into 1× solution prior to use.

2.2 Analysis of Membrane Fusion

High Ca^{2+} intracellular medium (HIM) 2× concentration: 440 mM glutamate (free acid), 1 M glycine, 10 mM NaCl, 20 mM PIPES (free acid), 49.9 mM $CaCl_2$, 38 mM EGTA, pH adjusted to 6.7 with KOH.

Low Ca^{2+} intracellular medium (LIM) 2× concentration: 440 mM glutamate (free acid), 1 M glycine, 10 mM NaCl, 20 mM PIPES (free acid), 1.9 mM $CaCl_2$, 38 mM EGTA, pH adjusted to 6.7 with KOH.

Ca^{2+} stocks for fusion assays, prepared according to the table below; final mixtures are adjusted to the pH indicated so that the CV consistently remain at pH 6.7 in the subsequent fusion assays. *See* **Note 1**.

Stock	$[Ca^{2+}]_{free}$ in final fusion assay (μM)	2× LIM (mL)	2× HIM (mL)	pH
1	0.3	188.6	61.4	6.71
2	2.4	109.1	140.9	6.72
3	5.6	94.7	155.3	6.73
4	14.0	84.5	165.5	6.78
5	25.2	77.4	172.6	6.82
6	41.7	72.6	177.4	6.84
7	63.4	68.2	181.8	6.86
8	105.0	57.8	192.2	6.90
9	188.9	45.3	204.7	6.93
10	475.6	0	250	6.95
11	1,000	2.7 (1 M CaCl2)	250	6.98

96-well and 384-well flat bottom microtiter plates.

Plate reader compatible with both 96- and 384-well microtiter plates and preferably able to scan at a number of wavelengths.

Coupled fast injector for the plate reader, to support kinetic assays; we have used both the Wallac Victor II microplate reader (Perkin Elmer, Boston, MA) and the POLARstar Omega (BMG Labtech, Offenburg, Germany).

Refrigerated centrifuge with a swinging bucket rotor.

2.3 Microscopy (Light, Electron, Confocal)

CellMask™ Deep Red Cell PM stain (Invitrogen, Molecular Probes); Nile Red lipophilic stain (Invitrogen, Molecular Probes); DiI (Invitrogen, Molecular Probes); Ca^{2+} ionophore (A23187) (Calbiochem); microscope dishes; plastic coverslips (Thermanox, USA); light, epifluorescence, or confocal microscopy system; automated freeze substitution system (Leica, Microsystems, Germany); ultramicrotome (Leica, Microsystems, Germany); electron microscope.

2.4 CV Membrane Protein Isolation for 2-Dimensional Gel Electrophoresis (2DE)

Hypotonic lysis buffer: 20 mM PIPES, pH 6.7.

Membrane protein solubilization buffer: 8 M urea, 2 M thiourea, 4 % (w/v) 3-[(3-cholamidopropyl)dimethylammonio]-1-propanesulfonate (CHAPS).

2.5 CV Membrane Neutral Lipid, Phospholipid, and Polyphosphoinositide Isolation

Acidic extraction buffer: 1.76 % (w/v) KCl, 100 mM citric acid, 100 mM disodium phosphate, 5 mM EDTA, pH 3.6.

Butylated hydroxytoluene (BHT) stock: 2.5 mg in 1 mL of chloroform:methanol (2:1, v/v); chloroform; methanol; 1 M NaCl/0.1 M HCl; chloroform:methanol (2:1, v/v); phosphate buffered saline (PBS), pH 7.4; water-saturated butanol; 0.88 % (w/v) KCl; supply of nitrogen.

3 Methods

3.1 Collecting and Maintaining Urchins for Experimentation

Depending on circumstances and urchins needed, researchers may collect their own or have gravid sea urchins delivered by commercial suppliers. Urchins can then be maintained in a minimal marine aquatic environment provided they have constant access to food (we have found lettuce and carrots suffice), and water conditions are kept within optimum marine parameters. Water quality (i.e., pH, salinity, nitrous waste) can be checked with standard kits available in aquatics stores, and occasional partial water changes with freshly prepared aquarium-grade artificial sea water can help maintain optimum conditions. Under these conditions, and with water temperature kept at ~2–5 °C below ocean ambient at the time of collection, gravid urchins can be maintained for several months at a time; we prefer to have urchins in these holding conditions for at least 1–2 weeks before using them for any experiments.

3.2 Harvesting and Handling of Eggs: Initial Quality Check via Fertilization

1. Use iris scissors to cut around the soft tissue surrounding the beak on the underside of the urchins. Lever the beak out, empty the intracoelomic cavity of fluid, and fill with 0.5 M KCl. Orange egg masses will appear at the crown of female urchins and white sperm at the crown of the males. See Note 2.
2. Place the urchins on beakers filled with sea water to collect eggs; most will settle to the bottom in <1 h. Work at the temperature at which the urchins are maintained. Decant seawater and pool egg suspensions. Only one male is needed, but do not let the collected sperm contact the eggs as this will trigger the cortical reaction and the eggs will no longer be usable. NB: Sperm can also be collected on a clean surface such as a petri dish or piece of parafilm.
3. Vortex 5 μL of sperm to activate (i.e., trigger capacitance reaction) and add it to 50 μL of egg suspension in 1 mL of ASW.

Incubate this mixture for 2–5 min at room temperature (RT) and then view under a light microscope; >90 % of eggs are expected to be fertilized (i.e., have a fertilization envelope; compare Fig. 1a, b). This serves as a measure of the viability of the urchin eggs and the cortical fusion reaction. See Note 3.

3.3 De-Jellying and Washing Eggs

Remove the jelly coat surrounding the eggs by gently passing the egg suspension through a ~100 μm nylon mesh a minimum of three times (rinse the mesh with CFASW after each pass). Examine under a microscope; the eggs will now contact each other if de-jellying was successful. The mesh size used should always be large enough to just enable passage of the eggs being washed. From this stage forward, all steps should be carried out on ice (i.e., ~4 °C).

Transfer the egg suspension to 50 mL tubes and centrifuge them at 700 × *g* for 2 min (make sure the pellet is not larger than ~20 mL as this ensures efficient washing). Gently resuspend the eggs in CFASW and repeat centrifugation thrice, replacing CFASW each time. See Note 4.

At this stage, the eggs can be used whole as described in Section 3.4, or processing to isolate CSC and then CV can proceed.

3.4 Ex Vivo Manipulation of Whole Eggs

The unfertilized sea urchin egg is a convenient system for testing the effect of pharmacological reagents that induce select metabolic changes [7]. For extended incubations it is necessary to collect the eggs in CFASW that has been filtered through a 2 μm mesh (this prevents bacterial growth during extended incubation periods); it is also best to rinse the urchin thoroughly with filtered CFSAW as well, prior to collecting gametes. Following washing and de-jellying as described above, the intact eggs are ready for long-term incubations (20–24 h); our experience to date has been to carry out these incubations at the temperature at which the urchins are maintained. Use 50 mL tubes for reagent incubations, ensuring continuous very gentle mixing over the time of incubation. To confirm the viability of eggs and the cortical reaction after the period of incubation, challenge an aliquot of the egg suspension with 50 μM Ca^{2+} and 50–100 μM of the Ca^{2+}-ionophore, A-23187; fertilization envelops should be visible in <1 min. However, see Note 5.

3.5 CSC Preparation via Homogenization

Resuspend the eggs in IM and wash with IM as described in Section 3.3.

Membrane fusion does not require cytosolic components [15]. Cytosol can be removed and CSC isolated via gentle dounce homogenization [31, 32] or simple solution shearing if the eggs are immobilized on a solid support (i.e., microscope slide) [4, 5, 19, 20].

For larger-scale isolations, as required for coupled fusion assays and molecular analyses, resuspend the washed eggs in a minimum volume of IM (with 1× PI and 2 mM DTT). Keeping the maximum volume of egg suspension to 10 mL ensures efficient homogenization in a 50 mL Potter-Elvehjem homogenizer; clearance between pestle and tube should be very close to the known egg diameter. *Important*: add 1 mL of 0.5 M EGTA/EDTA to this suspension before proceeding to the next step. Transfer the eggs to a Potter-Elvehjem homogenizer and disrupt them with 4–6 strokes of the Teflon pestle. In our hands, a twisting motion through the egg suspension while moving the pestle in and out has proven most efficient. Monitoring the progress using a standard light microscope is vital. Using a cutdown 100 μL pipette tip, transfer ~2–5 μL of the homogenized suspension to a glass slide and view under 10× magnification. If homogenization has been sufficient, ruptured eggs with cytosol leaking out (i.e., cell ghosts) should constitute ≥95 % of a field and very few whole eggs should be visible (Fig. 1c); over homogenization will result in extensive PM fragmentation (i.e., very small membrane fragments) from which it will be largely impossible to later isolate a high-purity CV preparation. See Note 6.

Transfer the homogenate to a 50 mL tube, fill to 45 mL with IM, and add 1 mL of 0.5 M EGTA/EDTA. Centrifuge this suspension at 700 × *g* for 4 min. Repeat this step until the supernatant appears clear of any cytosolic material.

The CSC can now be used to assess CV-PM fusion in vitro, using the various assays described in Sections 3.8–3.10; this is also an excellent preparation for microscopy and related analyses (Fig. 2). Membrane proteins and lipids can also be extracted as described in Sections 3.15 and 3.16, respectively.

3.6 CV Isolation

Full access to the fusion machinery is only enabled by working with isolated CV [1, 2, 4, 10–14, 18, 21, 33]. CV are released from the PM using the KCl-based chaotropic buffers KEA and PKME. Resuspend the CSC pellets in an equal volume of KEA supplemented with 1× PI and 2 mM DTT and incubate on ice for 10 min. Gently invert and then add an equal volume of PKME and incubate on ice for 1 h. Very gently invert the 50 mL tubes every 20 min to resuspend settled CSC. After an hour, most of the CV will have disassociated from the PM. NB: In most preparations, we are working with 2–6 tubes by this point.

Isolating and purifying the CV is now a process of differential centrifugation. First, neutralize the CV-chaotrope mixture by adding an equal volume of IM. Centrifuge this suspension, sequentially, at 525, 595, and 700 × *g* for 5 min each; there is no need to remove the supernatant between these steps. The supernatant now contains free-floating CV and small PM fragments. Transfer the supernatant to 15 mL tubes and centrifuge, up to four times, at 700 × *g* for 5 min to pellet the small PM fragments with any remaining attached CV. Again, this process should be regularly monitored using a light microscope. The supernatant should contain only CV before continuing to the next step; while this is a process of losses, the result will be a final CV isolate of very high purity (Fig. 1d).

CV can now be recovered from the supernatant by centrifugation at 2,000×*g* for 30 min at 4 °C (NB: it is critical to add 500 μL of 0.5 M EGTA/EDTA per 10 mL of CV suspension *before* this final centrifugation step). Remove the supernatant. The resulting pellet should be white or slightly off-white (depending on the species used) with a small orange layer on top containing residual yolk platelets; remove as much of this top layer as possible by gently aspirating using a 100 μL pipette tip. Make sure the CV pellet remains wet—that is, a small overlay of buffer must remain, as exposing the CV to air will result in lysis and thus loss of the preparation. See Note 6.

3.7 Analysis of Membrane Fusion

Before use in functional assays, CV are typically suspended in BIM spiked with 2.5 mM MgATP, at a working optical density of 1.0, measured at 405 nm. See Note 7. This is achieved by suspending the CV pellet in BIM, measuring the OD_{405} in a microtiter plate and adjusting the stock volume accordingly. This is the OD at which CV stock suspensions are routinely treated with reagents of interest (i.e., drugs, exogenous proteins, or other compounds). Once treatment is complete, the CV are recovered as described above and diluted to an OD_{405} between 0.3 and 0.4 for the functional assays. The most efficient way to achieve this is by pipetting 50 μL of the CV suspension into a well of a 96-well microtiter plate, reading the absorbance, and then adjusting the stock volume as required. It is advisable to move quickly through this stage in order to avoid affecting Ca^{2+} sensitivity; in our experience, incubations/handling that go beyond ~1 h result in a very gradual leftward shift in the Ca^{2+}-activity curve.

3.8 CV-CV Endpoint Fusion Assay: Establishing a Ca^{2+}-Activity Curve

This assay measures CV function by inducing fusion with different concentrations of Ca^{2+} and monitoring changes in OD_{405} [1, 29]. Essentially, this assay assesses the fundamental ability of CV to fuse. Adjust the OD_{405} to 0.3–0.4. Gently triturate the CV suspension with a pipette prior to dispensing into microplates; this serves to ensure that the initial suspension contains mainly free-floating CV monomers rather than aggregates (i.e., not pre-attached/docked, as that is also something that must be assessed; see below). Load 50 μL CV suspension into each of 12 wells of a 96-well microtiter plate. This represents an *n* of one using 11 different Ca^{2+} stocks and a water control. We advise preparing four replicates per experimental condition (total 48 wells). Centrifuge plates at 700×*g* for 10 min to ensure optimal contact/packing of CV and thus engagement of all available fusion machines; this does not contribute to subsequent fusion per se except for normalizing the inter-CV contacts [1]. Read the starting OD_{405} of each well that should now contain a CV "lawn." Add 50 μL of each Ca^{2+} stock to four replicate wells each, including replicate water controls. See Note 8. Centrifuge the plate(s) as described above in order to ensure full dispersal of CV content released after fusion and read the final OD_{405}. If using 384-well microplates, follow the protocol above, but use only 35 μL each of CV suspension and Ca^{2+} stock per well. NB: free-floating CSC can be assayed in the same manner (i.e., exocytosis in vitro). The final $[Ca^{2+}]_{free}$ assayed, determined from parallel mock samples not containing CV, will be required for the final data analysis.

3.9 CV-CV Settle Fusion Assay: Assessing Docking

As above, this assay measures CV function by inducing fusion with Ca^{2+} stocks. In this assay, however, CV are left to "settle" into contact thus assessing the capacity of the native machinery itself to establish effective intermembrane contact (i.e., docking) to support Ca^{2+}-triggered fusion. The CV are prepared and aliquoted into 96-well plates as described above, and the covered plate is set aside for 1 h in the dark without centrifugation to allow the CV to settle into contact. Read the initial OD_{405}. Add Ca^{2+} stocks as described above and incubate the plate at RT for 10 min. Centrifuge the plate(s) as described above and read the final OD_{405}.

3.10 CV-CV Fusion Kinetics: Assessing the Initial Rate of Fusion

This assay makes use of high-speed injectors that can be coupled to most currently available plate readers in order to measure changes in OD_{405} as fusion occurs. For each condition, four solutions will be injected: water, BIM, and two buffers yielding a high (i.e., saturating) and a low $[Ca^{2+}]_{free}$. The same CV suspension as was used in the other two assays should be dispensed in 35 μL aliquots into 32 wells in a 384-well microtiter plate per condition and centrifuged at $700 \times g$ for 10 min. This allows for eight replicates of the four different injection buffers. Take an initial reading of OD_{405}. For each well in turn, inject 35 μL of the Ca^{2+} stock solutions so that each runs down the wall of the microtiter plate well. Read the absorbance at the highest possible temporal resolution in order to measure progressive changes in the OD_{405} and thus fusion kinetics; as we are interested in initial kinetics to characterize mechanism, we routinely record at high temporal resolution for only ~2–3 s. Allow the plate to incubate for 10 min at RT after injection is completed. Read final absorbance at OD_{405}; these controls should match parallel samples in the standard endpoint assay. NB: kinetics in the settle assay and CSC preparations can be assessed in the same manner.

3.11 Assessing the Fusion Assays

Fusion as a percentage of maximal release is calculated using (1) for all of the assays described above (Fig. 3). The mean value of the final water blanks is first subtracted from all data collected in order to control for any sample absorbance that does not correspond to CV content.

$$\left[\left(OD_{initial} - OD_{final}\right)/OD_{initial}\right] \times 100 \tag{1}$$

Kinetics data is plotted as percentage of fusion vs. time (Fig. 4). However, once percentage of fusion has been calculated for each $[Ca^{2+}]_{free}$ in the endpoint and settle assays, the resulting activity curves are fit, yielding information concerning the extent and Ca^{2+} sensitivity of the triggered fusion reaction. A two-parameter variable function is thus used to fit control data [1, 34] (Fig. 3). Equations (2 and 3) are used to evaluate the data from experimental conditions.

$$\%\,Fusion = 50 \times erfc\left[\left(pCa + \log_{10}(-6) \times EC_{50}\right)\right]/\left(\sqrt{2} \times W\right) \tag{2}$$

In this equation, erfc = complimentary error function, $EC_{50} = Ca^{2+}$ concentration that induces 50 % fusion, and *W* = distribution width. Data from test conditions are fit using a three-parameter curve equation that is identical to that above (2) with the addition of extent as a variable parameter [1, 11, 34].

$$\%\,Fusion = 50 \times erfc\left[\left(pCa + \log_{10}(-6) \times EC_{50}\right)\right]/\left(\sqrt{2} \times W\right) \times Extent \tag{3}$$

We use TableCurve (Systat Software) to determine the parameters of each of the fusion assays and SigmaPlot (Systat Software) to present the graphical analyses.

3.12 Electron Microscopy

Native membrane structures are highly sensitive to chemical fixation, resulting in artifacts that confound attempts to understand ultrastructural arrangements [9, 35]. By slam freezing and carrying out fixation and dehydration at very low temperatures, rapid-freeze/freeze substitution preserves native structure, enabling assessments at nanometer resolution.

Oocytes have been processed using two different protocols for either ultrastructural examination or immunocytochemical studies. In both cases, the initial fixation of the eggs is by physical freezing. A plunge-freeze fixation system is used (made in house, according to [36]). The system uses liquid nitrogen to cool gaseous propane passing through a cooled aluminum block. Plunging a specimen into the liquid propane yields an extremely high freezing rate on the order of 10,000°/s [37]. Small plastic coverslips (Thermanox, USA) coated with poly-D-lysine are then seeded with several hundred eggs. The coverslip is then plunged into the liquefied propane. See Note 4.

For ultrastructural examination, transfer the plunge-frozen specimen to precooled osmium tetroxide (2 % (w/v)) solution [38] in an automated freeze substitution system (Leica Microsystems, Germany). Leave the specimen in substitution solution for 3 days before allowing the temperature to very gradually rise to RT. Now embed the specimen in epoxy resin and let it cure at 70 °C for 15 h. Specimen blocks are sectioned routinely on an ultramicrotome (Leica UCT, Germany). Cut semi-thin sections (500 nm) and examine under the light microscope to check for suitable oocyte profiles and then cut ultrathin sections (90 nm). Mount these on 300 mesh grids stained with 2 % (w/v) uranyl acetate and 2 % (w/v) lead citrate and let them air dry. Grids can then be viewed (Fig. 2); we are currently using a Morgagni transmission electron microscope (FEI, The Netherlands) at 80 kV, with images acquired using an integrated digital CCD camera (MegaView III, Olympus Soft Imaging Solutions, Germany).

For immunocytochemical studies the plunge-frozen specimen is freeze substituted without osmium tetroxide followed by embedding in LR White acrylic resin [39]. Ultrathin sections from these blocks are subjected to antigen retrieval, followed by immunostaining in a moist chamber at RT [40]. Positive immunostaining is detected by antibody conjugated affinity probes based on colloidal gold particles or quantum dot nanocrystals [41]; following a light counterstain with 2 % (w/v) uranyl acetate, the grids are viewed normally. Images are acquired using the camera set in manual exposure mode with gamma adjustment optimized for visualization of the electron dense affinity probes. Probe density is quantified on segmented images to give quantitative data and determination of labelling significance.

3.13 Fluorescent and Confocal Microscopy

Various light microscopy techniques allow for the examination of egg structures in a live system, both in terms of space and real time [33, 42]. The best results are achieved when the eggs are gently adhered to the slide surface. We favor the use of microscope dishes that have been pretreated with poly-D-lysine. Eggs that settle and adhere to this surface can be gently stained and washed provided pipetting is not too vigorous. Settled eggs can also be sheared to yield a CSC preparation of adhered PM and associated CV [33]. Both applications have proven quite informative. See Note 4.

To prepare the dishes, dispense a poly-D-lysine solution (15.2 μg/mL) onto the recessed section and incubate at 4 °C overnight. Once the poly-D-lysine coating has been completed, remove excess poly-D-lysine, rinse with high-purity water, and store the dishes at 4 °C until required. It is important that the egg suspension used is as pure as possible; immature eggs and other debris will confound imaging. See Note 3. It will be necessary to dilute the resulting egg suspension for imaging; if too concentrated, the eggs will not settle evenly or adhere well. Dilute eggs 1:10 (v/v) in CFASW to allow for uncluttered imaging with a single layer of settled eggs. A Ca^{2+} ionophore, A23187, is also used in such imaging experiments to stimulate CV fusion and thus fertilization envelope elevation; this is a positive control for viability of the triggered fusion pathway. See Note 6.

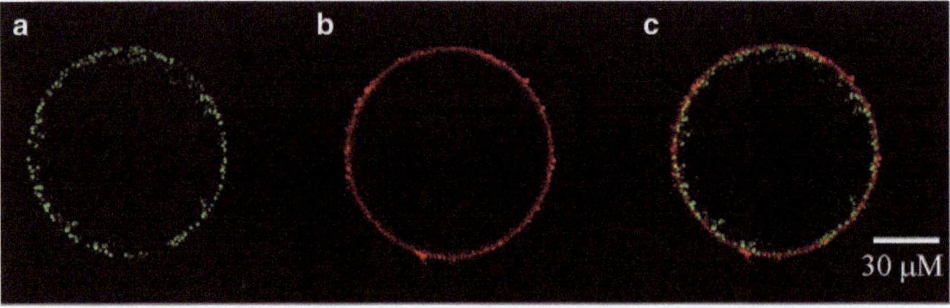

Fig. 5 Confocal images of (**a**) an intact egg with Nile Red stained CV, (**b**) the PM of the same egg stained with CellMask™ Deep Red, and (**c**) an overlay of A and B confirming CV stably docked at the PM

To make a CSC preparation for microscopy, coat a microscope slide with poly-D-lysine as described for the dishes above. Allow eggs to settle and adhere to the surface. Using a 19-gauge syringe, apply a 2 mL stream of CFASW to the slide at an angle no greater than 45°. A fine balance of the angle and pressure will shear the eggs and result in the adhered layer of PM and associated CV. PM sheets stripped of CV can be prepared by further shearing [33].

There are a number of fluorescent dyes that can be used to label various structures in sea urchin eggs. CellMask™ Deep Red PM stain (Ex 649 nm/Em 666 nm) can be used to stably label the PM, with little or no leakage into other structures for up to ~1 h. To stain, incubate eggs in 5 μg mL^{-1} CellMask™ Deep Red PM stain diluted in CFASW for 5 min. Remove the stain and wash the eggs gently with CFASW. Nile Red (Ex 552 nm/Em 636 nm) passes readily through the lipid bilayer and can be used to stain vesicles; to achieve optimum staining, incubate the eggs in 40 μM Nile Red solution (diluted in CFASW from a DMSO stock) for 2 min. This association is not stable, however, and Nile Red can leach into other areas of the egg if the staining duration is prolonged more than ~2 min. It is also possible to dual stain provided washing is sufficient to remove excess fluorescent stain (Fig. 5). 1,1′-dioctadecyl-3,3,3′,3′-tetramethylindocarbocyanine perchlorate (DiI) (Ex 549 nm/Em 565 nm) is another reagent that can be used to stably stain the PM, thus providing an alternative to CellMask™ Deep Red [43]. Lipidic structures can also be targeted with fluorescent dyes; cholesteryl BODIPY FL-C12 has been used to label cholesterol-rich membrane domains [33].

3.14 CV Membrane Protein Isolation for 2DE

1. Resuspend the CV pellet with 3× volume of hypotonic lysis buffer and incubate on ice for 90 s [7, 11, 33]. Restore isotonicity with an equal volume of 2× BIM (with 2× PI).
2. Ultracentrifuge the suspension at ~150,000 × *g* for a minimum of 3 h at 4 °C to pellet the CV membrane. Remove the supernatant and wash the pellet in a 3× volume of PKME to remove

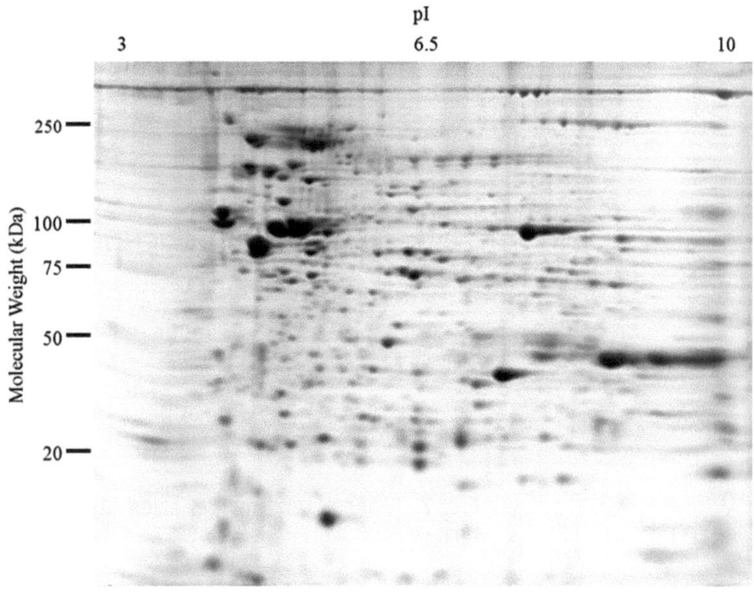

Fig. 6 CV membrane proteins resolved using 2DE

any content protein contamination [2, 7, 11, 33]. Neutralize with an equal volume of 2× BIM (with 2× PI) and repeat ultracentrifugation.

3. Resuspend the pellet in membrane protein solubilization buffer (with 1× PI). Store the protein sample at −80 °C.
4. 2DE is carried out as previously detailed (Fig. 6) [7, 44–48].

3.15 CV Membrane Neutral Lipid and Phospholipid Isolation (Modified Bligh and Dyer Method [7, 21, 49])

Protein-lipid interplay is a crucial component of membrane fusion. Though research in the field has been dominated by studies of the proteinaceous machinery, there is now an increased appreciation of the involvement of diverse lipid species in the fusion mechanism [6, 7, 21, 26, 27, 30, 33, 50].

1. Suspend CV membranes (from step 3 of Section 3.14) in 400 μL PBS and transfer to a silanized glass tube. Add 30 μL of BHT. Add 990 μL methanol and then 480 μL chloroform (0.8:2:1 ($H_2O/CH_3OH/CHCl_3$, v/v/v)). Now add 1 mL 0.1 M NaCl in 0.1 M HCl, followed by 500 μL chloroform (final ratio: 1.8:2:2 ($H_2O/CH_3OH/CHCl_3$, v/v/v)). Vortex vigorously after each addition.
2. Transfer the bottom chloroform layer to a silanized glass test tube using a glass Pasteur pipette and thoroughly dry under a stream of nitrogen. Alternatively, centrifuge lipid samples in a vacuum centrifuge for 2 h to evaporate chloroform. Fill the headspace above the resultant lipid film with nitrogen. Thoroughly seal the tube with parafilm to prevent entry of oxygen; store at −80 °C.

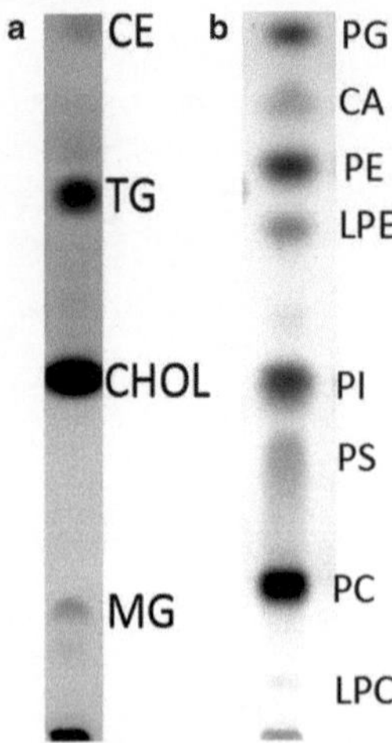

Fig. 7 CV membrane neutral lipids (**a**) and phospholipids (**b**) resolved using HPTLC

3.16 CV Membrane Polyphosphoinositide Extraction (Acidic Extraction [51])

(a) Solubilize CV membrane in 400 μL of PBS and transfer to a silanized glass tube. Add 30 μL of BHT stock, 990 μL methanol, 480 μL of chloroform, and vortex thoroughly. Incubate on ice for 15 min. Now add 500 μL of acidic extraction buffer, vortex thoroughly, and then 500 μL of chloroform. Vortex thoroughly and incubate on ice for 10 min. Collect the resultant chloroform phase and transfer to a silanized glass tube.

(b) Add 1 mL 0.88 % (w/v) KCl to the remaining aqueous phase, vortex vigorously, and then add 1 mL of water-saturated butanol.

(c) Incubate the suspension on ice for 30 min. Collect the upper butanol-rich phase and mix with the previous chloroform phase. Dry this solution under nitrogen and store as described above.

Neutral lipids, phospholipids, and polyphosphoinositides can now be resolved and quantified by a variety of methods; as previously described, we favor the high resolution achieved using automated multiple development high-performance thin-layer chromatography (Fig. 7) [7, 10, 21, 26, 27, 33].

4 Notes

Note 1: These stocks can be aliquoted and frozen (or sealed and stored at 4 °C) for gradual use over several weeks. Using a Ca^{2+} electrode (Orion Electron Products, Beverly, MA, USA) and voltmeter (usually integral to most pH meters), all final free Ca^{2+} concentrations ($[Ca^{2+}]_{free}$) used in fusion assays (see below) are verified in mock samples prepared in parallel during every experiment; Ca^{2+} standards are used to establish a calibration curve (supplier: World Precision Instruments, Inc., Sarasota, FL, USA). If necessary, the range of final $[Ca^{2+}]_{free}$ can be adjusted by altering the proportions of LIM and HIM used to prepare the Ca^{2+} stocks.

Note 2: Avoid cutting too deeply into the urchin as you can disrupt the ovaries and reduce your egg yield.

Note 3: When handling whole eggs, cut the pipette tip back with scissors to widen the bore and thus avoid shearing the eggs.

Note 4: Due to seasonal variability, it may be necessary to selectively remove immature or irregularly shaped egg-like structures. This can be done using repeated lower speed centrifugation washes; we recommend as low as $100 \times g$ for a clean preparation of mature eggs.

Note 5: Keep in mind that should eggs fail to raise a fertilization envelop after ex vivo treatment, this could also mean that the reagent used has caused CV undocking from the PM; this could obviously be important and would require further assays to validate.

Note 6: Failure to add EGTA/EDTA prior to homogenization or prior to isolating CV may well result in fusion and thus loss of the entire preparation.

Note 7: We typically determine OD at 405 nm, but as this is simple forward light scattering, another wavelength can be used should any compound being tested also absorb at 405 nm.

Note 8: When adding Ca^{2+} stocks to CV lawns, be sure to pipette down the side of the well wall rather than directly into the center of the well—disturbing the CV layer will yield erratic OD readings from well to well.

5 Conclusion

For nearly 5 decades the sea urchin cortical reaction has been an invaluable tool for the study of fast, Ca^{2+}-triggered exocytosis. Primed, fusion-ready CV enable detailed analysis of the final, triggered steps of the exocytotic pathway using tightly coupled functional, molecular, and ultrastructural assays. The purity of the CV isolates and the straightforward nature of the fusion assays has yielded many first reports of critical features of triggered release that have since proven to be highly conserved even in the mammalian central nervous system. Vacquier was the first to demonstrate that triggered fusion of fully docked secretory vesicles occurs in the absence of ATP and other cytosolic components [15]; this was soon confirmed by others using the urchin egg preparations [16] but only years later in mammalian neuroendocrine cells and then neurons (reviewed in [17]). Taking the confirmation of ATP as a priming component into consideration, the urchin work indicated that a mechanism that was already prepared to effect membrane merger was "released" by Ca^{2+}. Essentially, the CV of unfertilized sea urchin eggs have provided direct access to the fundamental or minimal essential fusion machine (i.e., FFM; [6, 7]). Thus, the CV model system and the work from a myriad of other labs has aided in the initial identification of critical components of the PFM that modulates the many fusion/fusion pore phenotypes that define the function of different secretory vesicles and cells [6, 10, 11, 52, 53]. This system also confirmed that, while essential to the exocytotic pathway as a whole, intermembrane SNARE protein interactions alone are insufficient to drive the final, Ca^{2+}-triggered membrane fusion step [2, 18]; this conclusion is now more widely recognized [54]. Instead, these SNARE interactions appear to play roles in priming and modulating the Ca^{2+} sensitivity of the fusion process [1–3, 17, 18, 54]; the role in priming has now also been confirmed in neuroendocrine cells [55, 56]. Moving forward, CV have since proven instrumental in establishing cholesterol as a critical component of the fusion mechanism. Through its structural configuration (i.e., intrinsic negative curvature) and ability to form microdomains that sequester other critical molecular components to the docking/fusion site, cholesterol has proven to be an essential part of the FFM (and thus also the PFM [6, 21, 33]). These critical functions have now been confirmed in a number of other secretory cells [27, 57–60]. The current working model arising from an analysis of CV fusion and substantial related research posits vesicles to each contain several fusion complexes, the distribution of which follows a Poisson process. These fusion complexes are differentially activated upon exposure to increasing $[Ca^{2+}]_{free}$. One such active fusion complex is sufficient to drive membrane fusion, and the rate of fusion is increased as more fusion complexes per vesicle are recruited with increasing $[Ca^{2+}]_{free}$ [19, 20, 34, 61]. This model remains to be fully tested in neurons and other secretory cells [23, 24].

By sharing a detailed outline of our methods, we hope that other researchers better appreciate the critical and quantitative questions we seek to address in order to most fully understand these fundamental cellular processes of regulated exocytosis and triggered membrane fusion. There is still much to learn about the fundamental triggered fusion pathway, and this minimal system provides direct access to these highly conserved mechanisms.

Acknowledgements

P.S.A. would like to thank the UWS School of Medicine for scholarship funding. EPW would like to thank Bellberry Limited for their donation to the UWS School of Medicine that provided a postdoctoral fellowship. J.R.C. acknowledges the support of the UWS School of Medicine and an anonymous private Australian foundation, as well as past support from the Canadian Institutes of

Health Research, the Natural Sciences and Engineering Research Council of Canada, and the Alberta Heritage Foundation for Medical Research that made much of this work possible. The introduction was largely reproduced from a text box that appeared in one of our recent publications [30], and was reproduced here with the permission of the publisher (Elsevier).

References

1. Coorssen JR et al (1998) Biochemical and functional studies of cortical vesicle fusion: the SNARE complex and Ca^{2+} sensitivity. J Cell Biol 143:1845–1857
2. Coorssen JR et al (2003) Regulated secretion: SNARE density, vesicle fusion and calcium dependence. J Cell Sci 116:2087–2097
3. Tahara M et al (1998) Calcium can disrupt the SNARE protein complex on sea urchin egg secretory vesicles without irreversibly blocking fusion. J Biol Chem 273:33667–33673
4. Zimmerberg J et al (2000) A stage-specific preparation to study the $Ca^{(2+)}$-triggered fusion steps of exocytosis: rationale and perspectives. Biochimie 82:303–314
5. Zimmerberg J et al (1999) Sea urchin egg preparations as systems for the study of calcium-triggered exocytosis. J Physiol 520 (Pt 1):15–21
6. Churchward MA, Coorssen JR (2009) Cholesterol, regulated exocytosis and the physiological fusion machine. Biochem J 423:1–14
7. Rogasevskaia TP, Coorssen JR (2011) A new approach to the molecular analysis of docking, priming, and regulated membrane fusion. J Chem Biol 4:117–136
8. Vogel SS et al (1992) Calcium-triggered fusion of exocytotic granules requires proteins in only one membrane. J Biol Chem 267: 25640–25643
9. Chandler DE, Heuser J (1979) Membrane fusion during secretion: cortical granule exocytosis in sea urchin eggs as studied by quick-freezing and freeze-fracture. J Cell Biol 83:91–108
10. Rogasevskaia T, Coorssen JR (2006) Sphingomyelin-enriched microdomains define the efficiency of native $Ca^{(2+)}$-triggered membrane fusion. J Cell Sci 119:2688–2694
11. Hibbert JE et al (2006) Actin is not an essential component in the mechanism of calcium-triggered vesicle fusion. Int J Biochem Cell Biol 38:461–471
12. Raveh A et al (2012) Observations of calcium dynamics in cortical secretory vesicles. Cell Calcium 52:217–225
13. Furber KL et al (2009) Enhancement of the $Ca^{(2+)}$-triggering steps of native membrane fusion *via* thiol-reactivity. J Chem Biol 2:27–37
14. Furber KL et al (2010) Dissecting the mechanism of Ca^{2+}-triggered membrane fusion: probing protein function using thiol reactivity. Clin Exp Pharmacol Physiol 37:208–217
15. Vacquier VD (1975) The isolation of intact cortical granules from sea urchin eggs: calcium ions trigger granule discharge. Dev Biol 43:62–74
16. Whitaker MJ, Baker PF (1983) Calcium-dependent exocytosis in an *in vitro* secretory granule plasma-membrane preparation from sea-urchin eggs and the effects of some inhibitors of cytoskeletal function. Proc R Soc Lond B 218:397–413
17. Szule JA, Coorssen JR (2003) Revisiting the role of SNAREs in exocytosis and membrane fusion. Biochim Biophys Acta 1641:121–135
18. Szule JA et al (2003) Calcium-triggered membrane fusion proceeds independently of specific presynaptic proteins. J Biol Chem 278: 24251–24254
19. Vogel SS et al (1996) Poisson-distributed active fusion complexes underlie the control of the rate and extent of exocytosis by calcium. J Cell Biol 134:329–338
20. Blank PS et al (2001) A kinetic analysis of calcium-triggered exocytosis. J Gen Physiol 118:145–156
21. Churchward MA et al (2005) Cholesterol facilitates the native mechanism of Ca^{2+}-triggered membrane fusion. J Cell Sci 118:4833–4848
22. Knight DE, Scrutton MC (1986) Gaining access to the cytosol: the technique and some applications of electropermeabilization. Biochem J 234:497–506
23. Lou X et al (2005) Allosteric modulation of the presynaptic Ca^{2+} sensor for vesicle fusion. Nature 435:497–501
24. Schneggenburger R et al (2012) $Ca^{(2+)}$ channels and transmitter release at the active zone. Cell Calcium 52:199–207

25. Coorssen JR et al (2002) Quantitative femto- to attomole immunodetection of regulated secretory vesicle proteins critical to exocytosis. Anal Biochem 307:54–62
26. Churchward MA et al (2008) Copper (II) sulfate charring for high sensitivity on-plate fluorescent detection of lipids and sterols: quantitative analyses of the composition of functional secretory vesicles. J Chem Biol 1: 79–87
27. Ormerod KG et al (2012) Cholesterol-independent effects of methyl-β-cyclodextrin on chemical synapses. PloS one 7:e36395
28. Vogel SS et al (1991) The sea urchin cortical reaction. A model system for studying the final steps of calcium-triggered vesicle fusion. Ann N Y Acad Sci 635:35–44
29. Vogel SS, Zimmerberg J (1992) Proteins on exocytic vesicles mediate calcium-triggered fusion. Proc Natl Acad Sci U S A 89:4749–4753
30. Rogasevskaia TP et al (2012) Anionic lipids in $Ca^{(2+)}$-triggered fusion. Cell Calcium 52:259–269
31. Sasaki H, Epel D (1983) Cortical vesicle exocytosis in isolated cortices of sea urchin eggs: description of a turbidometric assay and its utilization in studying effects of different media on discharge. Dev Biol 98:327–337
32. Haggerty JG, Jackson RC (1983) Release of granule contents from sea urchin egg cortices. New assay procedures and inhibition by sulfhydryl-modifying reagents. J Biol Chem 258:1819–1825
33. Churchward MA et al (2008) Specific lipids supply critical negative spontaneous curvature—an essential component of native Ca2+-triggered membrane fusion. Biophys J 94:3976–3986
34. Blank PS et al (1998) Submaximal responses in calcium-triggered exocytosis are explained by differences in the calcium sensitivity of individual secretory vesicles. J Gen Physiol 112:559–567
35. Parsons TD et al (1995) Docked granules, the exocytic burst, and the need for ATP hydrolysis in endocrine cells. Neuron 15:1085–1096
36. Sitte H et al (1994) A new versatile system for freeze-substitution, freeze-drying and low temperature embedding of biological specimens. Scanning Microsc Suppl 8:47–64
37. Tivol WF et al (2008) An improved cryogen for plunge freezing. Microsc Microanal 14:375–379
38. Franzini-Armstrong C et al (1978) T-tubule swelling in hypertonic solutions: a freeze substitution study. J Physiol 283:133–140
39. Sobol M et al (2011) A method for preserving ultrastructural properties of mitotic cells for subsequent immunogold labeling using low-temperature embedding in LR White resin. Histochem Cell Biol 135:103–110
40. Yamashita S et al (2009) Establishment of a standardized post-embedding method for immunoelectron microscopy by applying heat-induced antigen retrieval. J Electron Microsc (Tokyo) 58:267–279
41. Killingsworth MC et al (2012) Quantum dot immunocytochemical localization of somatostatin in somatostatinoma by widefield epifluorescence. Super-resolution light and immunoelectron microscopy. J Histochem Cytochem 60(11):832–843. doi:10.1369/0022155412459856
42. Terasaki M (1995) Visualization of exocytosis during sea urchin egg fertilization using confocal microscopy. J Cell Sci 108:2293–2300
43. Terasaki M et al (1991) Characterization of sea urchin egg endoplasmic reticulum in cortical preparations. Dev Biol 148:398–401
44. Butt RH, Coorssen JR (2006) Pre-extraction sample handling by automated frozen disruption significantly improves subsequent proteomic analyses. J Proteome Res 5:437–448
45. Butt RH et al (2006) An initial proteomic analysis of human preterm labor: placental membranes. J Proteome Res 5:3161–3172
46. Taylor RC, Coorssen JR (2006) Proteome resolution by two-dimensional gel electrophoresis varies with the commercial source of IPG strips. J Proteome Res 5:2919–2927
47. Butt RH et al (2007) Enabling coupled quantitative genomics and proteomics analyses from rat spinal cord samples. Mol Cell Proteomics 6:1574–1588
48. Harris LR et al (2007) Assessing detection methods for gel-based proteomic analyses. J Proteome Res 6:1418–1425
49. Bligh EG, Dyer WJ (1959) A rapid method of total lipid extraction and purification. Can J Biochem Physiol 37:911–917
50. Rituper B et al (2012) Cholesterol and regulated exocytosis: a requirement for unitary exocytotic events. Cell Calcium 52:250–258
51. Pettitt TR et al (2006) Analysis of intact phosphoinositides in biological samples. J Lipid Res 47:1588–1596

52. Thorn P et al (2004) Zymogen granule exocytosis is characterized by long fusion pore openings and preservation of vesicle lipid identity. Proc Natl Acad Sci U S A 101: 6774–6779
53. Larina O et al (2007) Dynamic regulation of the large exocytotic fusion pore in pancreatic acinar cells. Mol Biol Cell 18:3502–3511
54. Rizo J, Sudhof TC (2012) The membrane fusion enigma: SNAREs, Sec1/Munc18 proteins, and their accomplices-guilty as charged? Annu Rev Cell Dev Biol 28: 279–308
55. Becherer U, Rettig J (2006) Vesicle pools, docking, priming, and release. Cell Tissue Res 326:393–407
56. Borisovska M et al (2005) v-SNAREs control exocytosis of vesicles from priming to fusion. EMBO J 24:2114–2126
57. Linetti A et al (2010) Cholesterol reduction impairs exocytosis of synaptic vesicles. J Cell Sci 123:595–605
58. Waseem TV et al (2006) Influence of cholesterol depletion in plasma membrane of rat brain synaptosomes on calcium-dependent and calcium-independent exocytosis. Neurosci Lett 405:106–110
59. Wang N et al (2010) Influence of cholesterol on catecholamine release from the fusion pore of large dense core chromaffin granules. J Neurosci 30:3904–3911
60. Zhang J et al (2009) Roles of cholesterol in vesicle fusion and motion. Biophys J 97: 1371–1380
61. Blank PS et al (1998) The calcium sensitivity of individual secretory vesicles is invariant with the rate of calcium delivery. J Gen Physiol 112:569–576

Index

Peter Thorn (ed.), *Exocytosis Methods*, Neuromethods, vol. 83,
DOI 10.1007/978-1-62703-676-4, © Springer Science+Business Media New York 2014

D

E

M

N

O

P

MIX
Papier aus verantwortungsvollen Quellen
Paper from responsible sources
FSC® C105338

If you have any concerns about our products, you can contact us on
ProductSafety@springernature.com

In case Publisher is established outside the EU, the EU authorized representative is:
Springer Nature Customer Service Center GmbH
Europaplatz 3, 69115 Heidelberg, Germany

Printed by Libri Plureos GmbH
in Hamburg, Germany